International Conference Baltic Polymer Symposium 2018

AA006886

IOP Conference Series: Materials Science and Engineering
Volume 500

Jurmala, Latvia
12 – 14 September 2018

ISBN: 978-1-5108-8538-7
ISSN: 1757-8981

Printed from e-media with permission by:

Curran Associates, Inc.
57 Morehouse Lane
Red Hook, NY 12571

Some format issues inherent in the e-media version may also appear in this print version.

Printed by Curran Associates, Inc. (2019)

Institute of Physics
Dirac House, Temple Back
Bristol BS1 6BE UK

Phone: 44 1 17 929 7481
Fax: 44 1 17 920 0979

techtracking@iop.org

Additional copies of this publication are available from:

Curran Associates, Inc.
57 Morehouse Lane
Red Hook, NY 12571 USA
Phone: 845-758-0400
Fax: 845-758-2633
Email: curran@proceedings.com
Web: www.proceedings.com

Table of contents

Volume 500

International Conference Baltic Polymer Symposium 2018

12–14 September 2018, Jurmala, Latvia

Accepted papers received: 25 February 2019
Published online: 5 April 2019

Preface

Papers

IOP Conf. Series: Materials Science and Engineering **500** (2019) 011001 doi:10.1088/1757-899X/500/1/011001

Preface

The issue contains selected papers of XVIII international conference Baltic Polymer Symposium (BPS), which was held in Jurmala, Latvia, from 12th of September until 14th of September of 2018.

The conference is annually organized by the technical universities of the three Baltic States. Since the first event in 2001, range of the topics, covered by the conference, has comprised all aspects of modern **Polymer Science**, including synthesis, processing, recycling, development and characterization of polymer composites as well as revealing the possibilities of nanotechnologies. During its eighteen years of course BPS has gained considerable popularity and at the moment it is an awaited event not only for the polymer scientists but also among the researchers working in the related fields of **Materials Science** and **Nanotechnologies** due to increasing multidisciplinary of the research area. Participation in BPS offers great possibilities of showing yourself, learning, getting new experience and ideas, as well as founding new contacts. Symposium is an excellent chance for young researchers to improve their ability to present the achievements of their research to the international community.

Baltic Polymer Symposium 2018 in Jurmala was visited by participants not only from the three Baltic States, but also by participants from Sweden, Spain, Taiwan, Germany, Czech Republic, Poland, Ukraine, Belarus and Russia. During the conference **5 plenary lectures** were given by world-recognized scientists as well as the representative of SKELOTON technologies, global leader in ultracapacitor technology. The audience was introduced with most recent novelties in the research field of polymer materials science during **28 oral** and **71 poster presentations.**

Peer review statement

All papers published in this volume of *IOP Conference Series: Materials Science and Engineering* have been peer reviewed through processes administered by the proceedings Editors. Reviews were conducted by expert referees to the professional and scientific standards expected of a proceedings journal published by IOP Publishing.

IOP Conf. Series: Materials Science and Engineering **500** (2019) 012001 doi:10.1088/1757-899X/500/1/012001

Mechanical and rheological properties of wood plastic composites based on polypropylene and birch plywood sanding dust

J Kajaks[1] K Kalnins[2] and J Matvejs[3]

[1] Assoc.professor, Institute of Polymer Materials, Faculty of Materials Science and Applied Chemistry, Riga Technical University, P. Valdena str. 3/7 Riga, LV-1048, Latvia.

[2] Troja Ltd , Riga, Bauskas str. 143, and [3] Latvija Finieris, Bauskas str.59, Riga, Lv-1004, Latvia

E mail: janis.kajaks@rtu.lv

Abstract: Article summarizes investigation results of thermal, mechanical and rheological properties of industrially prepared two new wood plastic composites, based on a virgin polypropylene (PP) and birch plywood production residues-plywood sanding dust (PSD). The studies revealed that the tensile strength of composites increased about 25-30% and the modulus for 4.0-4.5 times, but the lightness, the whiteness degree and the common colour changes are smaller in comparison with previously investigated similar composites [5,6].The softening temperatures by VIKA of approbated composites are 154.4 and 158.6^{o}C. The starting temperatures of thermal oxidation are 250.5 and 253.5^{o} C and the thermal expansion of samples at 90° C is only 3%. The rheological experiments showed the decrease of composite melts viscosity with the increase of temperature and the shear rate, but the fluidity index is smaller than 1, that is typical phenomenon for pseudoplastic polymer liquids. The SEM pictures of the fractured surface replicas confirmed that the composites failure has mixed character: partially fracture the polymer matrix and the border surface of polymer matrix and wood fibre.

Introduction

Over the past 20-30 years numerous researchers have paid attention to the studies of different properties of wood plastic composites (WPCs) [1-4]. The most useful from the wide selection of thermoplastic polymers is polypropylene (PP), but as a reinforcements different waste products comprising wood lignocellulose fibres often are used. Such materials also are different residues that arises in plywood production industry [5-10]. Our previous studies [7-10] have showed that plywood sanding dust can be seen as a promising reinforcement of PP. The additives of the plywood sanding dust increase the tensile and the flexural modulus and the microhardness up to 40-50 wt. %, but decrease the deformation ability of the PP matrix and the impact strength, the water resistance as well as the fluidity of the composite melts. Modifying composites with the coupling agent maleated polypropylene (MAPP) (up to 3-5 wt. %) significantly improve all of the above-mentioned properties: the tensile and the flexural strength and the modulus, the microhardness, the impact strength and the water resistance increase to compare with the unmodified composites. The investigated WPCs practically does not swell in a water and the total amount of the absorbed water after 600 h water uptake is no more than 2.2 % at the presence of the coupling agents. The similar results were gained by authors [11-13]. The studies of the thermal, the mechanical, the aging [6] and the rheological properties [5] of the polypropylene (PP) with 40 wt.% content of the plywood sanding dust (PSD) also showed the positive influence of the PSD on exploitation and processing properties of the PP matrix. Due to the results of these studies, two composites which presented the best exploitation properties were chosen. The composition of these composites contains lubricants and different stabilizers [5, 6], but for increasing of the colour intensity of the selected composites, the composition was supplemented with the additions (2.5 and 3.0 wt.%) of the pigment concentrate based on low density polyethylene (PIG). The aim of this work was for both composites to do tensile, flexural and impact strength tests, as well as to check microhardness or the surface. The thermal properties like softening temperature (by VIKA), melting and thermal oxidation temperatures also were noted.The changes of the colour of materials were evalueated by lightness, common colour changes and whiteness degree measurements. The failure mechanism of composites was attempted to clarify with SEM studies of the fractured surface of specimens. The processing possibilities, rheology of the composite melts were examined by capillary viscosimetry tests.

Materials and methods

Polypropylene type PPC 3650 with MFI=3.4g/10min. serve as a polymer matrix, but as a reinforcement – the birch plywood sanding dust (PSD) with the noted fibre fractions [5] were used. In our work two PP+ 40 wt.% PSD composites with the following compositions: 1-A(1) PP+40 wt.% PSD+3 wt.% Struktol TWP+0.33 wt. % stab.108+0.33wt.% stab.1010+0.5 wt.% stab.770+ 2.5 wt.% PIG and 1-A(2) PP+40 wt.% PSD+3 wt.% Struktol TWP+0.33 wt. % stab.108+0.33 wt.% stab.1010+1.0 wt.% stab.770+ 3.0 wt.% PIG were approbated. The colour changes were evaluated by Colour colorimeter anlyzer (Easy Color QA+Software CQA Easy Color V3.0 light origin D_{65} angle of observation 10^0 1964). The whiteness degree of specimens was observed with equipment NOVO-SHADE DUO™ $45^0/0$ opacity/shade Meter). According to the standard ASTM 638 M (EN ISO 527-2:2012) the tensile strength experiments of the composites were performed. The microhardness measurements by Vickers M-41 at the load 200 g were done. Scanning electronic microscope TESCAN TS 5136 MM and VEGA TC computer software were used for visual inspection of the fractured surface of the specimens, as well as Sputter coater EMITECH K 550X for applying an Au layer on the analysed surface. The thermal properties of the composites were noted by Differential Scanning Calorimeter (DSC) measurements on Metler Toledo DSC-1 apparatus at the heating rate 10°C/min. by 300°C. Rheological properties were evaluated as in [5].

Results and discussion

Our studies of the physical mechanical properties (see Table 1) showed that the tensile strength of the new composites increase about 25-30% and the modulus for 4.0-4.5 times to compare with the results of the similar systems given in [6]. The microhardness by Vikers M-41 at the load 200 g values maintain at the same level about 92 - 113 MPa. The gained results revealed the positive influence of additions of the PIG on mechanical properties of the new composites. DSC curves of the composites 1-A(1) and 1-A(2) showed two melting temperatures: one correspond with the melting point of Struktol TWP (156.7 °C) and second with the melting point of the polypropylene (162.07 °C and 161.25 °C) respectively. The widen of the melting limits of the composites: 1-A(1) from 75.7 °C up to 191.7 °C and 1-A(2) from 34.1° C till 189.7 °C are observed that is more enlarge to compare with results gained in [6]. This broadening obviously is connected with the presence of PIG concentrate based on LDPE addittions 2.5 and 3.0 wt.% respectively. The beginning of thermal oxidation temperatures are 250.5 °C and 253.5 °C. The degree of crystallinity that is derectly proportional the enthalpy of the melting ΔH_m differs very significantly 72.75 J/g (1-A(1)) and 104.48 J/g (1-A(2)). It is impossible to explain this fact that we do not know the technological history of the industrially prepared composites. The thermal-mechanical properties of the composites are high: softening temperatures by VIKA are 154.4 °C (1-A(1)) and 158.6 °C (1-A(2)) and thermal expansion of the samples at 90 °C is only 3%. Density of the composites is 1.115 g/cm^3 and 1.092 g/cm^3 respectively.

Table 1. Results of mechanical (tensile, flexural, impact strengths and microhardness) tests

Comp.	Tensile strength, MPa			Flexural strength, MPa			A	MH
	σ_t	E_t	e_{t},%	σ_f	Ef	e_{f}.,%	kJ/m^2	MPa
1-A (1)	25.56	2998.54	1.78	43.38	2930.27	2.26	7.99	92.01
1-A (2)	25.31	2511.80	1.27	56.88	3605.38	2.28	9.71	113.14

In the Table 2 are given numerical values of the lightness (L*), colour difference in lightness (ΔL*), satiation (C_{ab}*), colour difference in satiation (ΔC_{ab}*) colour tone (h_{ab}*), colour difference in tone (ΔH_{ab}*) and the common colour changes as well as the whiteness degree (W). Test results show that the lightness (L*), the colour difference in satiation (ΔC_{ab}*) and the whiteness degree (W) differ very little from a standard specimen (A). The similar composite 1-A was checked in [6]. One of the aim of this tests was to obtain the colour characteristics of the new materials closer to the standard specimen (A).

IOP Conf. Series: Materials Science and Engineering **500** (2019) 012001 doi:10.1088/1757-899X/500/1/012001

Seems that the composite 1-A(1) colour is very close to A because the numerical values of the lightness (L*) and the whiteness degree (W) of the composite 1-A(1) differs from corresponding values of A smaller than from 1-A(2).

Table 2. Result of the measurements the colour parameters and whiteness degree.

Comp.	Colorimetry parameters and whiteness degree (W) values (medium from 10 paralell measurements)							
	L*	ΔL*	C_{ab}*	ΔC_{ab}*	h_{ab}*	ΔH_{ab}*	ΔE_{ab}*	W
A (stand)	40.30	-	5.37	-	-0.92	-	-	12.2
1-A(1)	42.77	-1.53	4.09	-1.28	0.31	5.35	5.71	12.5
1-A(2)	43.28	-1.02	4.61	-0.76	0.74	7.26	7.37	13.2

The rheological experiments presented in Fig.1-3 showed that the melts behaviour of the both materials indeed is very similar. Before the melt flow index test (MFI) was done and the gained values of MFI were 0.728 g/10min. for 1-A(1) and 0.525 g/10min. for 1-A(2) and that also differs a little. From Fig.1 can be seen that the shear stress increase lower than the shear rate for both composites. That indicates on the character of typical pseudo-plastic liquids for which viscosity not only depend on the temperature and the shear rate, but also decreases with the increase of the shear deformation rate (Fig.2).

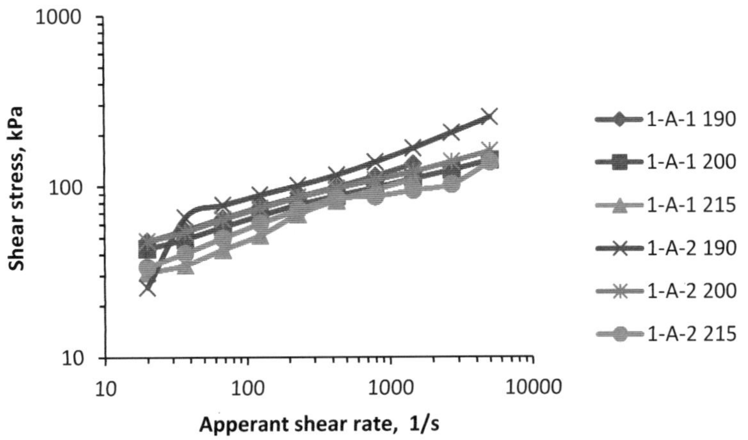

Figure.1. Dependence of the shear stress of the composites based on PP+40 wt. % PSD with different additives on the shear deformation rate:: 1-A (1) and 1-A (2).

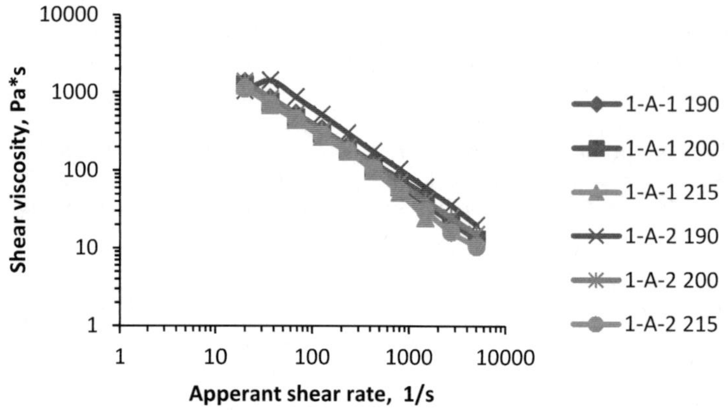

Figure 2. Dependence of the shear viscosity of the composites based on PP+40 wt. % PSD with different additives on the shear deformation rate:: 1-A (1) and 1-A (2).

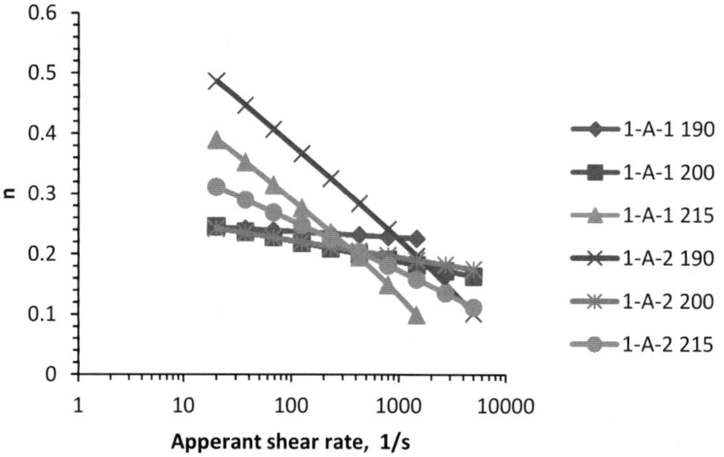

Figure 3. Depedence of the fluidity index (n) of the composites based on PP+40 wt. % PSD with different additives on the shear deformation rate: 1-A (1) and 1-A (2).

Besides, the shear viscosity can decrease up to 10 times with the increase of the shear rate till 8.000-9.000 1/s. Fluidity index n which shows degree of the distinction of the investigated melts from Nuton liquids (n=1) is smaller than 1 (Fig. 3) and decrease with the increase of the shear rate. This fact and results in Fig. 2 and Fig. 3 confirms our previous viewpoint of the pseudo-plasticity of the both composite melts. Similar conclusions were made by the authors [18-20, 22]. All SEM pictures (see Table 3) of the replicas of the fractured surface showed that composites failure has mixed character: partially fracture the polymer matrix and the border surface between the polymer matrix and the wood fibre. It means that the both composites necessary to strengthen with additions of the coupling agents like the maleated polypropylene waxes as in [7].

Table 3. SEM images of the fractured surface of WF/PP composites taken at 1000 and 2000 times of magnification: A, B, C for composite 1-A(1) and D, E, F for composite 1-A(2).

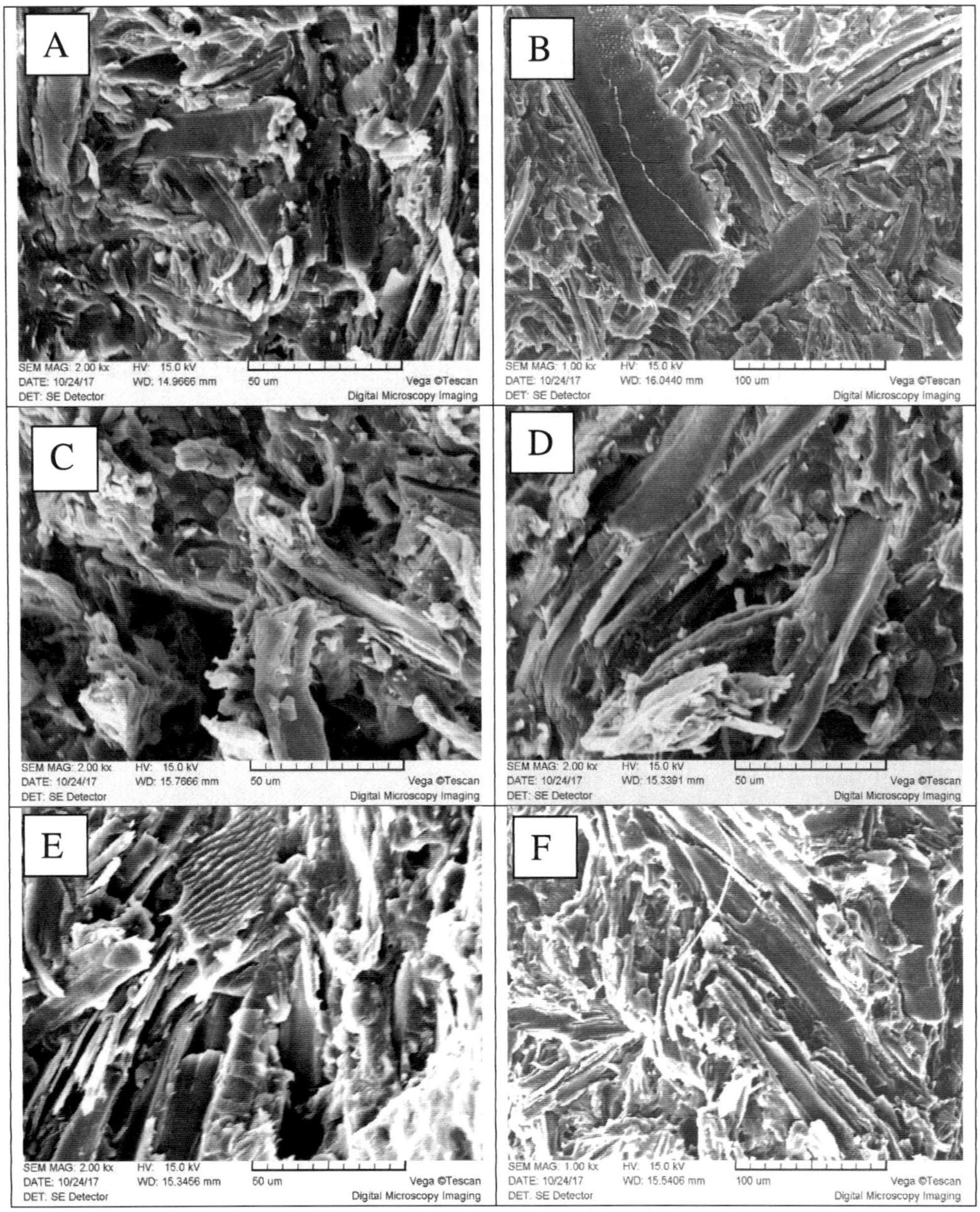

Conclusions

The additons of pigment concentrate (PIG) up to 3 wt.% amplified the physical mechanical properties of the investigated composites and materials colour parameters approached to the colour characteristics of standard specimen. The rheological test showed that the both composite melts viscosity decreases with the increase of temperature and shear rate, fluidity index is smaller than 1, that is typical for pseudoplastic polymer liquids (melts). SEM studies demonstrated that the failure of composites has diverse characteristics: partially fractured polymer matrix and the border surface between the polymer matrix and wood fibre. Undoubtedly, the both composites need to strengthen with additions of the coupling agents.

Acknowledgement: The authors gratefully acknowledge the financial support in accordance with the contract No. 1.2.1.1/16/A/009 between "Forest Sector Competence Centre" Ltd. and the Central Finance and Contracting Agency, concluded on 13th of October, 2016, the study is conducted by "Troja" LTD with support from the European Regional Development Fund (ERDF) within the framework of the project supervised by "Forest Sector Competence Centre".

References

[1] Raj R et al 1989 Journal of Applied Polymer Science **7** (4) 1089-110
[2] Kokta B et al 1989 Polymer Plastic Technology and Engineering **28** (3) 247-259
[3] Maldas D et al 1994 International Journal of Polymer Materials **27** (1-2) 77-88.
[4] Polyolefin Composites Ed. by Domasius Nwabunma and Thein Kun 3M Company, Wiley-Interscience A. John Wiley and Sons INC Publications, 2007, P. 3-82, 87-123, 150-201.
[5] Kajaks J et al 217 Key Engineering,Materials 762 226-232.
[6] Kajaks J et al Open Engineerig, 2018 (in press)
[7] Kajaks J et.al 2018 Proceedings of Estonian Academy of Science, Materials Science **67** (2)
[8] Kajaks J et al 2014 Central European Journal of Engineering, **4** (4) 385-390
[9] Kajaks J et al 2015 Open Engineering **5** (1) 457-464
[10] Kajaks J et al 2017 Key Engineering Materials. ISSN: 1662-9795, 721:48-52. doi: 104028 www.Scientific Net./KEM 721.48
[11] Sobczak L et al 2012 Composites Science and Technology **72** (5) 550-557
[12] Bulylina S et al 2011 Applied Composite Materials **18** (2) 101-11.
[13] Ashori A 2010 Advanced Materials Research **123-12**, 1195-1198
[14] Azwa Z et a..2013 Materials and Design **47** 424-442
[15] Xie L et al 2010 Journal of Reinforced Plastics and Composites **29** (13) 1940-1951
[16] Pereira L et al 2015 Journal of Composite Materials **49** (30) 3727-3740.
[17]. Areas A 2010 Journal of Polymers and Enviroment **18** (3) 318-325
[18] Maiti S et al 2014 J. of Applied Pol. Sci. **91** (1) 644-650
[19] Mazzanti V et al 2015 Polymer Composites Doi: 10.1002/pc. 232546 1-15.
[20] Lewandovski K et al 2016 Polymer Testing, **51** 58-62
[21] Bettini S et al Carbohydrate Polymers **94** 800-806.
[22] Ramzy A et al 2014 J. of Appl. Pol. Sci. Doi: 10.1002/AP 39861.

IOP Conf. Series: Materials Science and Engineering **500** (2019) 012002 doi:10.1088/1757-899X/500/1/012002

Mechanical properties of composite material based on scrap tyres and polyurethane binder with different reactivity

R Kurme and L Malers

Institute of Polymer Materials, Faculty of Material Science and Applied Chemistry, Riga Technical University, Latvia,

E-mail: laimonis.malers@rtu.lv

Abstract. Some mechanical properties of composite material based on scrap tyres and polyurethane type polymer binder were investigated to underline the significance of binder reactivity on selected mechanical properties of the composite material. Compressive stress and compressive modulus of elasticity E at 10 % deformation and Shore C hardness were investigated. Connection was observed between the above-mentioned selected mechanical properties of composite material and reactivity of the polymer binder. It was confirmed previously that the crosslinking degree of polyurethane type binder separately and therefore at the same time mechanical properties of binder were strongly connected with water as a significant factor during the hardening of the polymer. The special objective of this work is to investigate connection between selected mechanical properties of the composite material, content and reactivity of the polymer binder under invariable composite material moulding and hardening conditions.

1. Introduction

Utilization and recycling of used vehicle tyres still must be considered as an important way to unload the environment from non-degradable waste [1]. Production of composite materials [2] is one of the most perspective ways for the reuse of scrap tires, and the produced composite material can be used in a wide range of applications, for example, in the production industry as an absorbent of mechanical vibrations, in the road industry as protective barriers, in the building industry as sound insulating and damping material [3].

In our previous investigations, [4,5] optimization of composition and technology of the composite material production from mechanically grinded scrap tyres and polyurethane type binder were realized. It was also elucidated that mechanical properties of the composite material are strongly dependent not only on the composition of the material, but also on technological parameters (moulding pressure, temperature T). Correlation between mechanical properties of polyurethane type binder and selected crosslinking conditions of the polymer was demonstrated in [5].

The special objective of this work is to investigate the connection between selected mechanical properties of the composite material, reactivity and content of the polymer binder.

2. Experimental part.

Grinded at room temperature rubber crumbs (particle sizes ranging from 0.2 to 5.0 mm) from scrap tires and polyurethane type polymer binder with different reactivity (C_{-CNO} 2.4, 5.5 and 7.4%) were used. Composite material samples were prepared by mechanical mixing of required components ($C_{pol.}$= 8,13,18,23 mass %), moulding of samples in uniform moulds and hardening of samples under defined conditions (T = 18-23^0 C; RH = 28-34%; P = 0.04 kg/cm^2). Shore C hardness (ISO 7619-1; ISO 868), compressive stress at 10% deformation σ_{10} in different loading modes (static and cyclic) and compression modulus of elasticity E (EN 826) in the connection with composition of the composite material and reactivity of polymer were investigated.

3. Results and Discussion

Influence of polymer binder content and reactivity on the investigated parameters were observed. It was established that the existing polyurethane-type binder becomes hard in moisture cure reactions between isocyanate functional groups in polymer and air humidity [5].

It is shown (Table 1) that values of Shore C hardness increase with the increase of polymer binder content in the composite material independently from the isocyanate functional group content in polyurethane type binder.

It was previously demonstrated that a higher content of isocyanate groups in the polyurethane binder leads to a higher degree of cross-linking of polymer [6] and therefore may provoke an increase of the tested in compressive mode mechanical properties as well as higher Shore C hardness of composite material samples. Results presented in present work (Table 1) confirm this presumption .

Connection between reactivity of the polymer binder and mechanical properties of the composite material in compressive mode at 10% deformation is demonstrated also in cyclic mode of deformation for the tested composite materials (figures 1a- 1d).

Table 1. Compressive stress ($\sigma_{10\ \%}$) and compressive modulus of elasticity (E) at 10 % deformation, Shore C hardness of the composite material in dependence of activity and content of the polymer binder.

Nr.	Activity of polymer binder (content of NCO groups, %)	Polymer binder content, Wt., %	Shore C Hardness scale	$\sigma_{10\ \%}$ kPa	E, kPa
1	2,4	8	40	22	280
		13	43	31	490
		18	47	41	780
		23	51	68	1030
2	5,5	8	41	27	305
		13	44	41	500
		18	48	50	795
		23	53	56	1095
3	7,4	8	45	29	365
		13	52	51	600
		18	53	62	725
		23	61	70	1280

In the same time connection between hardness and the compressive stress at 10% deformation, as well the compressive modulus of elasticity of the composite material were observed (see Table 1).

The results presented in Figures 1a; 1b; 1c and 1d demonstrated changes in compressive stress at 10% deformation with the number of deformation cycles (till 20 cycles) of composite material with different activity of polymer binder (CNO group content - 2,42; 5,3 and 7,4 %) and binder content (8,13,13 and 23 mas.%). It was demonstrated that a more essential decrease of compressive stress at 10% deformation is ascertained during the first cycles of deformation independently from activity and the content of polymer binder. It can be explained with degradation of weaker intermolecular bonds between the rubber particles and polymer binder in the material during the first cycles of deformation. After the first three deformation cycles changes in compressive stress are less intrinsic. Material remains practically with similar and invariable compressive stress values. Relatively higher initial and also compressive stress values during all cycles of deformation were observed for samples with higher content and higher activity of polymer binder. It can be explained with the direct influence of polymer binder activity and content on crosslinking degree of polyurethane binder and therefore higher compressive stress of composite material.

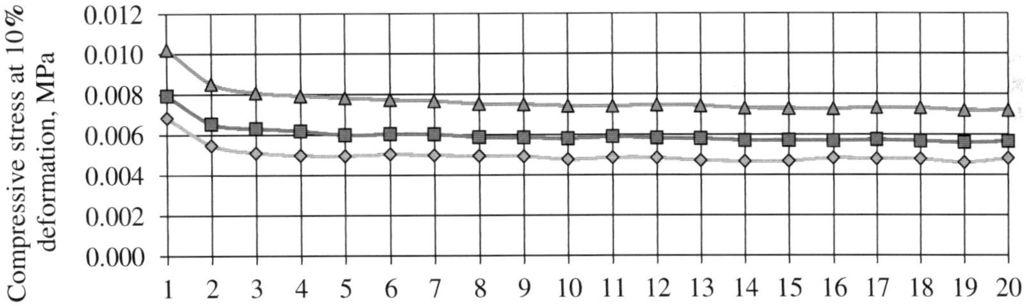

1(a)

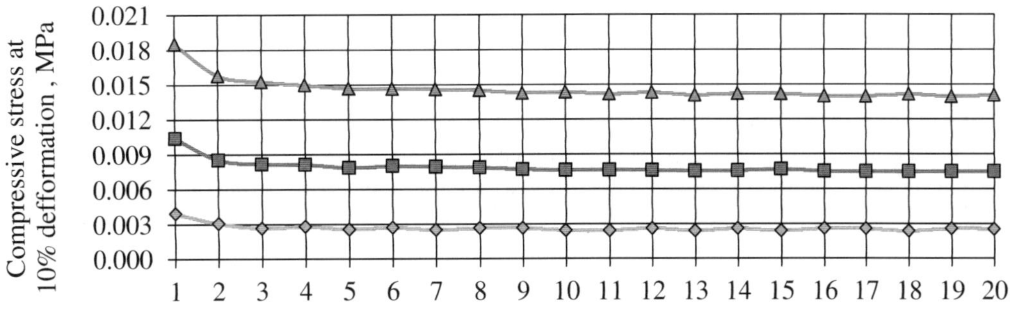

1(b)

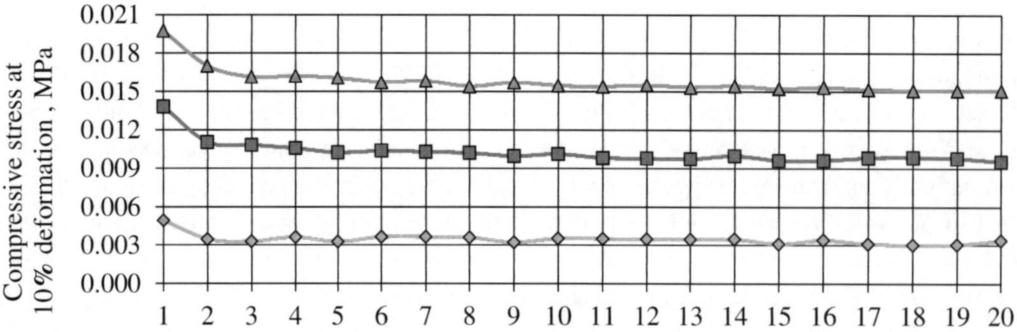

1(c)

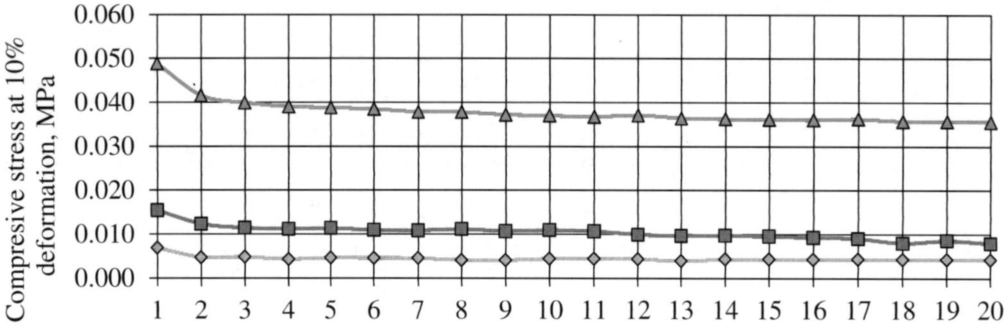

1(d)

Figures 1(a) to 1(d) Dependence of the compressive stress at 10% deformation with the deformation cycles (till 20 cycles) for composite materials with different binder content (mas.%) (**1(a)** -8; **1(b)** -13;**1(c)**-18; **1(d)** – 23) and polymer activity (C_{-NCO} :♦ - 2,24%, ■ – 5,5%, ▲- 7,4%)

Purposeful change of reactivity of the polymer binder during the formation of composite material samples, as well as controlled composition of the composite material provoke changes of the selected mechanical properties.

4. Conclusions

The obtained results show that variation of composition of the composite material and reactivity of the polymer binder in defined intervals have influence on the selected mechanical properties of the material.

Increase of the polymer activity through variation of the CNO groups content from 2,4 till 7,4 % and polymer binder content from 8 till 23 mas.% leads to increase of selected mechanical properties of the composite material.

Thus purposeful selection and mutual combination of composition of the material and activity of polymer binder offers possibility to ensure predictable mechanical properties of composite material- Shore C hardness, compressive stress and compressive modulus of elasticity at 10% deformation .

References

[1].Mark J E, .Erman B ,.Erich R The Science and Technology of Rubber 3 d. Ed. Elsvier Inc. USA 2006

IOP Conf. Series: Materials Science and Engineering **500** (2019) 012002 doi:10.1088/1757-899X/500/1/012002

[2] Hughes A H , Pennington S, Precoated rubber crumb for composites GB Patent 2364708
[3] Shulman V L 2004 Tyre Recycling, e.d. Smithers Rapra Shrewsbury GBR
[4] Plesuma R ,.Megne A ,.,Mateusa-Krukle I.,.Malers L 2015 Progress in Rubber, Plastics and Recycling Technology Nr.2 vol.31 pp 69-75
[5] Plesuma R. , Malers L 2016 Key Engineering Materials ISSN:1662-9795 vol.721 pp.3-7
[6].Malers L ,.Plesuma R, .Locmele L. 2011 Scientific Journal of Riga Technical University, Material Science and Applied Chemistry vol.23 pp 103-106.

IOP Conf. Series: Materials Science and Engineering **500** (2019) 012003 doi:10.1088/1757-899X/500/1/012003

Water Absorption and Tensile Properties of Hemp Waste and Polyethylene Composites

Zane Zelca[1,a *], Silvija Kukle[2, b], Janis Kajaks[3,c]

[1,2] Institute of Design Technology, Faculty of Material Science and Applied Chemistry, Riga Technical University, Riga, Latvia
[3] Institute of Polymer Materials, Faculty of Material Science and Applied Chemistry, Riga Technical University, Riga, Latvia
[a]gusare@inbox.lv, [b]silvija.kukle@rtu.lv, [c]janis.kajaks@rtu.lv

Abstract. This research focuses on improving the mechanical properties of linear low density polyethylene (LLDPE) composites reinforced with hemp waste and reducing water absorption. Cross-linking is introduced between the filler and the polymeric matrix. Cleaning the filler surface of dusts with alkali pre-treatment promotes an increase of the aforementioned properties. Hemp waste pre-treatment with TEOS and modified sols decrease composite water absorption and elongation at break, but silanization increase elastic module. The choice of different silanes for the treatment of hemp waste and treatment method affects the properties of composite.

Introduction

The mechanical properties and water absorption are essential properties that limit the use of composites. The properties of the composites are influenced by the components and the manufacturing process.

The usage of hemp fiber producing waste helps to solve the waste disposal problem [1]. Replacing a part of the polymer with fossil resources with the enriched components of biological origin can reduce pollution and produce new materials. The combination of polymeric matrix and natural fibers provide flexibility and have relatively low density and low manufacturing prices [2].

The waste-reinforced composites can be applied in many fields such as automotive, aerospace, packaging, construction, and transportation industries. The choice of polymer as a matrix material is important for the kinds of natural fiber reinforcement. Polyethylene, polystyrene, polylactides, polyvinyl chloride and polypropylene are thermoplastic materials and can be used as matrix materials for composites [3].

This research focuses on mechanical properties and water absorption of composites consisting of linear, low-density polypropylene and hemp waste. The water absorption of composites could be controlled by waste content and treatment methods. Biodegradable fibers are modified by various chemical processes with variable results. Sources often publish data of silanization, mercerization, acetylation, sol-gel method and other treatments, with regard to the effects of natural reinforcing components on the properties of composites [4;5].

The partially degradable composites can be recycled by returning them to the production stream or separating the components and processing individually. The more the composites were recycled, the lower water absorption potential they exhibited by fiber encapsulation, and recycling could lead to the degradation of hemicellulose leaving natural fibers less hydrophilic [6].

Materials and methods

The hemp waste (W) used was supplied from the factory of "Zalers". Composites with waste content 40 wt% using LLDPE grade 6201XR as matrix (PE) were prepared by two rolls mill (T=150°C, t=10 min) and pressed for testing.

Table 1. Hemp waste composites

Designation	Waste modification					
	NaOH	ZnO	ZAD	TEOS or APTES	Dipping time	Heating
PE	-	-	-	-	-	-
PE/ZnO	-	-	-	-	-	-
PE/W-NaOH	x	-	-	-	30 min	-
PE/ZnO/W-NaOH	x	-	-	-	30 min	-
PE/W-NaOH-Sol_TEOS	x	-	-	0.14 mol	10 min	120°C
PE/W-NaOH-Sol_TEOS_ZAD	x	-	x	0.9 mol	10 min	120°C
PE/W-NaOH-Sol_TEOS_ZnO	x	x	-	0.9 mol	10 min	120°C
PE/W-NaOH-Sil_TEOS	x	-	-	2 mol	60 min	110°C
PE/W-NaOH-Sil_APTES	x	-	-	2 mol	60 min	110°C

Alkalized waste was prepared using 2 % Sodium hydroxide solution (NaOH). Silanized waste (Sil) was prepared by treated alkalized waste with 2 mol APTES - (3-Aminopropyl) triethoxysilane or TEOS - tetraethyl orthosilicate. Waste dipping time is seen in table 1. Nanosols (Sol) were prepared by a controlled hydrolysis, adding ethanol into TEOS, after adding water and hydrofluoric acid. Solutions were stirred for 30 minutes; the obtained nanosols were homogeneous and then modified with 7.5 wt% zinc acetate dehydrate (ZAD) or zinc oxide (ZnO). The humidity of air was 65 ± 2 % and room temperature was -20 ± 2 °C. The waste samples were then dried and heated.

In some samples the zinc oxide nanoparticles used in composite manufacturing process as nanofiller in viscous state. Used reagents: sodium hydroxide, ethanol, hydrofluoric acid, APTES and zinc acetate dehydrate were purchased from Sigma Aldrih, TEOS from Alfa Aesar and zinc oxide nanoparticles from Nanjing High Technology Nano Materials Co., Ltd .

Scanning electron microscopy (SEM) was used to examine the nature of the waste surface modification.

The tensile tests were measured on the material testing machine Tinius Olsen 25ST by standard ASTM D 638 M – 3.

Water exposure experiments (according to the standard ASTM D 570-98) were carried out at room temperature (+23°C).

Results and discussion

LLDPE elastic module is 69.8 MPa and elongation at break is 21.2 %. The flexibility of the matrix is high, both inorganic and organic ingredients are used to decrease it. ZnO 2wt% nanoparticle filler additive increased elastic module (E_t) to 233.4 MPa, but elongation at break decreases by 26 %, figure 1. The waste pretreated with alkali solution used to clean the waste particles surfaces are applied to all variants under discussion, figure 1. The alkali-treated hemp waste 40 wt% composite (LLDPE/W-NaOH) elastic module (451 MPa) is 7 times higher than the module of net LLDPE, but decreases by 7 % when compared with 40 wt% composite E_t without waste alkali treatment (483.3 MPa) [7]. At the same time elongation at break drops by almost 6 times ($\varepsilon_t = 4$ %), figure 1. Samples PE/ZnO/W-NaOH reach elastic module 576 MPa when alkali treated waste combined with ZnO nanofiller in viscous state.

Optional waste treatment with the TEOS based silica sol raises the composite module by 14 % compared with the alkali pre-treated and by only 9 % compared with the waste non-treated variant.

Modules and elongation values of the next three variants are quite close, with an average improvement of 19 % E_t and decrease of 29 % elongation at break compared with the waste non-treated variant (ε_t = 4.2 %). The highest E_t obtained using APTES (647 MPa) exceeded the module of non-treated variant by 1.34 times.

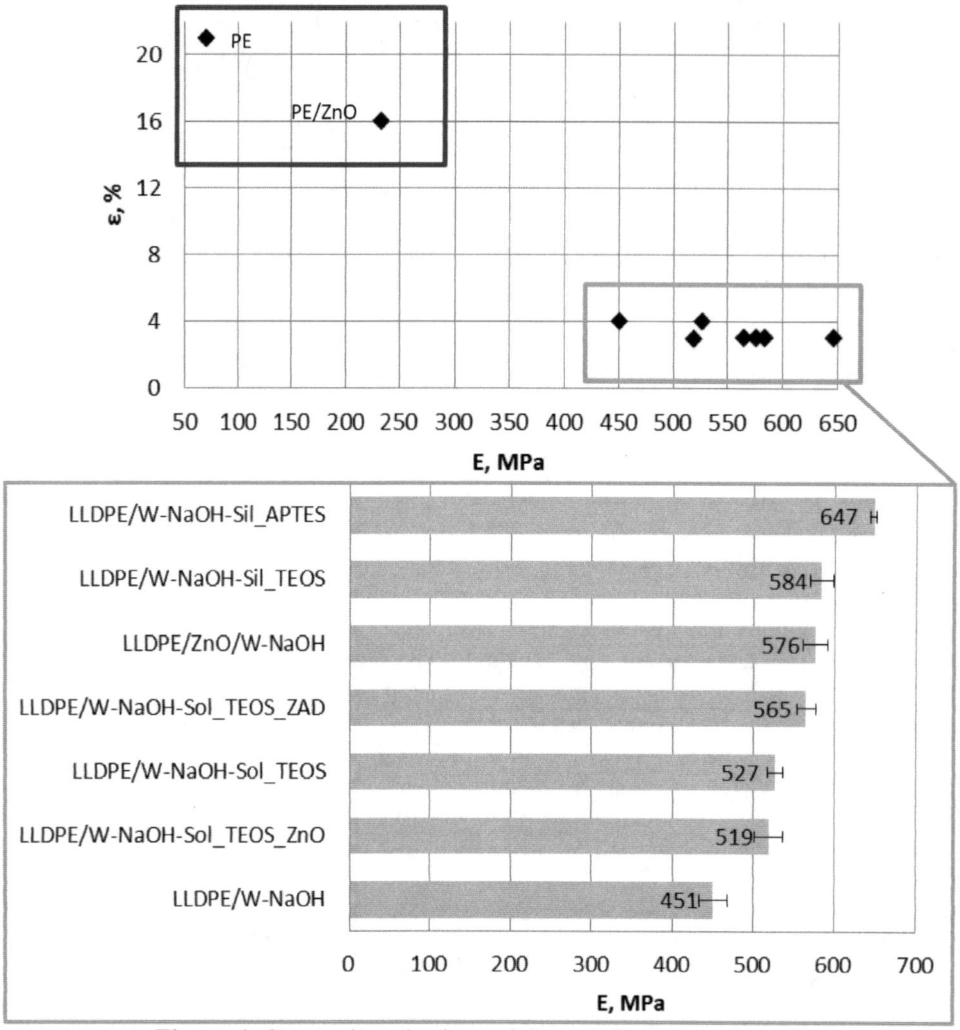

Figure 1. Composites elastic module and elongation at break

LLDPE does not absorb water, LLDPE/ZnO composite absorbs 0.6% water due to air inclusions occuring in the manufacturing process, which serves as a way of penetrating water. Untreated hemp waste composite water absorption is 13.6 % [8]. Water absorption of 40 wt% alkali-treated hemp waste composite is 13.7 %. It is substantially (by 26%) increased if the waste is subsequently subjected to the several pre-treatments as in variant LLDPE/W-NaOH-sil_TEOS (figure 2) indicating that the fibers are damaged and silanisation with TEOS is not suitable for hemp waste.

It seems that in the case of several pre-treatments, it is very important to choose the proper combination. The combination of when TEOS based sol treatment follows after 2 wt% alkali treatment (composite LLDPE/W-NaOH-sol_TEOS) with very low saturation moisture content (5.2 %) works excellently. Composite samples of variants LLDPE/W-NaOH- TEOS_ZAD and LLDPE/W-NaOH-Sol_TEOS_ZnO (7.5 % and 7.4 % respectivily) show equal absorption disposition in waste pre-treated with sol containing Zn compounds (figure 2).

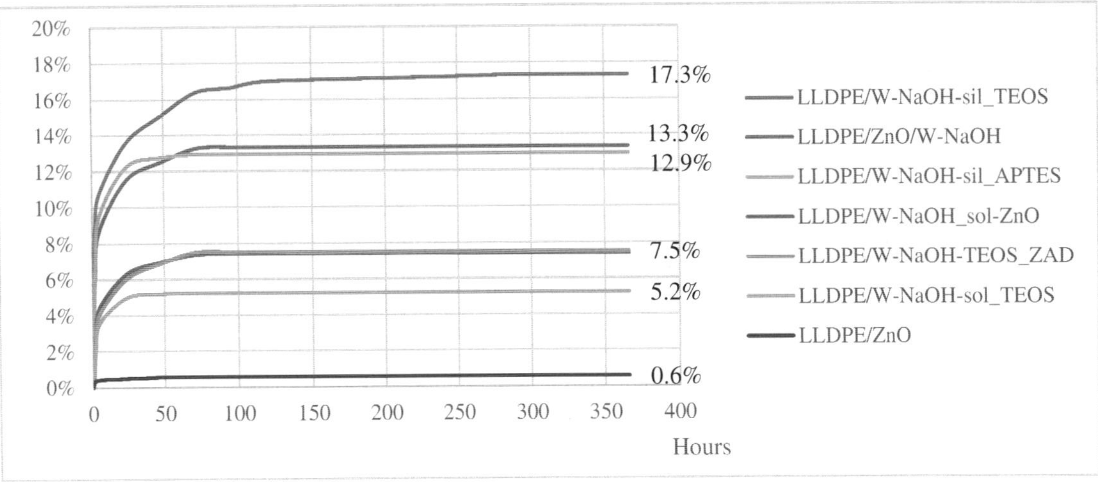

Figure 2. Composites water absorption kinetics

Composites rapidly absorb water in the first 24 hours and reach a point of equilibrium within 72 hours except composite LLDPE/W-NaOH-Sil_TEOS.

As we can see, the untreated hemp waste surface is rough with small dust particles, but alkali treatment cleans the surface which means that a part of the hemicelulose and another water soluble components are washed out together with dust, figures 3 and 4. Clean waste surfaces promote a more durable coating after treatment with silane.

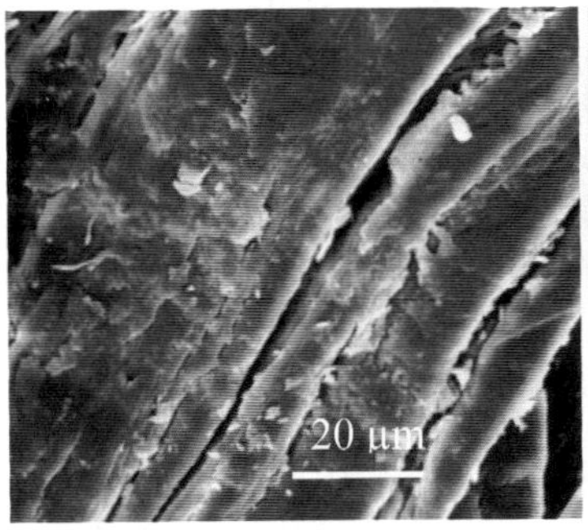

Figure 3. Untreated hemp waste

Figure 4. Hemp waste with 2 % NaOH solution treatment

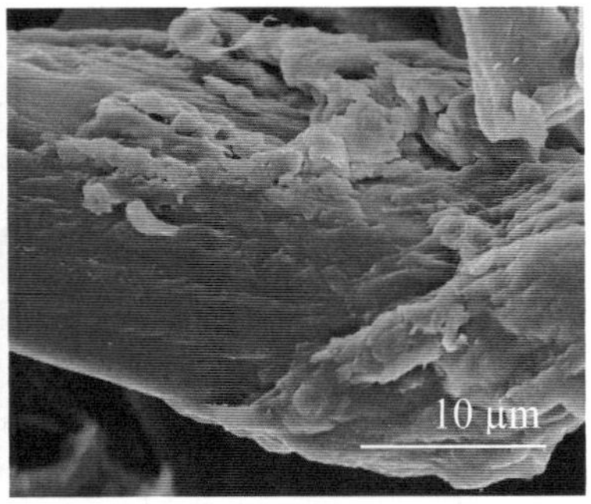

Figure 5. Hemp waste with 2 % NaOH treatment and silanized with 2 mol APTES

The hemp waste surface modified with sol-gel method forms a silica coating on the surface, but silanization also penetrates into some deeper layers and creates a smooth surface (figure 5).

Summary

The highest elastic module and the lowest elongation at break, from all eight composites, shows that composites with alkali pre-treatment and waste silanization are effective for hemp waste treatment to increase composite mechanical properties. The high elastic module and low elongation at break also show composites with alkali treated waste and ZnO 2 wt% additive.

To decrease hemp waste reinforced composite water absorption, alkali pre-treatment could be combined with the sol-gel technology. The composites with the lowest water absorption are made from hemp waste pre-treated sequentially with 2 wt% alkali and 0.14 mol. TEOS sol. Waste pre-

IOP Conf. Series: Materials Science and Engineering **500** (2019) 012003 doi:10.1088/1757-899X/500/1/012003

treatment with 0.09 mol. TEOS and 7.5 wt% Zn acetate dehydrate precursor sol moderate increase water sorption still not exceeding 8 %, at the same time elastic module of composite increase too; these composites are suitable for outdoor use. Waste silanization leads to the water absorption capability increase above 12 wt% of LLDPE composites with the hemp waste content 40 wt%, and this treatment can't be recommended to reduce moisture absorption.

As we can see, it is not possible to achieve high mechanical composite properties and low water absorption with the same waste treatment methods, therefore, the purpose of using composites and the use of the most suitable processing combinations should be considered.

References

[1] Karahan M, Masood Z, Nawab Y and Karahan N 2018 Development and Characterization of Hybrid Green Composites from Textile Waste *Advances in Ergonomics of Manufacturing: Managing the Enterprise of the Future* (Springer International Publisher) ed S. Trzcielinski , pp 37-49

[2] Todor M P, Bulei C, Heput T and Kiss I 2018 Researches on the development of new composite materials complete / partially biodegradable using natural textile fibers of new vegetable origin and those recovered from textile waste *IOP Conf. Ser.: Mater. Sci. Eng.* **294** 012021

[3] Vu N D, Tran H T and Nguyen T D 2018 Characterization of Polypropylene Green Composites Reinforced by Cellulose Fibers Extracted from Rice Straw *International J. of Polymer Sci.* **2018** 1813847

[4] Ngo T T, Kohl J G, Paradise T, Khalily A and Simonson D L 2015 Improving Mechanical Properties of Thermoset Biocomposites by Fiber Coating or Organic Oil Addition *International J. of Polymer Sci.* **2015** 840823

[5] Zelca Z, Vihodceva S and Kukle S 2017 Sol-Gel Coating Processing Optimisation for Natural Fibers *Proc. of the Estonian Academy of Sci.* (Estonian Academy Publishers) **66(4)** 467-472

[6] Nadalia E, Layeghib M, Ebrahimib G, Naghdia R, Jonoobib M, Khorasanic M M and Mirbagherid Y 2018 Effects of Multiple Extrusions on Structure-property Performance of Natural Fiber High-density Polyethylene Biocomposites *Mat. Res.* **21(2)** e20170301

[7] Zelca Z, Kukle S and Kajaks J 2016 The Hemp Primary Processing Waste and Linear Low Density Polyethylene Composite *16th Intern. Multidisciplinary Sci. GeoConf. SGEM 2016*, (Austria, Sofia: STEF92 Technology Ltd.) pp 55-60

[8] Zelca Z, Kukle S, Kajaks J and Geikina-Geimane M 2017 Hemp Fibers Waste and Linear Low Density Polyethylene Composite Properties *Key Eng. Mater.* **721**

IOP Conf. Series: Materials Science and Engineering **500** (2019) 012004 doi:10.1088/1757-899X/500/1/012004

Properties of polyurethane coatings based on linseed oil phosphate ester polyols and expandable graphite

Vladimir Yakushin, Arnis Abolins, Dzintra Vilsone, Irina Sevastyanova

Latvian State Institute of Wood Chemistry, 27 Dzerbenes, LV-1006, Riga, Latvia

E-mail: yakushin@edi.lv

Abstract. The effect of two grades of expandable graphite (EG 096 and EG 290) on the thermal decomposition and flammability of polyurethane (PU) coatings, based on two kinds of linseed oil phosphate ester polyols, was studied. Thermogravimetric analysis (TGA) and the cone calorimeter test at a heat flux of 35 kW/m^2 were used for this purpose. The chemical structure of the synthesized polyols was identified by Fourier Transform Infrared (FTIR) and Nuclear Magnetic Resonance (^{31}P NMR) spectroscopy. The content of EG was varied from 0 up to 25%. For all characteristics of PU coatings, the maximum effect was reached at 25% of EG. It was found that PU coatings filled with EG 096 had higher char residue at thermal decomposition under nitrogen. The flame retardancy of the PU coating filled with EG 290, which generated more blowing gases at thermal decomposition and therefore had a higher expansion volume, was higher than the flame retardancy of the coatings filled with EG 096. The best characteristics were shown by filled PU coatings based on the polyol, which was phosphorylated in the presence of isopropyl alcohol and which, besides phosphate mono- and diesters, had phosphate triester and pyrophosphate monoester.

1. Introduction

To decrease the flammability of bio-based polyurethanes (PUs), phosphorus-containing polyols are used due to the strong carbonization effect of phosphorus in the condensed phase and the free radical quenching effect in the gaseous phase of polymer combustion [1]. For the synthesis of such polyols, different methods can be used [2, 3]. One of them is the phosphorylation of epoxidized vegetable oils, when firstly, the oil is oxidized with peroxyacids, and then ring-opening of the epoxy group is performed using phosphoric acid. The phosphorylation of epoxidized oils was investigated mainly for epoxidized soybean oil (ESO). Zhong, Guo et al. showed that PU coatings and foams based on phosphorylated ESO had not only good adhesive and mechanical characteristics [4–7], but also some decrease in the burning rate [8] and increase in the oxygen index [9] of PU foam samples.

To achieve a potentially greater effect on the flammability characteristics of PU coatings, epoxidation and phosphorylation of linseed oil (LO) has been investigated in a previous study [10]. Due to the higher content of double bonds in LO, it was possible to obtain epoxidized LO (ELO) with a higher value of epoxy equivalent to that of ESO and then, by phosphorylation, to incorporate more phosphate groups in the ester polyol structure. The main parameters of the flame retardancy of coatings, based on the synthesized polyols, were increased, but only up to 15–30%. To further enhance the flame retardancy of coatings, additional flame retardants would be used.

Among non-halogen flame retardants, expandable graphite (EG) is one of the most effective intumescing additives. It has an exfoliated structure intercalated with sulphuric acid into a graphite crystal structure. At high temperatures, the intercalated agent releases gases and graphite, expanded in a perpendicular direction to the carbon layers in the crystal structure [11]. Most of the studies on the use of EG in urethane systems refer to PU foams. The effects of the particle sizes, expansion volume and percentage of EG on PU and PIR foams' thermal degradation and flammability have been investigated in numerous works [12–15]. Duquesne et al. [16, 17] investigated the thermal degradation of PU coatings with EG and the mechanism of flame retardancy induced by EG in PU. Based on these studies, it has been concluded that the retardancy of EG results from the physical action rather than from the chemical one.

At the same time, the combination of EG with phosphorus-containing compounds may demonstrate the synergistic effect on polyurethane flame retardancy [18, 19]. In the present work, we will investigate the effect of two grades of EG, which differed in particle size and expansion volume, on the thermal degradation and flame retardancy of the PU coatings based on linseed oil phosphate ester polyols.

2. Experimental

2.1. Materials

For the synthesis of phosphorus-containing polyols, refined LO from Alberdingk Boley GmbH, Germany, and glacial acetic acid (CH_3COOH), 35% hydrogen peroxide (H_2O_2), ion exchange resin Amberlite IR-120 H, 85% phosphoric acid (H_3PO_4), isopropyl alcohol (IPA), diethylene glycol butyl ether (DGBE), and ethyl acetate (EtOAc), purchased from Sigma-Aldrich Chemie GmbH, Germany, were used.

For PU coating preparation, polymeric diphenylmethane diisocyanate (p-MDI), grade Voratec SD 100 (NCO content 31.5%, functionality 2.7) from Dow Deutschland GmbH, Germany; Dabco T-12 Catalyst (dibutyltin dilaurate) from Fluka Analytical, Switzerland; toluene puriss ($\geq$ 99.7%) from Sigma-Aldrich Chemie GmbH were used. Two grades of EG, EG 290 and EG 096, supplied by SINOGRAF SA, Poland, were used as the flame retardant.

2.2. Polyols synthesis and characterization

For comparison purposes, two kinds of polyols were synthesized in two stages. First, LO was epoxidized using 35% H_2O_2, CH_3COOH and ion exchange resin Amberlite IR-120 H as the catalyst. Then, phosphorylation of ELO with 2 wt% of H_3PO_4 was carried out in the presence of IPA (IPA polyol) or DGBE (DGBE polyol). The completion of the synthesis stages was monitored by the changes in the epoxy equivalent of the reaction products. The synthesis of polyols as well as the procedure of polyols washing, drying and separation from the water and solvents have been described in detail in a previous work [12]. It was shown that PU coatings based on IPA and DGBE polyols had different mechanical and thermal stability characteristics, but approximately the same flame retardancy.

The epoxy equivalent, hydroxyl (OH) number, and acid value of the polyols were determined according to ISO 3001:1999, ISO 14900:2001, and ISO 2114:2000 standards, respectively. The dynamic viscosity of polyols at 20°C was measured using a rotation HAAKE Viscotester 6L/R plus (Thermo Electron GmbH, Karlsruhe, Germany). The content of phosphorus was determined according to the AOCS Official Method Ca 12-55.

The chemical structure of the synthesized polyols was identified by FTIR spectroscopy using a Nicolet iS50 FTIR Spectrometer (Thermo Scientific Inc.) and ^{31}P NMR spectroscopy using Varian 400-MR (161.9 MHz), and software Mnova 11.0 (Mestrelab Research).

2.3. Preparation and testing of polyurethane coating

Preparing PU coatings, EG was preliminary mixed in a 50% toluene solution of polyol. The content of EG was varied from 5% to 25% of the total coating mass. The premixed polyol composition was then mixed with p-MDI at a NCO/OH molar ratio of 1.1:1. For the thermogravimetric test, PU coatings were prepared in the form of free films, ~300 μm in thickness. For the flammability test, PU coatings with the same thickness were applied on standard wood (pine) samples (100 × 100 × 16 mm). The consumption of the PU composition upon its applying to wood samples was controlled by the weight method (400 g/m^2). All PU coatings were hardened at a temperature of 21 ± 2°C for 7 days.

The thermal decomposition of the coatings was studied by TGA under a nitrogen atmosphere using TGA/SDTA 851e Mettler Toledo. The samples (about 8 mg) were heated in aluminium oxide crucibles from room temperature up to 800°C at a heating rate of 10°C/min and a gas flow rate of 20 cm^3/min. The fire behaviour of wood samples with PU coatings was studied using a FTT Cone Calorimeter (Fire Testing Technology Ltd.). Testing was performed at a heat flux of 35 kW/m^2, which is recommended in ISO 5660-1:2015 for exploratory testing. Test duration was 30 min.

3. Results and discussion

3.1. Characteristics of polyols

Both synthesized polyols were light yellow liquids. The OH number of DGBE polyol (295 mg KOH/g) was slightly higher than that of IPA polyol (283 mg KOH/g) due to the fact that the acid value of DGBE polyol also was higher (23.2 mg KOH/g in contrast to 11.5 mg KOH/g). The viscosity of DGBE polyol was 1200 mPa·s, while that of IPA polyol was much higher (26900 mPa·s). Such a big difference in viscosity was explained later owing to spectroscopic data. The content of phosphorus in the DGBE polyol (0.45%) was slightly lower than that in the IPA polyol (0.50%).

FTIR spectra of both IPA and DGBE polyols had the absorption bands of the P=O stretching vibration at 1140 cm^{-1} and the P–O–C stretching vibration at 1024 cm^{-1} for IPA polyol and 1029 cm^{-1} for DGBE polyol. In other ranges of spectra, absorption bands spectra of both polyols were similar. The difference in the polyols' structure was identified only by the ^{31}P NMR spectra:

DGBE polyol ^{31}P-NMR (162 MHz, δ, ppm): 15.90 [O=P(OR)(OH)$_2$]; 1.07 [O=P(OR)$_2$(OH)]; 0.021 [O=P(OH)$_3$].

IPA polyol ^{31}P-NMR (162 MHz, δ, ppm): 15.90 [O=P(OR)(OH)$_2$]; 1.07 [O=P(OR)$_2$(OH)]; 0.021 [O=P(OH)$_3$]; -0.91 [O=P(OR$_3$)]; -6.7 [O=P(OH)$_2$–O–P(OH)$_2$=O].

The ^{31}P NMR spectra of both polyols had shifts at 15.90 and 1.07 ppm, assigned to phosphate mono- and diesters, and the chemical shift at 0.021 ppm was assigned to the traces of unreacted free phosphoric acid. The IPA polyol ^{31}P NMR spectrum, besides these chemical shifts, had two more signals at -0.91 and -6.7 ppm, related to phosphate triester and pyrophosphate monoester. The reaction scheme of the formation and representative structure of the mentioned phosphate esters is presented in our previous work [12]. The difference in the polyols' composition can be explained by the different polarity of the solvents. Due to the presence of phosphate triester and pyrophosphate monoester, which are formed in the medium of the less polar IPA, the viscosity of the IPA polyol was much higher.

3.2. TGA of polyurethane coatings

Two grades of expandable additives, EG 096 (EG1) with a particle size of 0–0.2 mm and EG 290 (EG2) with a particle size of 0.2–0.6 mm, had a declared expansion volume of 50–90 ml/g and 200–300 ml/g, respectively. Camino et al. [20] have shown that the graphite expansion process at thermal decomposition takes place due to the redox reaction between H_2SO_4 and carbon, which generates the blowing gases CO_2, SO_2 and H_2O. These gases escape through the edges of the graphite particles, leading to their irreversible expansion.

The expandability of EG, as follows from the manufacturer's declared characteristics and literature data, depends mainly on the particle size. As the particle size of EG gets smaller, it is difficult to keep and seal significant amounts of sulphuric acid inside the particles [21]. The experimental TGA curves of EG1 and EG2 under nitrogen follow this rule. The EG1 with smaller particles at thermal decomposition generated a smaller quantity of blowing gases and at 800°C lost only 14.8% of mass (Fig. 1). The mass loss of EG2 at this temperature was 23.7%. The redox and expansion process of these kinds of EG occurs mainly in the temperature range of 210–450°C and presents the maximal rate of mass loss at 233°C for EG1 and 235°C for EG2 (Figs. 3, 4).

To compare the effect of EG1 and EG2 filled PU coatings based on DGBE polyol, polyols with a lower viscosity were prepared. Afterwards, taking into account the results of the combustibility test, EG2 also was used only for filling of IPA coatings. TGA curves for DGBE polyol based neat PU coatings and those filled with 25 wt% of EG are presented in Fig. 1. The IPA coating had analogues TGA curves.

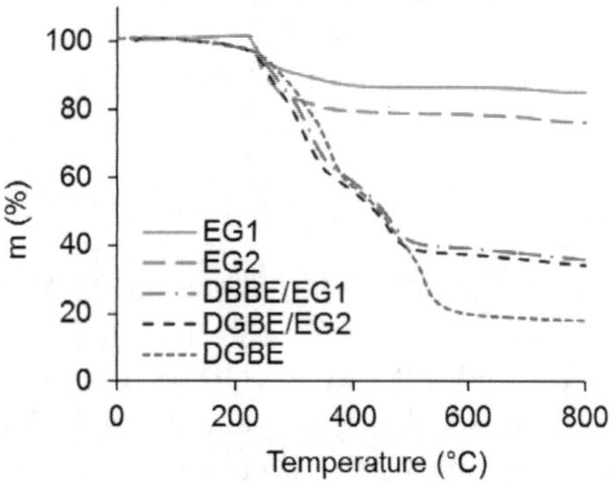

Figure 1. TGA curves for EG1, EG2, and filled with EGs and neat DGBE coatings.

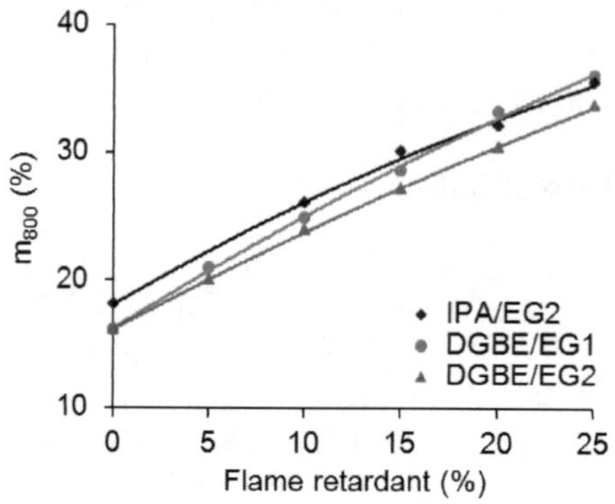

Figure 2. Effect of the EG content on the char residue at 800°C of PU coatings.

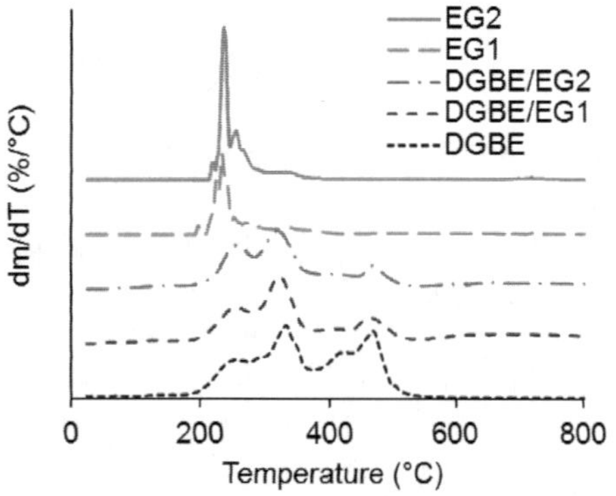

Figure 3. DTG curves for EG1, EG2, and filled with EGs and neat DGBE coatings.

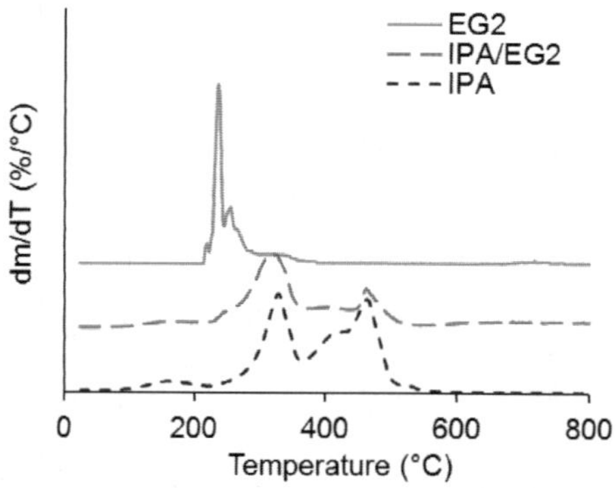

Figure 4. DTG curves for EG2, and filled with 25 wt% of EG and neat IPA coatings.

DTG curves for neat and filled coatings based on DGBE and IPA polyols are presented in Figs. 3 and 4, respectively. The thermal decomposition of neat PU coatings had three steps of mass loss. The first step took place in the temperature range of 215–370°C for DGBE and 230–370°C for the IPA coating, and presents the maximum rate of mass loss (T_{1max}) at 330°C and 326°C for DGBE and IPA coatings, respectively. This step is assigned to the primary scission of the urethane linkage and the phosphate ester bond [1]. In filled coatings in this temperature range, the redox and expansion process of EG also takes place. As a result of the summation of these processes at increasing EG content, T_{1max} gradually decreased and reached its minimum value at 25 wt% (318°C and 320°C for DGBE and IPA coatings, respectively).

The second and the third steps of PU thermal decomposition, attributed to the polyester degradation and char residue formation, take place in the temperature range of 370–440°C and 440–540°C [10]. The char residue gradually increased with the addition of EG, reaching the maximum values at their content of 25 wt% (Fig. 2). The char residue at 800°C of the DGBE coating filled with 25% of EG1 (36%) was higher than the residue of the coating with 25% of EG2 (34%) due to smaller mass loss of

EG1. The char residue at 800°C of IPA coatings with 25% of EG2, despite the greater mass loss of EG2, was practically the same (36%) due to the slightly higher content of phosphorus and the presence of phosphate triester and pyrophosphate monoester in IPA polyol.

3.3. Cone calorimeter test of polyurethane coatings

The fire behaviour of wood samples with applied PU coatings was studied in the cone calorimeter test and was characterized in terms of mass loss (ML), peak of heat release rate (PHRR), total heat release (THR), mean effective heat of combustion (EHC), maximum average rate of heat emission (MARHE) and total smoke release (TSR).

Figure 5 presents HRR curves of wood samples with a neat DGBE and IPA coatings and those filled with 25 wt% of EGs, at the initial stage of combustion. The first stage of samples' combustion (0–100 s) was attributed to the combustion of volatile products of thermal decomposition of PU coatings and wood. This stage of combustion was accompanied by the maximum HRR. With loading of EG, the PHRR of coatings gradually decreased (Fig. 6) and reached their minimum values at 25 wt% of EG. Time to the peak of HRR at increasing content of FR also decreased due to the thermal decomposition and high ability to absorb infrared radiation and thermal conductivity of EG [21]. The effect of EG1 loading on PHRR was lower than that of EG2. The PHRR of coatings with 25 wt% of EG2 decreased 2.5-fold, compared to the case of neat coatings.

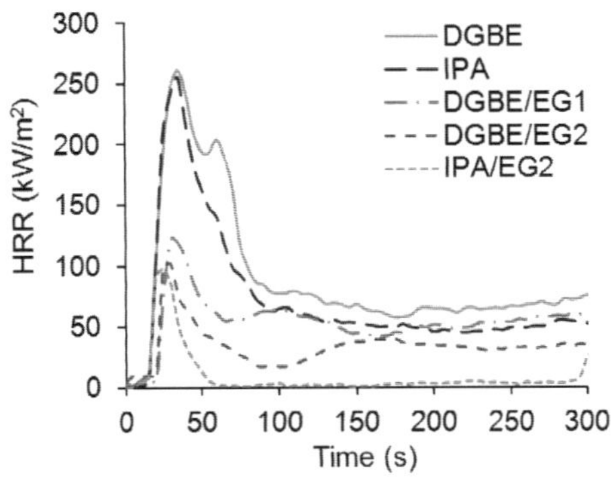

Figure 5. HRR curves of samples with neat and filled PU coatings. EG content 25 wt%.

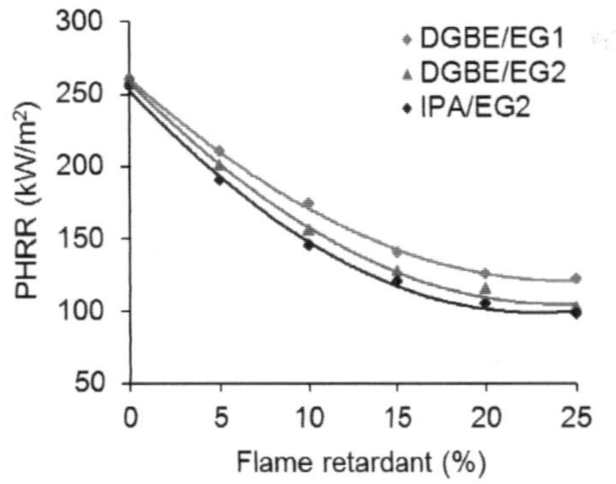

Figure 6. Effect of the EG content on the PHRR of samples with PU coatings.

With further exposure to heat flux, wood samples with the neat PU coating and the coatings with lower content of EG continued to burn, and extinguished only after 1300–1500 s. The coating with the EG content higher than 15 wt%, after ignition and the expansion of graphite, extinguished. However, after 300–400 s, the samples with the protective EG layer ignited again, because the pyrolysis of wood under the loose protective layer at constant heat flux exposition was continued, and flammable gases went out through this layer. Only some samples with the IPA coating filled with 25 wt% of EG2 did not ignite again.

Total and mean values of the main flammability parameters of wood samples with PU coatings for the whole period of the test time (0–1800 s) were decreased at increasing of EG content, like PHRR. All these parameters for samples with neat and filled coatings are listed in Table 1. The flammability parameters of samples filled with 25 wt% of EG were significantly lower than these parameters of wood samples without coatings. The effect of EG1 with lower expansion volume of flame retardancy was lower than that of EG2 with a higher expansion volume. Due to the presence of phosphate triester and pyrophosphate monoester in IPA polyol, the IPA coating filled with 25 wt% of EG2 had the lowest values of PHRR, mHRR, THR, mEHC, MARHE, ML and TSR, compared to filled DGBE coatings.

Table 1 Flammability parameters of wood samples with IPA and DGBE coatings

Flame retardant	0%		25% EG2		25% EG1	Wood
Coating	IPA	DGBE	IPA	DGBE	DGBE	-
ML (%)	87	88	68	77	81	92
PHRR (kW/m²)	255	260	98	102	122	190
THR (MJ/m²)	120	124	56	82	100	131
mEHC (MJ/kg)	13.7	13.6	8.7	10.8	12.7	14.5
MARHE (kW/m²)	112	124	42.8	52.5	71.7	104
TSR (m²/m²)	522	495	73.7	85	146	180

4. Conclusions

PU coatings reached the maximum char residue in the TGA test under nitrogen atmosphere at 25 wt% of EG. Due to a smaller mass loss of EG 096, the PU coating filled with it had a higher value of the char residue than the PU coating filled with EG 290. At the same time, EG 290, which at thermal decomposition generated more blowing gases and as a result had a higher expansion volume, had higher flame retardancy than EG 096. Minimum values of flammability parameters for all coatings were reached at the maximum content of EG (25 wt%). Due to the presence of phosphate triester and pyrophosphate monoester in the structure, IPA coatings filled with 25 wt% of EG 290 had the lowest values of all flammability parameters in the cone calorimeter test.

Acknowledgement

The present study was supported by the National Research Programme "Forest and earth entrails resources: research and sustainable utilization – new products and technologies" (ResProd).

References

[1] Chattopadhyay D K and Webster D C 2009 *Prog Polym Sci.* **34** 1068
[2] Illy N, Fache M, Ménard R, Negrell C, Caillol S and David G 2015 *Polym Chem.* **6** 6257
[3] Byard B, Wang K, Morgan A and Benin V 2018 *Fire Mater.* **42** 3
[4] Zhong B, Shaw C, Rahim M and Massingill J 2001 *J Coat Technol.* **73** 53
[5] Guo Y, Hardesty J H, Mannari V M and Massingill J L 2006 *J Coat Technol Res.* **3** 327
[6] Patel M M, Patel B P and Patel N K 2012 *Int J Plast Technol.* **16** 67
[7] Fan H, Tekeei A, Suppes G J and Hsieh F-H 2012 *Int J Polym Sci.* **907049** 1
[8] Norzali NRAA, Badri K H, Ping O S and Ahmad I 2014 *J Anal Sci.* **18** 456
[9] Heinen M, Engel A G and Petzhold C L 2014 *Polym Degrad Stab.* **108** 76
[10] Abolins A, Yakushin V and Vilsone D 2018 *J Renew Mater.* (In Print).
[11] Puri R G and Khanna A S 2017 *J Coat Technol Res.* **14** 1
[12] Modesti M, Lorenzetti A, Simioni F and Camino G 2002 *Polym Degrad Stab.* **77** 195
[13] Thirumal M, Khastgir D, Singha N K, Manjunath B S and Naik Y P 2008 *J Appl Polym Sci.* **110** 2586
[14] Kurańska M, Cabulis U, Auguścik M, Prociak A, Ryszkowska J and Kirpluks M 2016 *Polym Degrad Stab.* **127** 11
[15] Lorenzetti A, Dittrich B, Schartel B, Roso M and Modesti M 2017 *J Appl Polym Sci.* **134** 1
[16] Duquesne S, Le Bras M, Bourbigot S, Delobel R, Camino G, Eling B, Lindsay C and Roels T 2001 *Polym Degrad Stab.* **74** 493
[17] Duquesne S, Le Bras M, Bourbigot S, Delobel R, Vezin H, Camino G, Eling B, Lindsay C and Roels T 2003 *Fire Mater.* **27** 103
[18] Chen X-Y, Huang Z-H, Xi X-Q, Li J, Fan X-Y and Wang Z 2018 *Fire Mater.* **42** 447
[19] Yang S, Wang J, Huo S, Wang M, Wang J and Zhang B 2016 *Polym Degrad Stab.* **128** 89
[20] Camino G, Duquesne S, Delobel R, Eling B, Lindsay C and Roels T 2001 Mechanism of expandable graphite fire retardant action in polyurethanes *Fire and Polymers, American Chemical Society*, pp 90–109
[21] Shi L, Li Z, Xie B, Wang J, Tian C and Yang M 2006 *Polym. Int.* **55** 862

IOP Conf. Series: Materials Science and Engineering **500** (2019) 012005 doi:10.1088/1757-899X/500/1/012005

Effect of extraction on the coatings produced from nanoparticle gels obtained from hardwood and softwood bark

M Laka[1,*], M Skute[1], S Janceva[2], V Fridrihsone[1,3], I Filipova[1], L Vecbiskena[1] and U Grinfelds[1]

[1]Laboratory of cellulose, Latvian State Institute of Wood Chemistry, 27 Dzerbenes St., LV – 1006, Riga, Latvia
[2] Laboratory of lignin, Latvian State Institute of Wood Chemistry, 27 Dzerbenes St., LV – 1006, Riga, Latvia
[3]Institute of Polymer Materials, Faculty of Material Science and Applied Chemistry, Riga Technical University, 3 Paula Valdena St., Riga, LV-1048, Latvia

*lamar @edi.lv

Abstract. For coatings, non-extracted hardwood (birch) and softwood (pine) bark and that extracted in biorefinery were used. Lyophilic extractives (fats, waxes, etc.) were extracted with hexane and hydrophilic extractives (polyphenols, tannins, etc.) by means of successive extraction with solvents with increasing polarity, i.e. hexane, ethyl acetate, and ethanol water (1:1). The bark was destructed by the thermocatalytic destruction method and then dispersed in a water medium at a concentration of 8-10%. As a result, gel-like dispersions were obtained, which contained nanoparticles. It has been established that the coatings improve the Gurley air resistance and increase the mechanical properties (tensile strength, burst strength, stretch) of paper sheets. The coatings decrease the tensile strength in a wet state. The coatings made from extracted bark nanoparticle gels affect the properties of paper sheets to a greater extent. It is because extractive substances hinder the bond formation between the cellulose fibres and nanoparticles.

1. Introduction

Recently, various coatings have been used for improving the barrier and mechanical properties of papers. Thus, chitosan-caseinate bilayer coatings lead to a decrease in the water vapour permeability of paper sheets [1]. Chitosan coatings improve also the gloss, oxygen barrier properties and water absorption capacity [2, 3]. The use of microfibrillated cellulose as a surface coating on various papers considerably reduces the air permeability and improves the oil barrier properties of papers [4]. In our work [5], we also have established that the coatings made from bark nanoparticle gels and their mixtures with chitosan and sodium carboxymethylcellulose (Na-CMC) solutions improve the barrier and mechanical properties of paper sheets.

In this work, the effect of extraction of bark on the mechanical and barrier properties of paper sheets with coatings produced from hardwood and softwood bark nanoparticle gels was investigated.

2. Experimental

For coatings, nanoparticle gels from hardwood (birch) and softwood (pine) bark were obtained. Non-extracted bark and that extracted in biorefinery [6] were used. Lyophilic extractives (fats, waxes, etc.) were extracted with hexane in a Soxhlet apparatus and hydrophilic extractives (polyphenols, tannins, etc.) by means of successive extraction with solvents with increasing polarity, i.e. hexane, ethyl acetate, and ethanol water (1:1). The bark was destructed by the thermocatalytic destruction method developed at the Laboratory of Cellulose of the Latvian State Institute of Wood Chemistry [7]. The bark was impregnated with a thermocatalytic destruction catalyst – weak hydrochloric acid solution – and then thermally treated at a temperature of 120 °C until a dry state. The partially destructed bark was dispersed in a water medium in a ball mill at a concentration of 8-10%. As a result, gel-like dispersions were obtained, which contained nanoparticles, average size of ~300 nm. Size of particles was determined by Zetasizer Nano ZS90 (Malvern Panalytical Ltd, UK), where measurements were made using dynamic light scattering (DLS) method, and therefore were influenced more by longitudinal size of particle. Diameters of these particles usually are smaller than 100 nm, however it cannot be correctly measured by DLS.

Using the obtained nanoparticle gels, coatings were made on both sides of paper sheets (made from 100% recycled fibres, grammage 90 g/m^2, thickness 0.11 mm) produced by the Ligatne Paper Mill (Latvia). Coatings with area of 0.04725 m^2 were made using K-Control Coater 202 (RK Print Instruments Ltd., UK). Gel concentration was 4-10%. In the suspension form, the coatings' thickness was 24, 40, 60 and 100 μm. Owing to the partial diffusion of the gel into the paper pores and water evaporation, the coatings' thickness after drying decreased. Final thickness was determined as a difference (divided by two) in the thicknesses of uncoated and coated paper.

Air resistance and mechanical properties of paper sheets were investigated. Air resistance according to Gurley (SCAN-P) was determined using an L&W Air Permeance Tester. Tensile strength was determined on a Frank Tensile Tester (Frank-PTI, Austria) according to the International Standard ISO 1924-1:1992(E). Burst strength was determined on a Frank Burst Tester for Paper (Frank PTI, Austria) according to the International Standard ISO 2758-1983(E). Determination of the tensile strength in a wet state was based of the TAPPI Standard 220 om-87. The StD of mechanical indices was below ± 4%.

3. Results and Discussion

3.1. Effect of coatings obtained from bark nanoparticle gels on the air resistance of paper sheets

Figure 1 demonstrates the Gurley air resistance of paper sheets versus thickness of coatings made from 6% non-extracted and extracted birch and pine bark nanoparticle gels. In the case of non-extracted bark, air resistance slightly increases with increasing thickness of coatings. In the case of extracted bark, the air resistance considerably increases with increasing coating thickness. It increases especially greatly in the case of birch bark. Thus, at the coating thickness of 30 μm, the air resistance of the coatings made from extracted birch and pine bark nanoparticle gels increases by 177 and 28 %, respectively, in comparison with the case of uncoated paper.

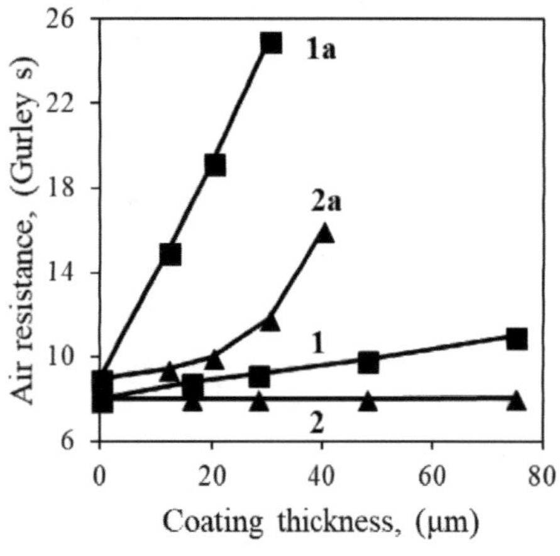

Figure 1. Gurley air resistance versus thickness of coatings made from 6% non-extracted and extracted (a) birch (1) and pine (2) bark nanoparticle gels.

Similarly, Gurley air resistance increases with increasing coating weight (Figure 2). At the coating weight of 0.45 g, Gurley air resistance increases for coatings made from non-extracted birch and pine bark nanoparticle gels by 20 and 9%, respectively, and for those from extracted ones – by 62 and 47%.

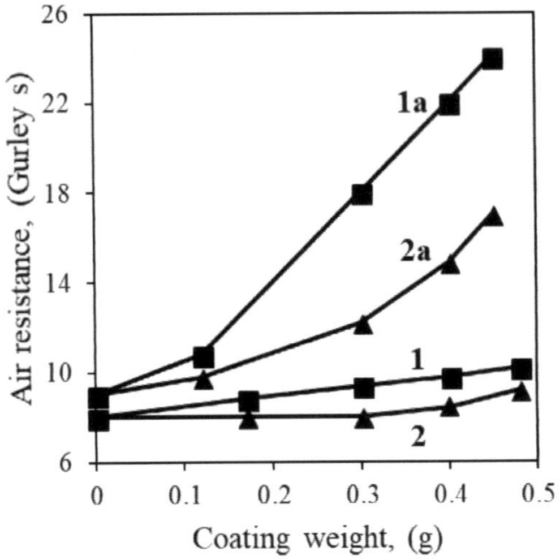

Figure 2. Gurley air resistance versus weight of coatings made from 6% non-extracted and extracted (a) birch (1) and pine (2) bark nanoparticle gels.

The higher Gurley air resistance of the paper sheets with coatings made from bark nanoparticle gels can be explained by the fact that nanoparticles cover the micro- and submicrovoids between the fibres and fibrils, and improve the barrier properties. The greater effect of the coatings made from extracted bark nanoparticle gels is because the extractive substances hinder the bond formation between the cellulose fibres and nanoparticles.

The effect of coatings on the Gurley air resistance increases with the nanoparticle gel concentration (Figure 3). Thus, for coatings made from extracted birch bark and pine bark nanoparticle gels (coating thickness in the suspension form was 100 μm) at a concentration of 4 %, Gurley air resistance increases by 117 % and 55 %, but at a concentration of 10 %, by 180 % and 120 %, respectively, in comparison with uncoated paper. For coatings made from non-extracted birch and pine bark nanoparticle gels, Gurley air resistance differs a little and, with increasing concentration, increases slightly. At a concentration of 10 %, it increases by 37 % on the average.

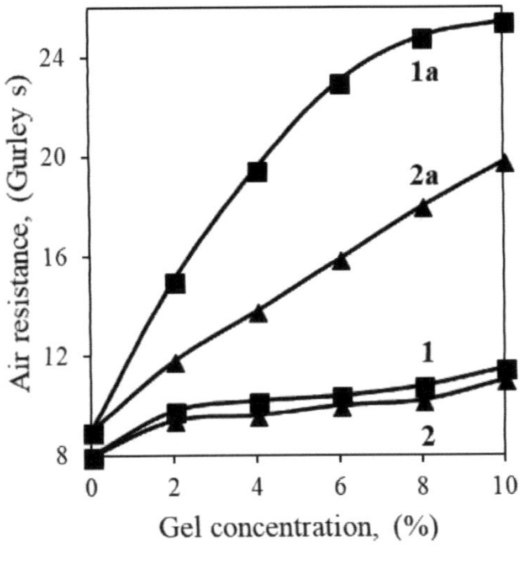

Figure 3. Gurley air resistance versus the concentration of the nanoparticle gel made from non-extracted and extracted (a) birch (1) and pine (2) bark.

3.2. Effect of coatings obtained from bark nanoparticle gels on the mechanical properties of paper sheets

The coatings made from extracted bark nanoparticle gels affect also the mechanical properties of paper sheets to a greater extent in comparison with those made from non-extracted bark nanoparticle gels. Figure 4 demonstrates the tensile strength in a dry state and burst strength of paper sheets versus coating thickness for coatings from birch bark (4a) and pine bark (4b) 6 % nanoparticle gels. The tensile strength of paper sheets with coatings from the gels obtained from non-extracted bark nanoparticle gels practically does not change with increasing coating thickness. In the case of coatings made from extracted bark nanoparticle gels, tensile strength increases with increasing coating thickness. At a coating thickness of 30 μm, tensile strength increases by 9.5 % for birch bark, and by 5.0 % for pine bark. Burst strength increases more with increasing coating thickness. At a coating thickness of 30 μm, burst strength for coatings made from non-extracted birch and pine bark nanoparticle gels increases by 21.4 and 11.5 %, respectively, and for coatings made from extracted bark by 46.2 and 22.8 %.

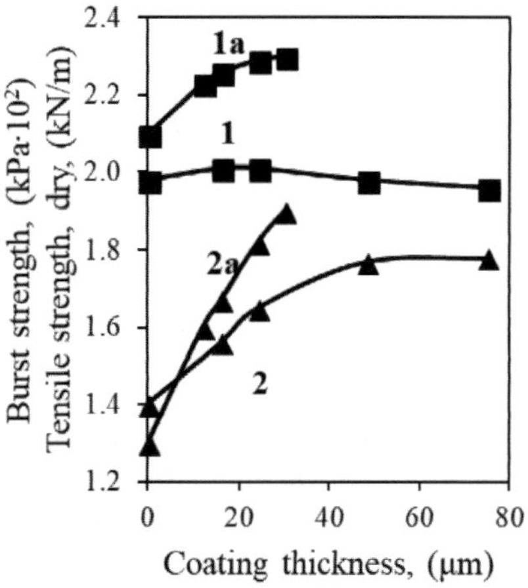

(a)

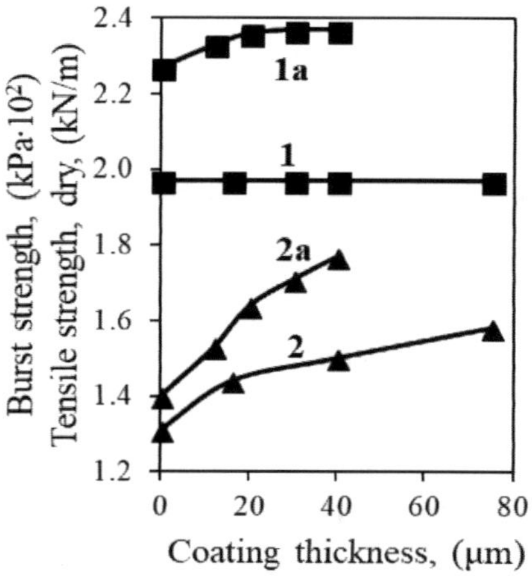

(b)

Figure 4. Tensile strength (1;1a) and burst strength (2;2a) versus thickness of coatings made from 6% non-extracted and extracted (a) birch (Fig. 4a) and pine (Fig.4b) bark nanoparticle gels.

Figure 5 demonstrates tensile strength and burst strength versus the concentration of non-extracted and extracted birch (5a) and pine (5b) bark nanoparticle gels (coating thickness in the suspension form was 100 μm). Tensile strength for paper sheets covered with coatings made from non-extracted and extracted bark nanoparticle gels practically does not change until 4-6 % and then decreases. Burst strength increases with increasing gel concentration until 6% and then also decreases (except for non-extracted pine bark). At a concentration of 6 %, burst strength increases in the case of non-extracted birch and pine bark by 25.0 and 15.0 %, respectively, and in the case of extracted bark by 44.6 and 28.6 %.

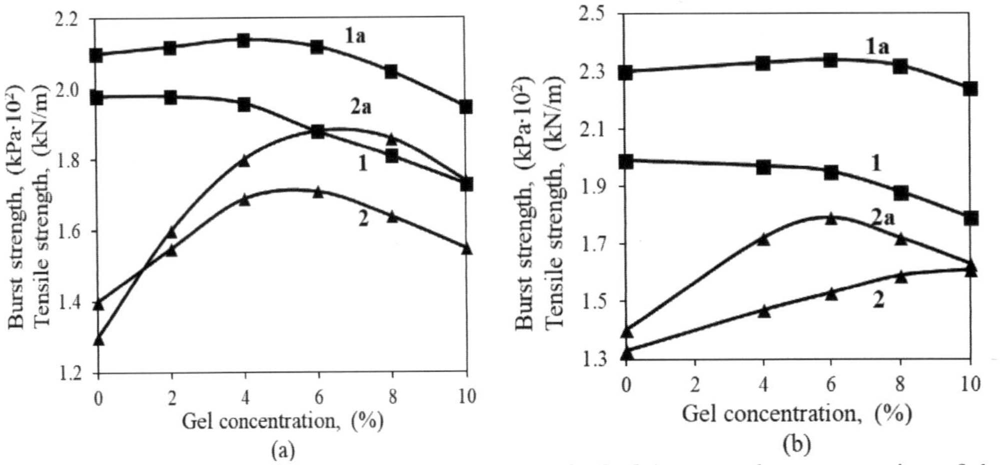

Figure 5. Tensile strength (1; 1a) and burst strength (2; 2a) versus the concentration of the nanoparticle gel made from non-extracted and extracted (a) birch (Fig. 5a) and pine (Fig. 5b) bark.

In a wet state, tensile strength decreases with increasing coating thickness (Figure 6) and gel concentration. This is probably connected with the greater hydrophilicity of the bark nanoparticle coatings in comparison with paper sheets.

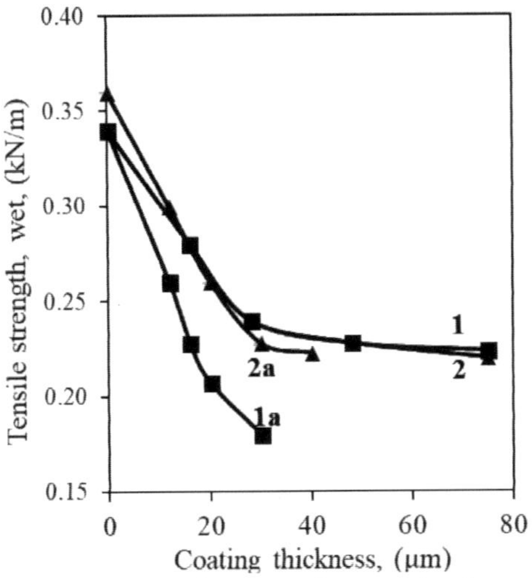

Figure 6. Tensile strength in a wet state versus thickness of coatings made from 6% non-extracted and extracted (a) birch (1) and pine (2) bark nanoparticle gels.

The coatings increase the stretch of paper sheets. Figure 7 demonstrates the stretch of the paper sheets covered with bark nanoparticle gel coatings versus coating thickness. Also, in this case, the nanoparticle coatings made from extracted bark increase the stretch of paper sheets to a greater extent. At a coating thickness of 30 μm, the coatings from non-extracted birch and pine bark 6% nanoparticle gels increase the stretch of paper sheets by 31.6 and 25.8 %, respectively, and from extracted bark by 79.2 and 37.5 %.

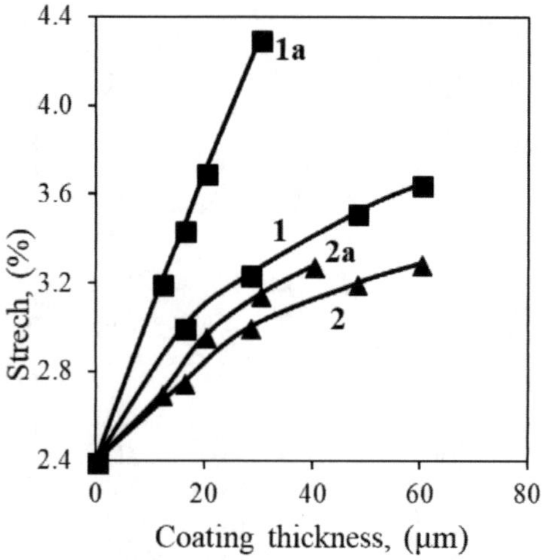

Figure 7. Stretch versus thickness of coatings made from 6% non-extracted and extracted (a) birch (1) and pine (2) bark nanoparticle gels.

4. Conclusion

Coatings were obtained from non-extracted and extracted hardwood and softwood bark nanoparticle gels, which were covered on both sides of paper sheets.

It was established that nanoparticle coatings improve the Gurley air resistance and mechanical properties (tensile strength in a dry state, burst strength) of paper sheets and increase their stretch. The effect of coatings depends on the coating thickness and gel concentration.

The greater effect of coatings, produced from extracted bark nanoparticle gels, can be explained by the fact that extractive substances hinder the bond formation between the cellulose fibres and nanoparticles.

For reinforcing of paper, the coatings from extracted bark nanoparticle gels can be used.

5. Acknowledgements

The study is financed by the Latvian State Programme “Res Prod” and Latvian State Institute of Wood Chemistry Bioeconomy grant „RenModCell”.

6. References

[1] Khwaldia K, Basta A H, Aloui H and El-Saied Chitosan-caseinate bilayer coatings for paper packaging materials 2014 *Carbohyd. Polym.* **99** 508-516.

[2] Vartiaine J, Motion R, Kulonen H, Ratto M, Skytta R and Ahvenainen R Chitosan-coated paper: Effects of nisin and different acids on the antimicrobial activity 2004 *J. Appl. Polym. Sci.* **94** 986-993.

[3] Reis A B, Yoshyda C M P, Reis A P C and Franco T T Application of chitosan emulsion as a coating on Kraft paper 2011 *Polym. Int.: Special Issue: Advances in Chitin and Chitosan Research* **6** 963-969.

[4] Aulin C and Gaelstedt M Oxygen and oil barrier properties of microfibrillated cellulose films and coatings 2010 *Cellulose* **17** 559-574.

[5] Laka M, Treimanis A, Chernyavskaya S, Skute M, Rozenberga L and Vikele L Micro-nanoparticles obtained from bark for their use alone and with chitosan and Na-CMC in paper coatings 2015 *Holzforschung* **69** 745-749.

[6] Telysheva G, Dizhbite T, Bikovens O, Ponomarenko J, Janceva S and Krasilnikova J Structure and antioxidant activity of diarylheptanoids extracted from bark of grey alder (*Alnus incana*) and potential of biorefinery-based bark processing of European trees 2011 *Holzforschung* **65** 623-629.

[7] Laka M and Chernyavskaya S, Latvian Republic Patent 11184 (1996).

Finite Element Simulation of Indentation Experiment on Branched Epoxy Novolac Resin

Andrejs Kovalovs[1], Andris Chate[1], Sergejs Gaidukovs[2], Arturs Medvids[3]

[1] Riga Technical University, Institute of Materials and Structures, Kipsalas St. 6A, LV-1048, Riga, Latvia

[2] Riga Technical University, Institute of Polymer Materials, Paula Valdena St. 3, LV-1048, Riga, Latvia

[3] Riga Technical University, Institute of Technical Physics, Paula Valdena St. 3, LV-1048, Riga, Latvia

andrejs.kovalovs@rtu.lv

Abstract. The main purpose of this study was to investigate the micro-indentation test method on the branched epoxy novolac resin (ENR) and to validate it by means of finite element simulation. At the first stage, the micro-indentation test was carried out with the purpose of establishment of the Young's modulus of ENR by the empirical relation to polymer materials. Experiments were performed by using the Vickers-type indenter. At the second stage, the finite element model was used to model the load-displacement curve of the material with the use of the initial values of elastic modulus and yield strength of material. The effectiveness of the use of these properties was investigated by using the planning of experiments and response surface technique. Full factorial design method was used to perform a series of numerical experiments. Then, a comparison of the results of micro-indentation test and finite element simulations was performed in order to compare and validate the values of Young's modulus obtained from two different techniques. Good correlation between the experimental and numerical Young's modulus was observed.

1. Introduction

The non-destructive experimental procedure is currently being widely studied by many researchers. The traditional test methods for measuring the mechanical properties of the materials, such us the Young's modulus or yield strength, are often destructive and need numerous sample materials.

The indentation test is an alternative method to establish the mechanical properties of the materials. This method is widely used to establish the hardness of the material, which is obtained by dividing maximum indentation load by the contact area. The method is non-destructive and inexpensive [1].

Many studies were carried out in order to find an empirical relationship between the hardness and mechanical properties of the materials [2]. Load-displacement curve, which was recorded during the test, showed sensitivity to the mechanical properties [3].

Some researchers are studying indention process by using Finite Element analysis. Application of FE analysis can be employed for investigation of the stress and strain field under indenter. The results obtained from numerical simulation can be compared with the experimental test to validate indentation process. The previous results show that the finite element simulation could be a good method for the establishment of the material hardness [4-6].

The main purpose of this study is to investigate the micro-indentation test method on branched epoxy novolac resin (ENR) and to validate it by finite element simulation. At the first stage, the micro-indentation test with Vickers indenter was carried out with the purpose of establishment of the

Young's modulus of ENR. At the second stage, the finite element model was used to model the load-displacement curve of the material with the initial values of elastic modulus. Then, a comparison between the results of micro-indentation test and FE simulation was performed to compare and validate the values of Young's modulus obtained from two different techniques.

2. Indentation Process

Microhardness was used as a very powerful applied technique in order to characterize the outcome of the finished coating photocuring process. Thereto, Vickers micro-indentation hardness (*HV*) of the polymer is related to the macroscopic mechanical properties, such as the empirical relation *HV*~E/10 for polymer materials [7]. For validation of this relation the finite element simulation of the micro-indentation test was performed.

According to [8], the Vickers hardness value was calculated by using the following equation:

$$HV = 0.189\frac{F}{d^2} \tag{1}$$

where F – the indentation force, N; d – average length of the two diagonal left by the indenter, mm.

Correspondingly, by knowing the Vickers micro-indentation hardness one can solve the inverse problem. The identification of mechanical properties of coating materials was performed by employing numerical experimental method based on experiment design method and approximation technique. 3D finite element simulation was carried out by using specimens and diamond indenter. The indenter was penetrating to the calculated depth with various values of elastic properties of the test material. The indentation forces obtained from numerical simulation were compared with those obtained from the experiments to minimize the gap between the experimental and numerical parameters of response.

The depth of contact of the indenter in the test material was established from the following equation:

$$h = \frac{d}{2\tan\frac{136°}{2}} \tag{2}$$

where d was calculated from the Vickers hardness value:

$$d = \sqrt{0.189\frac{F}{HV}} \tag{3}$$

3. Materials and Methods

3.1. Materials

The branched poly (phenyl glycidyl ether)-co-formaldehyde obtained from Advanced Polymer Coatings was used as ENR for the photopolymerization synthesis. The resin's viscosity is 2,000 cPs, and the specific gravity is 1.21 g/cc. The UV 1242 cationic photoinitiator (PI) bis (4-dodecylphenyl) iodonium hexafluorantimonate was obtained from Deuteron GmbH. Hereafter, PI for the photoinitiator will be used in the text. As a reactive diluent, C12-, C14-glycidilether was applied. Its content is maintained at approx. 1 part per 1 part of PI. Acetone (>99.5%) obtained from Aldrich. For reference, a commercial two-component ENR formulation was used. The commercial mixture of amine and imidazole obtained from Advanced Polymer Coatings was used as a curing agent.

3.2. Methods

3.2.1. Sample Preparation

Samples were prepared by solution casting and UV curing method. First, liquid mixtures of poly (phenyl glycidyl ether)-co-formaldehyde with a pre-calculated amount of bis (4-dodecylphenyl) iodonium hexaflurorantimonate and toluene were blended in a glass flask. The bis (4-dodecylphenyl)

IOP Conf. Series: Materials Science and Engineering **500** (2019) 012006 doi:10.1088/1757-899X/500/1/012006

iodonium hexafluorantimonate blend with 50 wt% C12-/C14-glycidilether was used in the preparations. The real weight content of the photoinitiator in the epoxy compound was adjusted to 0.5, 1.5 and 3 and 6 wt%. The mixtures of photoinitiator (PI), ENR were coated on a commercially available glass slides and steel plates. No special preparations of the substrate surface were applied to the specimens. Finally, the coating films were cured at room temperature for 1, 3, 5, 10, 20, and 50 min under UV irradiation with 96.5 W/cm^2 intensity and at a distance of 13 cm from the irradiation source. The coating thickness was in the range 120-170 µm. Accordingly, 1000 W Hg deep UV exposure lamp with broad emission spectra range from 200–600 nm was utilized for photopolymerization curing experiments in air environments.

3.2.2. Micro-Indentation Experiment

The micro-indentation tests was carried out by using a Vickers-type indenter according to the standard procedure (ISO 14577-4). The microhardness was measured on a Vickers M-17 1021 device equipped with optical microscope with a 50-g (0.49 N) load and loading time of 20 s. At least five indentations were made in order to check the repeatability of the measurements.

3.2.3. Response Surface Technique and Plan of the Experiment

The basic idea of this approach is that simple mathematical models (response surfaces method - RSM) are established only by using the finite element solutions in the reference points of the experimental design. At the first stage, a plan of experiments was produced depending on the number of design variables and number of the experiments. At the second stage, the numerical model was created in order to model the response of a structure, and then the finite element analysis was carried out in the reference points of the experimental design. At the third stage, the numerical data obtained by the finite element calculations in the sample points were used in order to build the approximating functions by using the response surface method.

The objective of the present study was to establish the functional relationships between the research parameters and indentation force. The numerical experiments were designed according to Full Factorial design. The 4^k factorial design was used to supply a relationship for indentation force as mathematical function of research parameters, where k is the number of parameters and the base 4 represents the level of treatment for each considered parameter [9]. The plan of experiments was formulated for 2 parameters, Young's modulus (X1), Yield strength (X2), and 16 experiment points. The levels are reported in Table 1.

Table 1. Research Parameters and Levels

Parameters	Levels			
Young's Modulus [MPa]	7000	7100	7200	7300
Yield Strength [MPa]	420	440	460	480

3.2.4. Finite Element Model

For comparison and validation of the micro-indentation test, 3D quarter-symmetry finite element (FE) model with Vickers indenter was used. In order correctly to simulate the indentation phenomenon, a contact analysis was conducted. 3D Finite element model was modelled by using SOLID185 elements. Target elements (TARGE170) were placed along the top surface of the resin and the contact elements (CONTA174) were used along the bottom surface of the indenter. The indentation was simulated by applying displacement boundary conditions in the *y*-direction to the nodes along the upper surface of the indenter. All nodes at the bottom of material were constrained to prevent them from moving in the x and z directions. Fragment of FE model is shown in Figure 1.

A pyramidal indenter with 68° face angle was used. This design provided the same projected area to depth as a Vickers indenter. The simulation of contact was carried out in two steps: loading and unloading. In the first step, the indenter was moved downwards along y-axis to calculate the depth (Figure 2). In the second step, the indenter was moved upwards. The contact between indenter and material was removed.

IOP Conf. Series: Materials Science and Engineering **500** (2019) 012006 doi:10.1088/1757-899X/500/1/012006

The material of indenter is the Diamond with the elastic modulus of 1140GPa and Poisson's ratio of 0.07. The Poisson's ratio of the research material is constant 0.2. The micro-indentation procedure was terminated at a calculated penetration depth of 2.31 μm. The friction coefficient between the indenter and the sample was equal to 0.15.

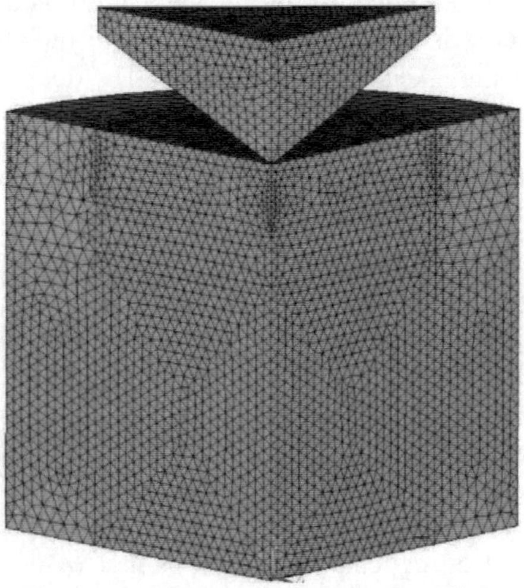

Figure 1. 3D Quarter-Symmetry Finite Element Model.

Figure 2. Distribution of Displacement.

4. Results and Discussions

The experiments showed that the highest hardness value equal to 712 MPa of coating was received with 1.5% of PI, and 3 min of UV-irradiation due to thee very dense cross-linking network structure was developed in the polymer. According to the empirical relation HV~E/10 for polymer materials, the Young's modulus of the ENR was 7120 MPa.

The numerical data obtained by the finite element calculations in the points of plan of experiments was used to build the approximating function by using the program EdaOpt [10]. A second-order polynomial equation is the following:

$$F(x) = 0.56 - 9.32\cdot10^{-5}x_1 + 1.09\cdot10^{-5}x_2 + 6.25\cdot10^{-9}x_1x_1 - 3.13\cdot10^{-7}x_2x_2 + 9.00\cdot10^{-8}x_1x_2 \qquad (4)$$

where x_1 - Young's modulus and x_2 – Yield strength.

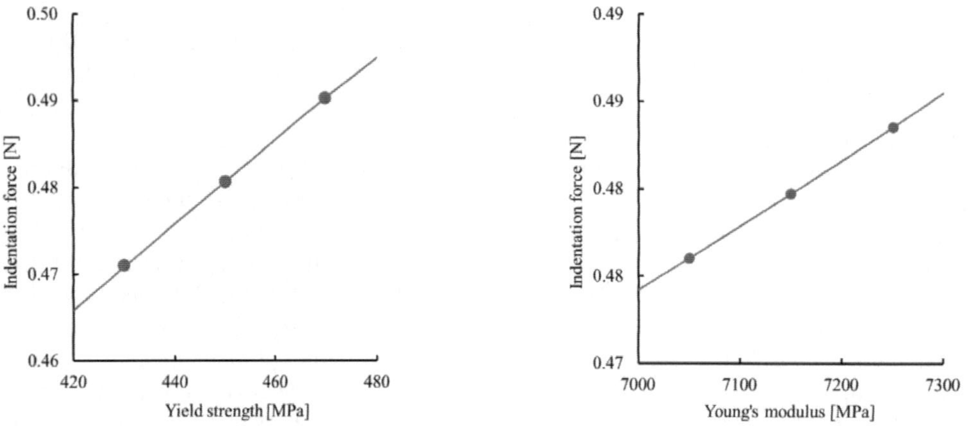

a) b)

Figure 3. Accordance Between Approximation Function and Control Points: a-Young's Modulus is 7300 MPa; b- Yield strength is 46 MPa

The approximating function was verified by the finite element solutions in the points different from the points taken in the plan of experiment. Examples of finite element verification of the response surfaces are presented in Figure 3, where a very good correlation is observed for the approximations and finite element solutions.

After selection of equation of regression the parametric studies were carried out to study the influence of the research parameters on the indentation force. The view of the response surfaces are shown in Figures 4. This figure shows the indentation force dependency on the Yield stress and Young's modulus. It can be seen that the value of the indentation force significantly increases with an increase of strength and modulus.

It is assumed that the indentation force does not exceed 0.49 N in order to detect the Young's modulus and Yield strength. It is presented in Figure 4 that there are many optimal solutions that satisfy the condition under a constant load of 0.49 N. Therefore, to establish the elasticity modulus of the material and to compare it with the results obtained from the experiment, the minimum and maximum values of the elasticity modulus from the possible range of the obtained solutions were selected.

The results of finite element simulations and the experimental test are presented in Table 2. It is seen that the minimal Young's modulus (7177 MPa) has the biggest Yield stress (48 MPa). Difference between the experimental and numerical results is 0.8%.

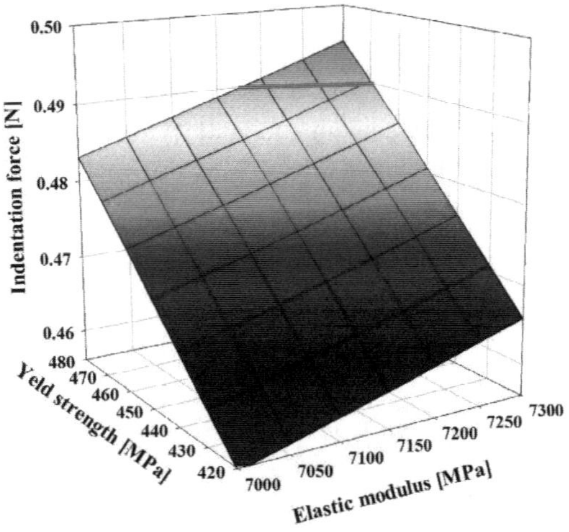

Figure 4. Dependency on the Indentation Force.

Similarly, the optimal result was obtained for minimal the Yield strength. The minimal Yield strength is 47 MPa, when the maximum Young's modulus is 7293 MPa. The difference between experimental and numerical modulus reaches 2.4%. It is seen that these differences are very little. Mostly residuals do not exceed 3%, which shows the good correlation between the approximating functions and numerical model. The load displacement curve obtained from the finite element simulation of the micro-hardness test with a Vickers indenter is shown in Figure 5.

Table 2. Young's modulus and Yield stress

	Experiment	RSM 1	Δ,%	RSM 2	Δ,%
Young's Modulus [MPa]	7120	7177	0.8	7293	2.4
Yield Strength [MPa]	–	480	–	470	–

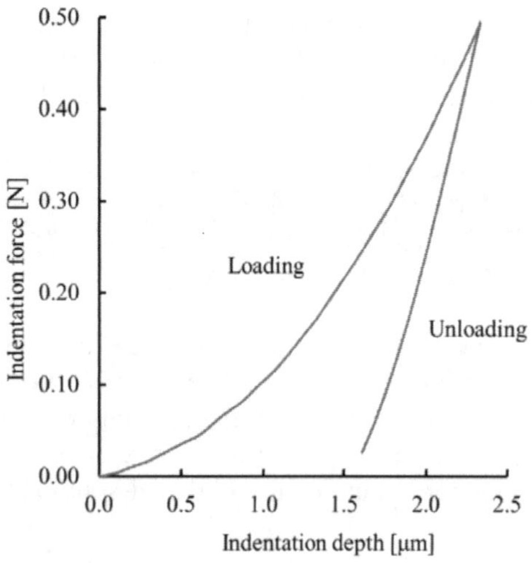

Figure 5. Load-Displacement Curve Obtained by Finite Element Simulations when Young's Modulus is 7177 MPa and Yield Strength is 480 MPa.

5. Conclusions

The micro-indentation experiment was performed on the branched epoxy novolac resin (ENR) by using Vickers indenter with the purpose of the establishment the Young's modulus of material. According to empirical relation for polymer materials, the value of Young's modulus was obtained. Then the finite element model was used to simulate the micro-indentation process in order to check the elastic properties.

Based on the indention load and displacement data extracted from the finite element simulation and from the micro-indentation experiment, the Young's modulus of branched epoxy novolac resin (ENR) was compared. As it can be seen, there is only a little difference between the Young's modulus obtained from finite element model and from the results of the experiments. The difference between the Young's modulus values obtained from the finite element simulation and from the micro-indentation experiment is negligible (0.8%). Additionally Yield strength of the material was defined.

References

[1] Z. Chen, K. Zhou, X. Lu, Y.C. Lam, "A review on the mechanical methods for evaluating coating adhesion," Acta Mech, Vol. 225, pp. 431–452, 2014.

[2] A.C. Fisher-Cripps, Introduction to Contact Mechanics. – Springer Verlag, New York, 2000.

[3] K. Komvopoulos, "An elastic–plastic analysis of spherical indenta- tion: Constitutive equations for single-indentation unloading and development of plasticity due to repeated indentation," Mech Mater, Vol. 76, pp. 93–101, 2014.

[4] L. Min, C. Wei-Min, U. Demirler, L. Kayali, W. Ling-Dong, "A numerical study of indentation using indenters of different geometry," J Mater Res, Vol. 19, pp. 73–78, 2014.

[5] X. Hernot, C. Moussa, O. Bartier, "Study of the concept of representative strain and constraint factor introduced by Vickers indentation," Mech Mater, Vol. 68, pp. 1–14, 2014.

[6] A. Karimzadeh, M.R. Ayatollahi, M. Alizadeh , "Finite element simulation of nano-indentation experiment on aluminum 1100," Comput Mater Sci, Vol. 81, pp. 595–600, 2014.

[7] G. H. Michler, F. J. Balta-Calleja, Mechanical properties of polymers based on nanostructure and morphology. – CRC Press, London, 2016.

[8] IS06507-1, Metallic materials – Vickers hardness test.

[9] R. H. Myers, D. C. Montgomery, Response surface methodology: Process and product optimisation using designed experiments. – John Wiley & Sons, New York, 1976. – 714 p.

[10] J . Auzins, A. Janushevskis, J. Janushevskis, E. Skukis, "Software EdaOpt for experimental design, analysis and multiobjective robust optimization" 2014 Proc. OPT-i Int. Conf. Engineering and Applied Science Optimization, pp 101-123

IOP Conf. Series: Materials Science and Engineering **500** (2019) 012007 doi:10.1088/1757-899X/500/1/012007

Compressive failure of quasi-static indented CFRP/aluminium honeycomb sandwich panels

Guntis Japins[1], Kaspars Kalnins[1], Olgerts Ozolins[1], Eduards Skukis[1]

[1]Institute of Materials and Structures, Riga Technical University, Riga, Latvia

guntis.japins@rtu.lv

Abstract. The present study examines sandwich panels, which are made of carbon fiber facesheets with aluminium honeycomb core and focuses on how varying factors (skin thickness, layup, core density and diameter of hemispherical indenter) are affecting collapse of artificially damaged panels in edgewise compression. Damage in the middle of sandwich specimens was introduced using quasi-static indentation. Hemispherical indenters of two diameters, namely, 20 mm and 150 mm, were applied. The residual strength was estimated through the edgewise compression test of damaged specimens. The study demonstrated that increase of the honeycomb density led to decrease in the safety margin in structure. Thicker skins may also lead to a catastrophic propagation of indent.

1. Introduction

Ultra lightweight sandwich construction materials are increasingly used in the aerospace structures due to their high specific stiffness and strength, corrosion resistance, and fatigue performance. Nevertheless, the major drawback of this structure is its susceptibility to localized impact damage caused by accidental out-of-plane load (tool dropping, hail stones, or runway debris) during the manufacturing process and/or service life [1-4]. This may lead to skin delamination and fracture, skin-core debonding, core crushing and shear failure [4]. Damage tolerance is affected by a multitude of factors, from panel construction to boundary conditions [3]. Current research focuses on failure characterization of in-plane loaded sandwich panels. The experimental set-up and testing were performed according to the ASTM standards. Artificial damage introduction in sandwich panels was carried out using quasi-static indentation. Although impact test is closer to real accidental dropping of an object on panel, quasi-static indentation is also a popular approach because it can be easier simulated with a numerical model.

2. Materials and Methods

2.1. Manufacturing of specimens

Composite skins were manufactured by laid up unidirectional pre-preg (Unipreg® 100) with a ply thickness of 0.1 mm and thermal consolidation in mechanical press under pressure of 0.3 MPa and temperature of 120 °C. Three different lay-ups (90/0), (-60/+60/0) and (0/-60/+60/0) were used. Also, three different 5052 aluminium honeycomb cores of 20 mm in thickness were selected for the study. More detailed information about density and cell size is shown in table 1. For interfacial bonding an epoxy-based paste adhesive Permabond E538 was used. To improve adhesion, the skins were processed with P120 sandpaper and degreased before bonding. 0° direction of the carbon fibres was aligned with the ribbon direction of the honeycomb core. Complete panel of 210 × 460 mm was machined into 6 specimens of 100 × 150 mm with the longer side aligned with the direction of edgewise compressive loading. After artificial damage introduction, specimens were dipped into an epoxy/sand mixture 20 mm from the ends. Cured resin forms an end-potting that prevent undesirable and premature failure of specimen ends. A complete specimen is shown in figure 1.

Table 1. Density and cell size of honeycombs used in the study.

Cell size (mm)	Density (kg/m^3)	Aluminium foil thickness (mm)
6.4	23.63	0.0178
3.2	49.66	0.0178
3.2	97.71	0.0381

2.2. Artificial damage introduction
Panels were damaged with quasi-static (QS) indentation by evenly pushing HS indenter into specimen skin with constant, 1 mm/min speed until predetermined displacement threshold from 0.5 mm to 2.5 mm was reached. Tests were performed in line with basic principles of ASTM D6264 and carried out on INSTRON E3000 universal testing machine. Each specimen was fixed with 4 rubber pads on steel frame with inner dimensions of 125 × 75 mm (figure 2). To compare the effect of indenter's diameter on damage propagation during in-plane compressive load, two different diameter (20 mm and 150 mm) hemispherical (HS) indenters were applied.

2.3.Edgewise compression test
After artificial damage is introduced in specimen, an appropriate and robust method for residual strength estimation must be applied. Edgewise compression test (popularly known as compression-after-impact (CAI) test [4]) was carried out by placing specimens between two steel plates in universal testing machine ZWICK Z100 and statically loaded with 0.5 mm/min rate. Lower compression plate was equipped with hemisphere to provide uniform distribution of the load. CAI test setup is shown in figure 3.

3. Results and Discussion
A summary of all tests is depicted in 3D graphs where simultaneous effect of three variables: honeycomb density, number of plies in facesheet (layup) and indent depth can be seen. Figure 4 shows the distribution of failure types for panels indented with 20 mm punch and figure 5 - with 150 mm punch respectively. Round markers represent panels with appropriate/satisfactory failure type, which means that these panels failed because of damage made by the indenter, while triangle markers show cases of inappropriate/unsatisfactory failures where damage was not affecting the panel enough to provoke failure. There is also a third group of cross shaped markers which are undamaged panels for a common reference. Data from 3D graphs is also shown in tables. Table 2 for intact panels, table 3 contains results of panels indented with 20 mm indenter and table 3 represents results of panels damaged with 150 mm indenter, plus and minus signs points to satisfactory or unsatisfactory failure of the panels. Schematic pictures of major failure types are shown in Figure 6. The very first attempts in practical testing showed, that there are several combinations of sandwich panel constructions where even visible damage under edgewise compression load does not cause premature failure. It can clearly be seen in 3D graphs (Figures 4 & 5) that many specimens did not fail appropriately. To be exact only 37% of panels damaged with 20 mm indenter had satisfactory failure type and the same index for 150 mm indenter was 42%. Not that much damage size, but panel construction is affecting the final failure

of the panel. For intact panels under edgewise compression load the most commonly observed failure type was skin crack and skin buckling simultaneously, about 38 % of the cases. Also, skin crack itself and skin buckling inside the panel with no cracks was observed. In case of indent provoked failures also most common was simultaneous skin crack and skin buckling. Under in-plane compressive load the indent was progressing from middle to panel sides making skin to buckle and when certain deformation is reached, in one moment skin cracks all way through the panel's width.

Figure 1. Complete sandwich panel specimen with reinforced ends for edgewise compression.

Figure 2. Steel frame for specimen fixation during quasi-static indentation test.

Figure 3. Edgewise compression test setup.

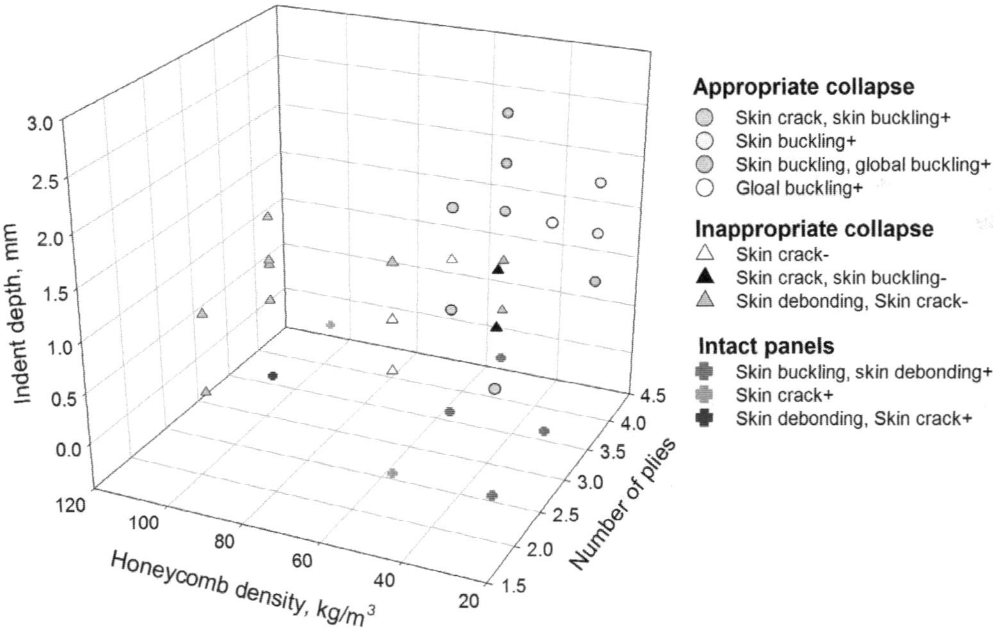

Figure 4. Distribution of failure types in edgewise compression test after out-of-plane quasi-static indentation with 20 mm indenter. Failure types distributed depending on honeycomb density, number of plies in facesheet and indent depth. All circle shaped markers show failures caused by artificial damage, triangles – inappropriate failures and crosses represents intact specimens.

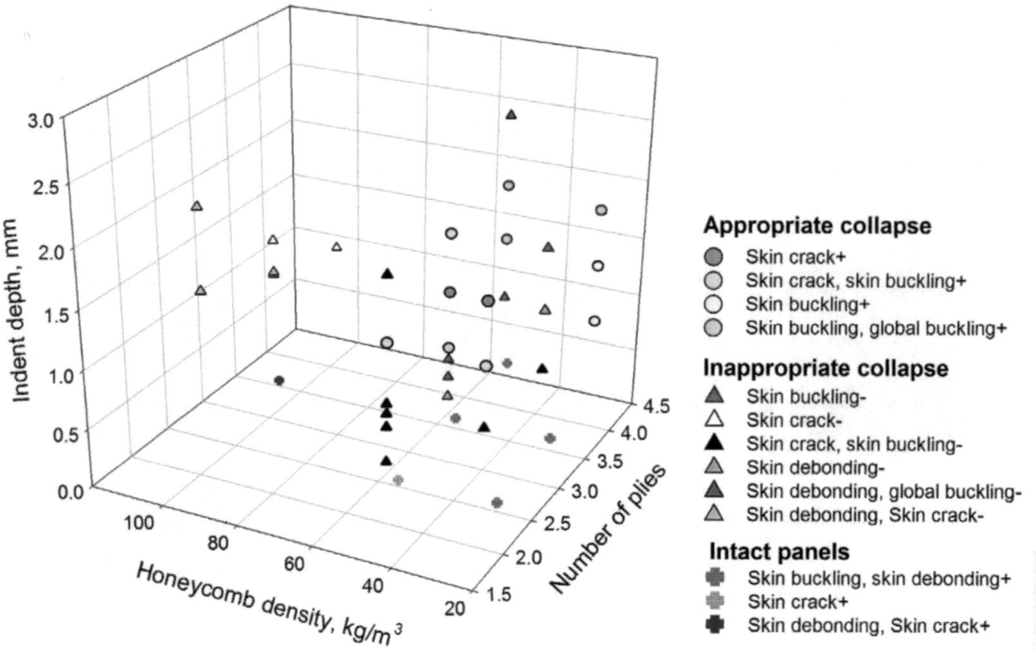

Figure 5. Distribution of failure types in edgewise compression test after out-of-plane quasi-static indentation with 150 mm indenter. Failure types distributed depending on honeycomb density, number of plies in facesheet and indent depth. All circle shaped markers show failures caused by artificial damage, triangles – inappropriate failures and crosses represents intact specimens.

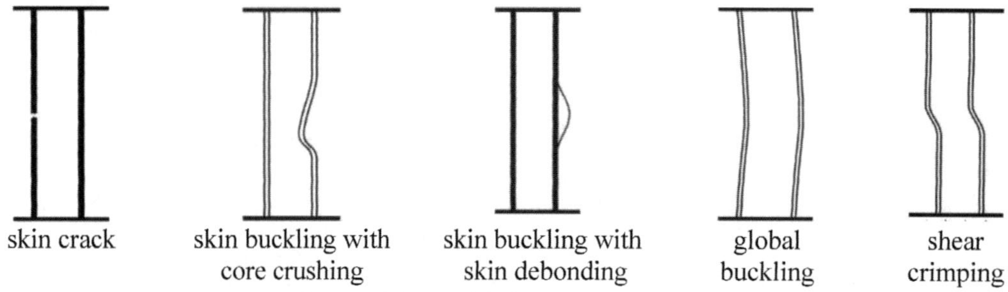

Figure 6. Major groups of sandwich panel failure types in edgewise compression test [3].

Table 2. Summary of failure types of undamaged specimens in edgewise compression.

No.	Facesheet lay-up	Honeycomb density (kg/m^3)	Failure
1	[90/0] 2 plies	25.63	Skin crack, skin buckling
2		49.66	Skin crack
3	[-60/+60/0] 3 plies	25.63	Skin buckling, skin debonding
4		49.66	Skin crack, skin buckling
5		97.71	Skin debonding, skin crack
6			Skin debonding, skin crack
7	[0/-60/+60/0] 4 plies	49.66	Skin crack, skin buckling
8		97.71	Skin crack

Table 3. Failure types in edgewise compression tests of panels damaged with 20 mm punch.

No.	Facesheet lay-up	Honeycomb density (kg/m^3)	Indentation depth (mm)	Failure
1	[90/0] 2 plies	25.63	1.0	Skin buckling, skin crack-
2			1.5	Skin crack, skin buckling+
3			2.0	Skin crack+
4		49.66	0.5	Skin buckling, skin crack-
5			0.8	Skin buckling, skin crack-
6			0.9	Skin buckling, skin crack-
7			1.0	Skin buckling, skin crack-
8			1.5	Skin crack, skin buckling+
9			2.0	Skin buckling, skin crack-
10		97.71	1.6	Skin debonding, skin crack-
11			2.3	Skin debonding, skin crack-
12	[-60/+60/0] 3 plies	25.63	1.0	Skin buckling, skin crack-
13			1.5	Skin debonding-
14			2.0	Skin buckling-
15		49.66	0.6	Skin debonding-
16			0.8	Skin buckling-
17			0.9	Skin buckling-
18			1.0	Skin crack, skin buckling+
19			1.5	Skin crack+
20			2.0	Skin crack, skin buckling+
21		97.71	1.3	Skin debonding, skin crack-
22			1.5	Skin crack-
23			1.5	Skin crack-
24	[0/-60/+60/0] 4 plies	25.63	1.0	Skin buckling+
25			1.5	Skin buckling+
26			2.0	Skin buckling, global buckling+
27		49.66	1.0	Skin buckling-
28			1.5	Skin buckling, global buckling+
29			2.0	Skin buckling, global buckling+
30		97.71	1.0	Skin crack-

Skin thickness also had significant influence on dent propagation during edgewise compression. Only 21% of all 2 ply facesheet panels failed due to damage, for 3 ply facesheets this index is quite higher – 30%, while panels with 4 ply skins in 73% of all tests failed due to propagation of indent. By decrease of density, the probability that damage will provoke collapse increases. 63% of panels with 25.63 kg/m^3 core had satisfactory failure types, in case with 49.66 kg/m^3 core only 42% of collapsed specimens belongs to the same group, but none of specimens with honeycomb density of 97.71 kg/m^3, made of thicker (0.0381 mm) foil, had satisfactory failure. This leads to a conclusion, that denser honeycomb has higher capacity to stabilize the facesheets even when its slightly crushed. Finally, it is obvious that larger dent depth increases the possibility of affecting specimen's fail in edgewise compression. Similarly, 29% of panels with 1 mm indent failed due to dent propagation, 47% with 1.5 mm indent and 62% with 2 mm indent.

Table 4. Failure types in edgewise compression tests of panels damaged with 150 mm punch.

No.	Facesheet lay-up	Honeycomb density (kg/m^3)	Indentation depth (mm)	Failure
1	[90/0] 2 plies	25.63	1.0	Skin crack, skin buckling+
2			1.5	Skin crack, skin buckling-
3			2.0	Skin crack, skin buckling-
4		49.66	1.0	Skin crack-
5			1.5	Skin crack-
6			2.0	Skin debonding, skin crack-
7		97.71	0.5	Skin debonding, skin crack-
8			1.0	Skin debonding, skin crack-
9	[-60/+60/0] 3 plies	25.63	2.0	Global buckling+
10		49.66	1.0	Skin crack, skin buckling+
11			1.5	Skin crack-
12			2.0	Skin crack, skin buckling+
13		97.71	0.8	Skin debonding, skin crack-
14			1.0	Skin debonding, skin crack-
15			1.0	Skin debonding, skin crack-
16			1.5	Skin debonding, skin crack-
17	[0/-60/+60/0] 4 plies	25.63	1.0	Skin buckling, global buckling+
18			1.5	Skin buckling+
19			2.0	Skin buckling+
20		49.66	0.5	Skin debonding, Skin crack-
21			1.0	Skin debonding, Skin crack-
22			1.5	Skin crack, skin buckling+
23			2.0	Skin buckling, global buckling+
24			2.5	Skin buckling, global buckling+

4. Conclusions

Aluminium honeycomb sandwich panels with 2 to 4 ply skins and core density between 25.63 and 97.71 kg/m^3 were manufactured, damaged with out-of-plane quasi-static indentation from 0.5 to 2.5 mm into panel's surface. The edgewise compression test showed that dominant failure types were skin cracking and skin buckling. Collapse of more than a half of panels was not provoked by artificial damage. Most likely this was due to excessive honeycomb density and insufficient penetration depth, as well as too thin facesheets. The sandwich panels with low core density and thick facesheets are more susceptible to barely-visible impact damage.

Acknowledgments

The study was done within European Space Agency study project (contract No: AO/1-7516/13/NL/KML) 'Development and validation of methodology for assessment of damage resistance properties of sandwich structures for European space sector'.

References

[1] Wei T, Falzon B G, Chiu L N S and Price M 2015 Predicting low velocity impact damage and Compression-After-Impact (CAI) behavior of composite laminates *Compos. Part A-Appl. S.* **71** 234

[2] Petit S, Bouvet C, Bergerot A and Barrau J J 2007 Impact and compression after impact experimental study of a composite laminate with a cork thermal shield *Compos. Sci. Technol.* **67** 3491

[3] Hill M D 2007 *Damage resistance and tolerance investigation of carbon / epoxy skinned honeycomb sandwich panels* (Leicestershire: Michelle Hill) p 390

[4] Zhou G and Hill M 2007 Damage characteristics and residual compressive strength of composite honeycomb sandwich panels (16th International conference on composite materials)

IOP Conf. Series: Materials Science and Engineering **500** (2019) 012008 doi:10.1088/1757-899X/500/1/012008

Mechanical Properties of Polypropylene Biocomposites Reinforced with Man-Made Cellulose Fibres and Cellulose Microfibres

Guntis Japins[1], Piotr Franciszczak[2], Kaspars Kalnins[1], Andrejs Kovalovs[1]
[1] Riga Technical University, Institute of Materials and Structures, Kipsalas St. 6A, LV-1048, Riga, Latvia
[2] West Pomeranian University of Technology Szczecin, Institute of Materials Science, Al Piastów 17, 70-310, Szczecin, Poland

guntis.japins@rtu.lv

Abstract. This paper presents the influence of man-made cellulose fibres and cellulose microfibres on the mechanical properties of reinforced composites based on polypropylene matrix. The hybrid combination of these reinforcements was also used. The man-made cellulose fibres and cellulose microfibres were compounded with polypropylene. The volume fractions of reinforcement and compatibiliser were maintained at the proportional levels to enable correlation of mechanical properties. The manufactured composite compounds were subsequently injection moulded into testing specimens. Mechanical properties such as tensile and flexural strength, tensile and flexural modulus were determined. The data obtained from experiments have shown that addition of the cellulose fibres and microfibres to the polypropylene matrix has reinforcing effect on its tensile and flexural properties, which depends mostly on volume fraction of reinforcement and its type. The hybrid combination has shown higher than expected properties, especially at higher volume fractions.

1. Introduction

Polymer structures and composites reinforced with materials of natural origin like wood and cellulose fibres are becoming significantly important for the production of a large variety of products owing to their advantages reflected in cost-effectiveness, mechanical properties, low weight and recyclability [1-3].

In recent years, composites reinforced with cellulosic fibres, such as man-made cellulose fibres, have become the focus of research. Application of cellulose fibres both of man-made and natural origin as reinforcement gives relatively good mechanical properties and higher elongation at break in comparison to glass reinforcement [4-6]. These fibres provide potential for lightweight applications in comparison to common glass fibre reinforcement on the one hand, while on the other they are obtained from raw-renewable resources and offer better recyclability and disposal [7-8].

The choice of polymer as a matrix material is important for the man-made cellulose fibres reinforcement. The use of polypropylene in the development of new composites is of great interest due to its low density, design flexibility and low price compared with other polymers [9]. It can be easily compatibilised with cellulosic fibres using polypropylene waxes with maleic anhydride functionality

[10]. Polypropylene matrix composites possess improved mechanical properties and offer new possibilities for technical and engineering applications.

The potential of short-cut rayon fibre (man-made cellulose) as reinforcements in polypropylene matrix was studied by Amash et al. [11-12]. Injection moulding compounds with high strength man-made cellulose fibres such as tyre cord yarn Cordenka (product name of tyre cord man-made cellulose) and polypropylene block copolymer were first reported by Weigel et al. [13]. Ganster et al. studied reinforcing effect of cellulose fibres in thermoplastic matrices in order to obtain a balanced property profile in terms of strength [14]. Khan et al. demonstrated that the property profile of Cordenka fibre reinforced polypropylene composites can be tailored by a partial replacement of the Cordenka by natural fibres such as jute. The jute/Cordenka hybrid composites were produced by a pultrusion process and subsequent injection moulding [15]. Recently, novel selective tailoring of dual interface in polypropylene reinforced with man-made cellulose fibres, as well hybridisation with polyethylene terephthalate fibres were developed by Franciszczak et al. [16-17].

The main purpose of this study was an investigation and comparison of the influence of man-made cellulose fibres and cellulose microfiller, as well as their hybrid combination on the mechanical properties of the reinforced composites based on polypropylene matrix. For this purpose injection moulding grade polypropylene HP400R was applied as a matrix polymer and man-made cellulose Cordenka fibres and HM400X cellulose microfibres were used as fibres and microfiller, respectively. The manufactured composite compounds were subsequently injection moulded into test specimens. Tensile and flexural characteristics of composites reinforced with different filling ratios of man-made cellulose Cordenka fibres, HM400X cellulose microfibres and their hybrid combination is presented. This research shows that the relatively costly man-made cellulose fibres may be partially substituted with HM400X, which results in a diminishing of tensile and flexural properties but to lesser extent than it would be expected by averaging of these properties of both Cordenka and HM400X composites. Especially elongation at break is maintained on the level of man-made cellulose reinforced composite after partial replacement these fibres with cellulose microfibres.

2. Materials and methods

2.1. Materials

Moplen HP400R was used as matrix material. This material is homopolymer polypropylene (PP) used for injection moulding applications in granulate form with a melt flow rate (MFR) of 25 g/10 min at 230 °C and 21.6 N.

For reinforcement, the man-made regenerated cellulose fibres Cordenka 610F having a linear density of 1.8 dtex (fibre diameter ~18 μm) and 1000 filaments were used. The initial length of fibres was approximately 1.7 mm [8, 17].

Cellulose microfibres Jelucel HM400X with a cellulose content of 99.5% and a bulk density of 55 g/l were added as a microfiller. They are irregular in shape with dia. ~21 μm and length of ~170 μm [8].

TP Licocene PP MA 6452, a maleic acid anhydride grafted polypropelene wax (MAH-g-PP) from Clariant was applied as an efficient compatibiliser between fibres/fillers and the matrix. 4 vol% of the compatibiliser for ~28 vol% of the reinforcement and 2 vol% of the compatibiliser for ~14 vol% of the reinforcement was applied in a composite respectively.

2.2. Methods

Due to their hydrophilic properties Cordenka fibres and HM400X microfibres were dried before compounding at temperature of 103°C in an oven for ~16 h [16]. Then materials were compounded at temperatures ranging from 150 to 200 °C to avoid decomposition of cellulose and at 40 RPM in order to reduce fibre shear damage. Native HP400R reference was processed without drying.

The pre-dried fibres and PP granulated were mixed with MAH-g-PP granulate. The extruded strands were cooled in a water tank and cut subsequently into granulate. Then granulate was dried before injection moulding at the same conditions as prior compounding.

Standard test specimens were prepared with these granulates according to DIN EN ISO 527 (Type A for tensile test) and DIN EN ISO 178 (Type B for 3-point bending test) using an injection moulding machine ARBURG ALLROUNDER 270 S 350–100 (clamping force 350 kN, screw diameter 25 mm, L/D = 20). The L/D ratio is the ratio of the flighted length of the screw to its outside diameter.
Static mechanical properties of manufactured test specimens were measured in a tensile and flexural test according to DIN EN ISO 527 and 178 using an Instron E3000 testing machine. All mechanical tests were performed at 23°C and 50% relative humidity after conditioning. Testing speed was 1 mm/min. Gauge length of extensometer in tensile testing was 50 mm. The presented values represent the averaged results of measurements performed on 10 samples for each type of composite.

3. Results and discussions

3.1. Volumetrict filling ratios

The filling ratios of the manufactured composites are collated as volumetric contents in Table 1. Filling volumes were calculated from the weight contents and densities.

For the composites the volume fraction of matrix was calculated according to:

$$M_{\text{vol.}} = \frac{M_{\text{wt}-\%} \cdot \rho_{\text{Tcomp}}}{\rho_{\text{Rmat.}} \cdot (1 - V_{\text{vol.}})} \quad (1)$$

while volume fraction of fibres was calculated according to:

$$F_{\text{vol.}} = \frac{F_{\text{wt}-\%} \cdot \rho_{\text{Tcomp}}}{\rho_{\text{Rfib.}} \cdot (1 - V_{\text{vol.}})} \quad (2)$$

where theoretical density of composite was calculated according to:

$$\rho_{\text{Tcomp.}} = \frac{100}{\frac{M_{\text{wt-\%}}}{\rho_{\text{R mat.}}} + \frac{F_{\text{wt-\%}}}{\rho_{\text{Rfib.}}}} \quad (3)$$

The void fractions were calculated from obtained real values of densities and theoretical densities:

$$V_{\text{vol.}} = 1 - \frac{\rho_{\text{Rcomp}}}{\rho_{\text{Tcomp}}} \quad (4)$$

where $M_{wt\text{-}\%}$ is weight content of matrix, $F_{wt\text{-}\%}$ is weight content of fibres, $\rho_{Rmat.}$ is measured density of matrix, $\rho_{Rfib.}$ is measured density of fibres, $\rho_{Tcomp.}$ and $\rho_{Rcomp.}$ is theoretical density and measured density of composite respectively.

Table 1. Volumetric filling ratios of the manufactured composite

Composite	Matrix [vol%]	Fibre [vol%]	Filler [vol%]	Fibre+Filler [vol%]	Voids [vol%]
HP400R	-	-			-
HP400R/HM400X	94.3	0.0	5.1	5.1	0.7
HP400R/HM400X	90.1	0.0	9.2	9.2	0.7
HP400R/HM400X	85.7	0.0	13.9	13.9	0.4
HP400R/HM400X	81.2	0.0	18.4	18.4	0.4
HP400R/HM400X	71.1	0.0	27.9	27.9	1.1
HP400R/Cordenka	94.7	4.5	0.0	4.5	0.8
HP400R/Cordenka	90.8	8.9	0.0	8.9	0.4
HP400R/Cordenka	86.3	13.0	0.0	13.0	0.8
HP400R/Cordenka	81.1	17.5	0.0	17.5	1.4
HP400R/Cordenka	69.4	27.4	0.0	27.4	3.2
HP400R/HM400X/Cordenka	91.0	4.3	4.4	8.8	0.2
HP400R/HM400X/Cordenka	84.7	6.9	7.1	14.0	1.3
HP400R/HM400X/Cordenka	81.4	8.7	8.9	17.5	1.0

HP400R/HM400X/Cordenka	71.6	13.2	13.5	26.6	1.8

3.2. Mechanical properties of the composite

3.2.1. Tensile and flexural strength

Tensile and flexural ultimate strength of composites reinforced with different filling ratios of Cordenka man-made cellulose fibres, HM400X cellulose microfibres and their combination are presented in Figure 1 and Figure 2, respectively.

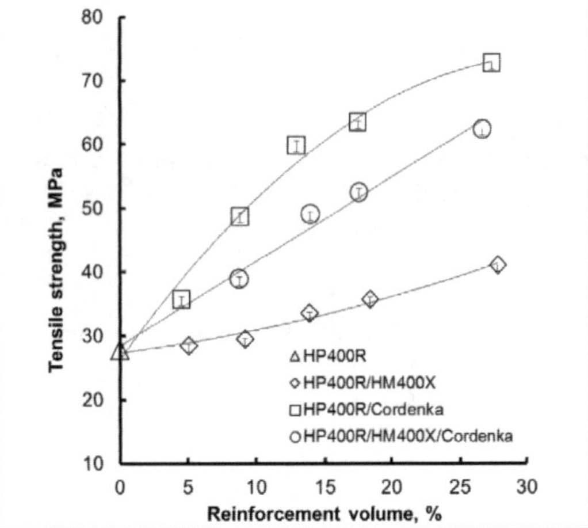

Figure 1. Tensile strength of composites with different filling ratios of cellulose reinforcements and their combination.

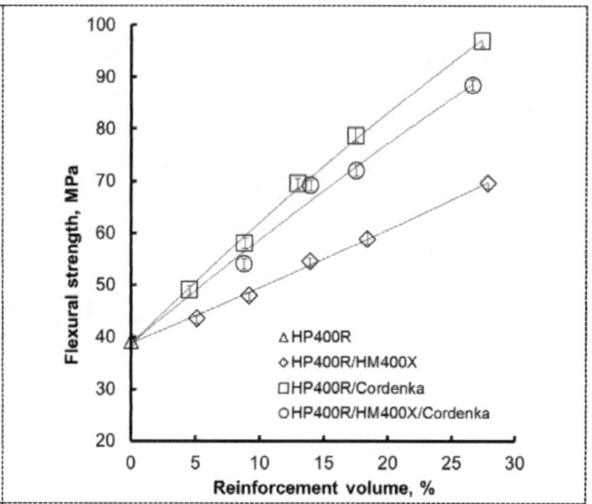

Figure 2. Flexural strength of composites with different filling ratios of cellulose reinforcements and their combination.

Tensile and flexural strength of composites increases with an increase of the reinforcement filling ratio. The greatest improvement was obtained with Cordenka fibres. The maximum stress reached 72.8 MPa and 96.9 MPa for tensile and flexural stress respectively, at ~27% reinforcement volume fraction. It can be explained by the significantly higher tensile strength of the Cordenka fibres, as well as their higher aspect ratio in comparison to HM400X microfiller [8, 17].

The composite with HM400X cellulose microfibres achieved lower strengths, which were 41 MPa and 70 MPa for tensile and flexural stress respectively at reinforcement volume of ~27%. The much lower improvement achieved in composites reinforced with microfibres is likely to be due to the fact that microfibres have lower aspect ratios than Cordenka fibres [6]. At this reinforcement volume fraction composite with Cordenka fibres had ultimate tensile strength higher by 45% and ultimate flexural strength by 28% in comparison to composite with HM400X microfibres. Lower strength of the composite with HM400X cellulose microfibres may be improved by using hybrid combination with Cordenka fibres. The differences of the tensile and flexural stress between composite with Cordenka fibres and composite with combination of man-made cellulose fibres and microfibres in ratio 1:1 was 14.3% (ultimate tensile stress) and 8.8% (ultimate flexural stress).

Averaged tensile stress-strain curves of manufactured composites with ~18% volume fraction of Cordenka man-made cellulose fibre, HM400X cellulose microfibres and their combination are presented in Figure 3.

The stress-strain characteristics of the composite with ~1:1 ratio of Cordenka man-made cellulose fibres and HM400X cellulose microfibres is more similar to the stress-strain characteristics of the composite reinforced with only Cordenka man-made cellulose than the composite reinforced with only HM400X microfibres . Comparing to the HM400X cellulose microfibres the ultimate tensile strength of composites increased from 36 MPa to 53 MPa for hybrid combination of man-made cellulose fibres and microfibres. The ultimate tensile strength of the reinforcement composite with Cordenka fibres achieved the maximum of 63 MPa.

The elongation at break had similar values for the composite reinforced with Cordenka fibres and hybrid composite with combination of Cordenka fibres and HM400X microfibres (8.5%). The composite with HM400X cellulose microfibres was breaking at lower elongation (4.4%).

Figure 4 shows representative flexural stress-strain curves of the manufactured composites with the same filling ratios as presented for tensile results. Comparing to the composite with HM400X cellulose microfibres the ultimate flexural strength of hybrid combination of Cordenka and HM400X increased from 59 MPa to 72 MPa, while the composite reinforced only with Cordenka fibres reached 79 MPa. In the same manner as for tensile testing, also in 3-point bending the stress-strain characteristics of hybrid was closer to stress-strain characteristics of composite reinforced with Cordenka man-made cellulose, as well as the elongation at break of Cordenka reinforced composite and hybrid reinforced with Cordenka and HM400X of 1:1 ratio, had similar values of approximately 13-15%, while the composite reinforced solely with HM400X microfibres broke at ~ 6%.

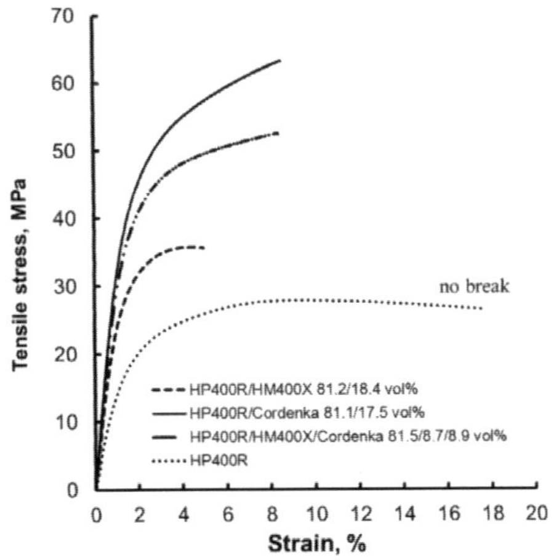

Figure 3. Tensile stress-strain curves of composites with similar filling ratios of cellulose reinforcements and their combination.

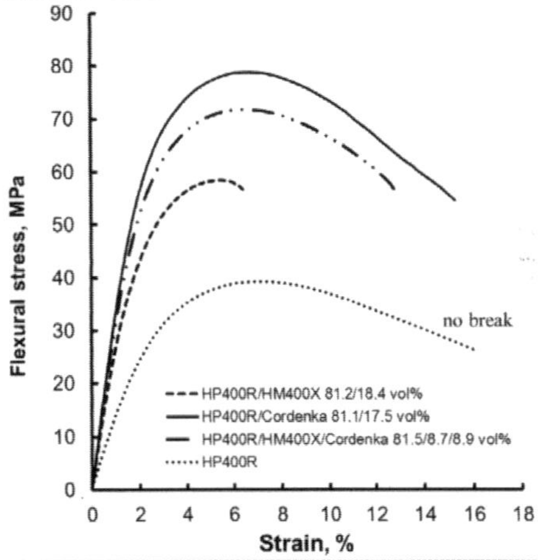

Figure 4. Flexural stress-strain curves of composites with similar filling ratios of cellulose reinforcements and their combination.

3.2.2. Tensile and flexural elastic modulus

Additional investigations were carried out to evaluate an influence of filling ratios on the tensile and flexural modulus. The results obtained for the tensile and flexural modulus with different filling ratio of the manufactured composites are presented in Figure 5 and Figure 6, respectively.

It can be seen that the values of elastic modulus significantly increase with an increase of the volume fraction of Cordenka fibres, HM400X microfibres and their combination. An addition of the cellulose microfibres in the composite increases tensile and flexural modulus for ~160% and ~130% respectively at reinforcement volume of ~5% compared to reinforcement volume of ~27%. The replacement of a cellulose microfibres by Cordenka fibres gives an increase of 61% in the tensile modulus and 58% in the flexural modulus. A similar increase is observed for the tensile and flexural modulus in the manufactured composites with combination of Cordenka man-made cellulose fibres and HM400X cellulose microfibres.

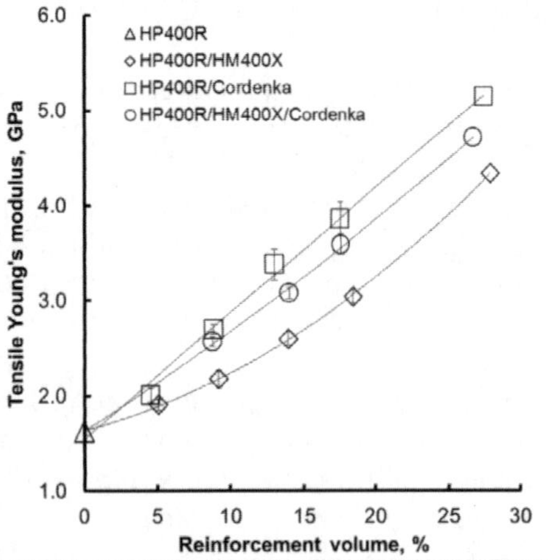

Figure 5. Tensile Young's modulus of composites with different filling ratios of cellulose reinforcements and their combination.

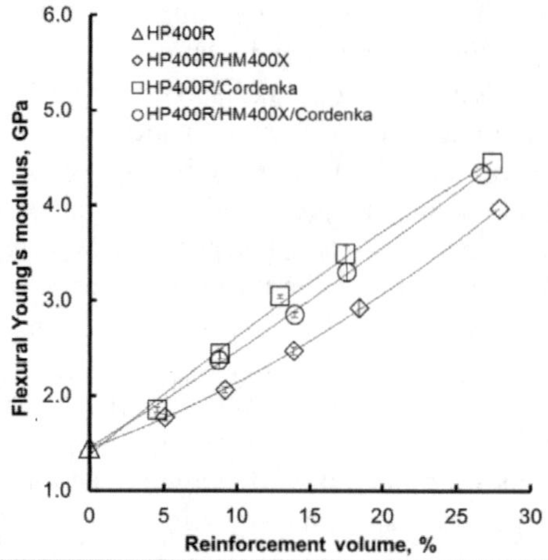

Figure 6. Flexural Young's modulus of composites with different filling ratios of cellulose reinforcements and their combination.

4. Conclusions

The effect of Cordenka man-made cellulose fibres, HM400X cellulose microfibres and their combination on the tensile and flexural strength, elastic and flexural modulus in the composites based on polypropylene matrix was investigated. Increasing volume fractions of Cordenka man-made cellulose fibres, HM400X cellulose microfibres and their combination were used in the research.

The reinforcing with HM400X cellulose microfibres had the smallest influence on the tensile and flexural strength. An addition of the man-made cellulose fibres together with cellulose microfibres increased tensile and flexural ultimate stress by 34% and 21%, respectively, at ~27%reinforcement volume. The maximal ultimate strength was obtained for the composite reinforced with highest filling ratio of Cordenka man-made cellulose fibres. Its tensile strength was higher by 45% and flexural strength by 28% in comparison to the composite reinforced with HM400X microfibres. It can be observed that hybridisation by partial substitution of more expensive man-made cellulose Cordenka fibres with cheaper HM400X cellulose microfibres is beneficial owing to maintaining the tensile and flexural characteristics closer to Cordenka fibre reinforced counterpart. The effect is more profound at higher volume fractions of reinforcement, which also give the highest performance in the investigated range of volume fractions. Especially elongation to break of hybrids is maintained close to Cordenka man-made cellulose composites of equivalent reinforcement volume. The manufactured hybrids have mechanical performance higher than WPC's and engineering plastics like polyamides, but are inferior to polypropylene reinforced with compatibilised glass fibres especially in terms of Young's modulus and strength, although their higher elongation to break might be desired in some particular applications.

Acknowledgements

The study was done partially within the framework of the projects:'Investigation on conditions for occurrence of positive hybrid effect in thermoplastic composites' 2014/15/N/ST8/03174 funded by the National Science Centre of Poland (NCN) and 'High performance short-fibre biobased hybrid composites for injection moulding (HyBiCo)' supported by the European Social Funded m-era.net network project No 4297.

References

[1] E. Labans, K. Kalnins, C. Bisagni, "Flexural behavior of sandwich panels with cellular wood, plywood stiffener/foam and thermoplastic composite core," Journal of Sandwich Structures and Materials, 2017. https://doi.org/10.1177/1099636217699587

[2] A.K. Bledzki, J. Gassan, "Composites reinforced with cellulose based fibres," Prog Polym Sci, Vol. 24, pp. 211–274, 1999. https://doi.org/10.1016/S0079-6700(98)00018-5

[3] R. Merijs-Meri, J. Zicans, T. Ivanova, I. Bochkov, M. Varkale, P. Franciszczak, A.K. Bledzki, J. Gravitis, "Some aspects of the development of oat husks containing polypropylene composites," AIP Conference Proceedings, Vol. 1981, 020128, 2018. https://doi.org/10.1063/1.5045990

[4] A.K. Bledzki, A. Jaszkiewicz, "Mechanical performance of biocomposites based on PLA and PHBV reinforced with natural fibers – a comparative study to PP," Compos Sci Technol, Vol. 70(12), pp. 1687–1696, 2010. https://doi.org/10.1016/j.compscitech.2010.06.005

[5] M Das, Chapter 2, Man-made cellulose fibre reinforcements (MMCFR). In: Biocomposites for High-Performance Applications, edited by D. Ray, Woodhead Publishing 2017, p. 336.

[6] J.K. Pandey, S.H. Ahn, C.S. Lee, A.K. Mohanty, M. Misra, "Recent advances in the application of natural fiber based composites," Macromol Mater Eng, Vol. 295(1), pp. 975–989, 2010. https://doi.org/10.1002/mame.201000095

[7] J. Ganster, H-P. Fink, "Novel cellulose fibre reinforced thermoplastic materials," Cellulose, Vol. 13, pp. 271–280, 2006. P. Franciszczak, K. Kalnins, A.K. Bledzki. "Hybridisation of man-made cellulose and glass reinforcement in short-fibre composites for injection moulding – effects on mechanical performance". Composites: Part B, Vol. 145, pp. 14-27, 2018. https://doi.org/10.1016/j.compositesb.2018.03.008.

[8] P. Franciszczak, K. Kalnins, A.K. Bledzki. "Hybridisation of man-made cellulose and glass reinforcement in short-fibre composites for injection moulding – effects on mechanical performance". Composites: Part B, Vol. 145, pp. 14-27, 2018. https://doi.org/10.1016/j.compositesb.2018.03.008

[9] J. Ganster, H-P. Fink, "Novel cellulose fibre reinforced thermoplastic materials," Cellulose, Vol. 13, pp. 271–280, 2006. https://doi.org/10.1007/s10570-005-9045-9

[10] A.K. Bledzki, P. Franciszczak, A. Mamun, "The utilization of biochemically modified microfibers from grain by-products as reinforcement for polypropylene biocomposite," eXPRESS Polymer Letters, Vol.8. pp. 767–778, 2014. Doi: 10.3144/expresspolymlett.2014.79

[11] A. Amash, P. Zugenmaier, "Study on cellulose and xylan filled polypropylene composites," Polym Bull, Vol. 40(1-2), pp. 251–258, 1998. https://doi.org/10.1007/s002890050

[12] A. Amash, P. Zugenmaier, "Morphology and properties of isotropic and oriented samples of cellulose fibre–polypropylene composites," Polymer, Vol. 41(4), pp. 1589–1596, 2000. https://doi.org/10.1016/S0032-3861(99)00273-6

[13] P. Weigel, J. Ganster, H-P. J. Fink, J. Gassan, K. Uihlein, "Polypropylene–cellulose compounds-high strength cellulose fiber strengthen injection moulded parts," Kunststoffe Plast Europe, Vol. 92, pp. 35–37, 2002.

[14] J. Ganster, H-P. J. Fink, M. Pinnow, "High-tanacity man-made cellulose fiber reinforced thermoplastics-injection moulding compounds with polypropylene and alternative materials," Composites: Part A, Vol. 37, pp. 1796–1804, 2006. https://doi.org/10.1016/j.compositesa.2005.09.005

[15] M. A. Khan, J. Ganster, H-P Fink, "Hybrid composites of jute and man-made cellulose fibers with polypropylene by injection moulding," Composites: Part A, Vol. 40(6-7), pp. 846–851, 2009. https://doi.org/10.1016/j.compositesa.2009.04.015

[16] P. Franciszczak, A.K. Bledzki, "Tailoring of dual-interface in high tenacity PP composites – Toughening with positive hybrid effect," Composites: Part A, Vol. 83, pp. 185-192, 2016. https://doi.org/10.1016/j.compositesa.2015.07.001

[17] P. Franciszczak, R. Merijs-Meri, K. Kalnins, A.K. Bledzki, J. Zicans, "Short-fibre hybrid polypropylene composites reinforced with PET and Rayon fibres – Effects of SSP and

interphase tailoring," Composite Structures, Vol. 181, pp. 121-137, 2017. https://doi.org/10.1016/j.compstruct.2017.08.075

Rigid Polyurethane Foams as External Tank Cryogenic Insulation for Space Launchers

U Cabulis[1], V Yakushin[1], W P P Fischer[2], M Rundans[1], I Sevastyanova[1], L Deme[1]

[1] Latvian State Institute of Wood Chemistry, Riga, Latvia
[2] ArianeGroup GmbH, Bremen, Germany

E-mail: cabulis@edi.lv

Abstract. In the framework of the present activity, work is carried out on the development of rigid polyurethane (PUR) foam material for external cryogenic insulation of a liquefied hydrogen (LH2) tank for the next generation launcher with a cryogenic upper stage. The main advantage of PUR foams in comparison with other thermal insulation materials is possibility to cover this material on the complicated shape metal surfaces by spraying method, which leads to significant cost savings. The spraying of PUR foams on metal constructions is chemical-technological process. As a result of work performed within the confidential European Space Agency contract, PUR foam material has been developed for the use as external insulation material for fuel tank structures. Raw ingredients from suppliers within EU have been used exclusively. The material has been tested for its performance in normal (room) and relevant (cryogenic conditions) environment and found to be mechanically durable and thermally excellent and therefore appropriate for use in space application. Accelerated ageing test indicate the material to be long-lasting. Safety coefficient, which characterizes the capability of insulation to withstand cryo-shocks, for optimal compositions reaches 4.7.

1. Introduction

The cryogenic insulation is very important for the space launchers with liquefied hydrogen (LH2) and liquefied oxygen (LOX) as the propellants. The basic requirements for the insulation materials are good thermal-insulating performance, especially at low temperatures, light weight and at the same time it must be non-flammable material, since during scent phase significant thermal loads are acting on the surface of the launcher. Although some of these characteristics are quite contradictory. For external thermal insulation (ETI) material additional important characteristics are UV-stability and the material's ability to resist cryo-pumping effects, and minimal outgassing at vacuum conditions. The leading contributors in space activities – USA [1], China [2] and also EU [3] are using cryogenic propellants and therefore, space research centres there are carrying out in-depth studies in the field of cryogenics.

Ever since the development of the vacuum flask, the vacuum is still the primary player in most insulation schemes on ground, where the weight of the system does not play a big role or in stationary conditions, where damage due to loads or vibration is not possible. In space industry there are two main solutions for cryogenic insulation. One common radiation barrier used in cryogenic applications is known as Multilayer Insulation (MLI). The MLI generally contains multiple layers of reflective material also called radiation shields (low emissivity foils, also called screens) separated by spacers having low conductivity (typically 0.25 mm thick) [4]. But, drawback is the poor thermal performance on ground or ascent conditions. The other solution is polymeric foams, which possess many advantageous properties, such as low thermal conductivity, light weight, low water absorption/permeability, and dimensional stability and can satisfy thermal requirements for phases where the pressure is above vacuum. In particular, polymeric foams do not change in volume along temperature gradients, in other words, they exhibit significant size stability [5]. Rigid polyurethane (PUR) foams have outstanding properties [6], and some of them are successfully used even today as cryogenic insulation. A combination of both above mentioned materials is also developed for cryogenic insulation [7]. In the future, PUR foams and other plastic foams are regarded as the main material for liquefied gas tank cryogenic insulation in different projects of liquefied natural gas (LNG) carriers [8] and space vehicles [6]. The main advantage of PUR foams in comparison with other thermal insulation materials is possibility to cover this material on the complicated shape metal surfaces by spraying method [9]. PUR foams obtained by spraying method were first developed in the 1960s and included applications for the Saturn V moon rockets and subsequently the Space Shuttle [6].

The properties of PUR foams and their adhesion to substrate materials depend not only on the chemical structure and macromolecule architecture of the polymeric matrix, but also on the technological factors of PUR foam production. PUR foam composition was developed in the frame of the confidential contract with European Space Agency. The tasks of this research were to develop and modify necessary polyol composition for production of spraying cryogenic insulation with optimal technological parameters. The used PU foam composition was calculated and optimised using previously described [10] approach for development of cryogenic insulation. Influence of technological parameters on the physical and mechanical characteristics of the ETI and their modifications was investigated. Also such important properties as flammability and UV-stability of ETI foams were measured.

2. Experimental

2.1 Materials

Cryogenic PUR foam composition (Cryo_PUR) was created combining Lupranol type (BASF) polyether type polyols with functionality 3 and 4 and OH-values 400 and 740 respectively as main ingredients in the recipe; also, reactive and additive flame retardants were introduced in system, and combination between physical (Solkane™ 365/227 (Solvay)) and chemical (water) blowing agents were used. Catalyst package included amine and tin containing catalysts. Desmodur® 44V20L (Bayer) was used as an isocyanate component for all rigid PUR foam samples. It is a solvent-free product based on 4,4'-diphenylmethane diisocyanate (MDI) and contains oligomers of high functionality. The average functionality was 2.8 to 2.9 and the NCO content was 30.5 to 32.5 wt.%. The calculated isocyanate index of composition was 110.

2.2 Sample preparation

Cryo_PUR foam panels were made with the high-pressure spraying machine Glass Craft VR. Cryo_PUR composition was spray-applied in one pass on 4 mm thick aluminium sheets, coated with a wax-based release agent. Aluminium sheets were heated until the desired temperature using an infrared heater. The temperature of aluminium sheets and small aluminium plates for the adhesion test was measured using a special surface thermocouple. Generally, samples for different tests were cut out from the panels' core. For production of Cryo_PUR foams by spraying method, the following parameters were controlled and maintained: A-component (polyol) and B-component (isocyanate) temperatures and viscosities, A and B components' volume ratios, pressure at A and B components' supply lines and also room temperature and relative air moisture.

2.3 Test methods

A Static Materials Testing Machine Z010 TN (Zwick GmbH & Co) for foam testing at 295 K was used. Compression tests were performed in two directions: parallel (Z) and perpendicular (X) to foam rise. To taking into account thickness limitation of sprayed layer, samples of 30×30×30 mm size were used. The testing of the foams' mechanical properties at 77 K was carried out with a Testing Machine Z100 (Zwick

GmbH & Co). For both tests the basic program testXpert V11.02 was used. A compact cryogenic facility and specific appliances for foam testing at cryogenic temperatures are shown in figure 1. Ring samples, 13-14 mm in width, with the inner diameter 43 mm and the outer diameter 53 mm, were used for tensile characteristics determination. A detailed analysis of this method compared to standard method for determination of tensile strength for foam materials (ISO 1926:2009) has been given in [11]. The adhesion of the Cryo_PUR foam to aluminium (tensile bond strength) was determined after the cryo-shock test (instant immersion in liquefied nitrogen, exposure therein for 60 min and warming at room temperature). The tests were performed according to EN 1607 "Determination of tensile strength perpendicular to faces of Thermal Insulation Products". The thickness of the foam material between the substrate (aluminium sheet) and the glued pad was 10 mm.

a) **b)** **c)**

Figure 1. a) Testing Machine Z100 with a cryogenic testing facility; **b)** appliance for the rings tensile test; **c)** appliance for the bond strength determination.

The ETI's possibility to withstand thermal strains can be characterized by safety coefficient K_S, which can be calculated from following equation:

$$K_S = \frac{\varepsilon_{77}}{\Delta l_{77-295}} = \frac{\varepsilon_{77}}{\alpha_x \cdot \Delta T \cdot 100} \qquad (1)$$

where:
ε_{77} – PUR foam elongation at 77K perpendicular to foaming direction, %;
$\Delta l_{77\text{-}295}$ – the contraction of PUR foam perpendicular to foaming direction freezing it from 295 K to 77 K, %;
α_x – coefficient of thermal expansion;
ΔT – temperatures difference.

The data for calculation of the average coefficient of thermal expansion (CTE) of the foam were obtained by the means of dilatometric measurements (3 samples) in the range of temperatures from 295K to 110 K using TMA Linseis PT1600 equipment. Only small specimens of 4×4×20 mm size can be used when using this device.

Thermal conductivity coefficient λ_{10} of foams was determined using a Linseis HFM 200 thermal analyser. The size of samples was 200×200×50 mm, top plate temperature 20°C and bottom plate temperature 0°C.

The flammability test of Cryo_PUR foams according to EN ISO 11925-2:2010 "Reaction to fire tests – Ignitability of products subjected to direct impingement of flame – Part 2: Single-flame source test" was performed in an apparatus produced by FTT-Fire Testing Technologies. Foam samples of 250×90×40-50 mm size were cut-out from the core of panels. Flame application time was 30 s.

Artificial ageing has been done using UV accelerated weathering chamber held at 60°C temperature and ambient pressure. A fluorescent lamp UVA-340 (295–420 nm, max 340 nm) with the intensity 0.89 W/m^2 was used. The degradation experiment was conducted for 5 weeks. Samples were periodically

IOP Conf. Series: Materials Science and Engineering **500** (2019) 012009 doi:10.1088/1757-899X/500/1/012009

photographed with a photo camera and SEM. SEM image analysis has been performed using Tescan TS 5136 MM equipment in secondary electron detection mode.

3. Results and discussions

The optimum spraying parameters for developed Cryo_PUR foams and ambient conditions are given in table 1.

Table 1. Main parameters of Cryo_PUR sample spraying

Volume ratio of components A:B	1.0 : 1.0
A and B components' temperature, °C	40
A-component viscosity at 25°C, mPa·s	450 - 500
B-component viscosity at 25°C, mPa·s	200 - 230
Hydraulic oil pressure, bar	40
A-component working pressure, bar	110 - 130
B-component working pressure, bar	130 - 150
Ambient temperature, °C	18 - 20
Air relative moisture, %	40 - 50

Three different aluminium sheet temperatures (20; 25 and 30°C) were chosen, since it is known, that at higher substrate temperature the bond strength to surface increases. The basic physical-mechanical characteristics and calculated safety coefficients are given in table 2.

Table 2. Test results and safety coefficient of spray applied Cryo_PUR foam

Al sheet temp., °C	ρ, kg/m^3	σ_Z, MPa	σ_X, MPa	Bond strength, MPa	σ_{77}, MPa	ε_{77}, %	$\Delta l_{77\text{-}295}$, %	K_S
20	48.7	0.238	0.184	0.46	1.21	5.66	1.23	4.61
25	48.4	0.220	0.187	0.74	-	-	-	-
30	48.8	0.219	0.176	0.62	1.20	5.92	1.29	4.66

It is confirmed that aluminium substrate sheet temperature does not have a significant influence on the safety coefficient and basic physical-mechanical characteristics. The obtained K_S above 4 is very high, and obtained Cryo_PUR foams from physical-mechanical point of view can be characterised as potential cryogenic insulation material [10].

The obtained dilatometric curves are shown in figure 2. Experimental data can be well enough approximated with a polynomial trend and extrapolated. Contraction level of foams and an average CTE value in the range of temperatures from 295 K to 77 K are calculated to be 1.49% and $68.4 \cdot 10^{-6}$ K^{-1} respectively. When comparing the data obtained from TMA PT1600 with the data from cryostat tests, it should be noted that data for calculations are obtained by the means of extrapolation of dilatometric test data. Moreover, the sample limitations for the TMA PT1600 device did not allow to assess the scale effect on the results of measurements as the maximum sample size limit for this device (4×4×20 mm) is obviously too small for cellular materials. Nevertheless, the obtained K_S using TMA analyser is 3.9, which is also high in comparison with conventional foams [10].

Physical-mechanical properties at different temperatures and capability to withstand cryo-shocks are not the only important characteristics for cryogenic insulation; material must be stable in long-term storage and also non-flammable and UV-stable.

The thermal insulation capacity of foam materials isn`t only influenced by chemical structure but also depends on the environmental factors such as temperature, humidity and radiation. In addition, the gas diffusion through the foam struts will result in the thermal insulation performance degradation with time and the aging will be affected by the operating environment [12]. Before long term storage at room temperature, λ_{10} for initial foams was 17.6 mW/m·K, after half a year – λ_{10} increases up to 20.7

mW/m·K (figure 3). The λ_{10} for aged conventional PUR thermal insulation materials reaches 26- 28 mW/m·K [13], the significantly lower λ_{10} of Cryo_PUR was achieved due to the fine cell structure (figure 5b) and chemical composition of foam matrix.

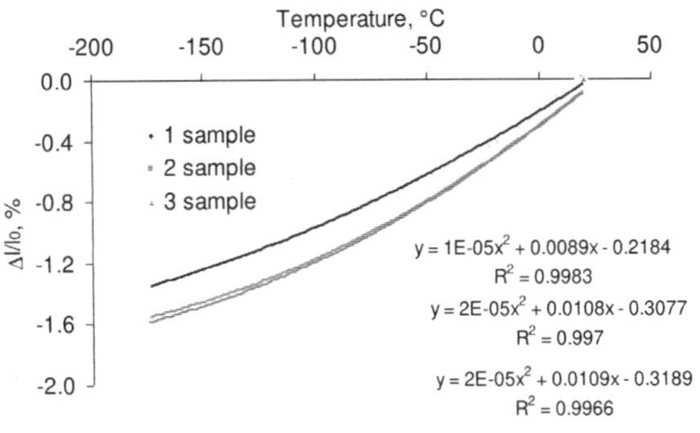

Figure 2. Changes of specimen length versus temperature for the Cryo_PUR foam

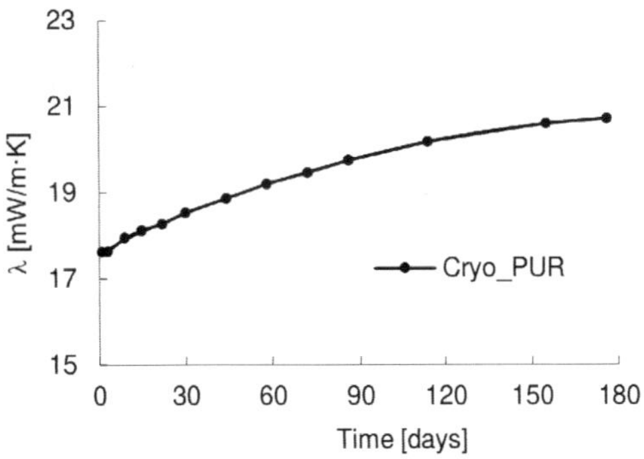

Figure 3. Changes of thermal conductivity coefficient (λ_{10}) at long term storage

Flammability test according to ISO 11925-2:2010 has been conducted and the results can be seen in table 3 below. Six samples in total were tested from the front and back sides (figure 4). In 3 out of 12 cases flame tip reached the mark of 150 mm, but in all cases the height of the damaged area did not exceed the 150 mm mark. Currently two types of flame retardants, additive and reactive, are being used to prepare polyurethane foams with low flammability [14]. Combination of both was used in Cryo_PUR composition and obtained flammability result turned out to be suitable for space launcher parts, which are subjected to aerodynamic friction.

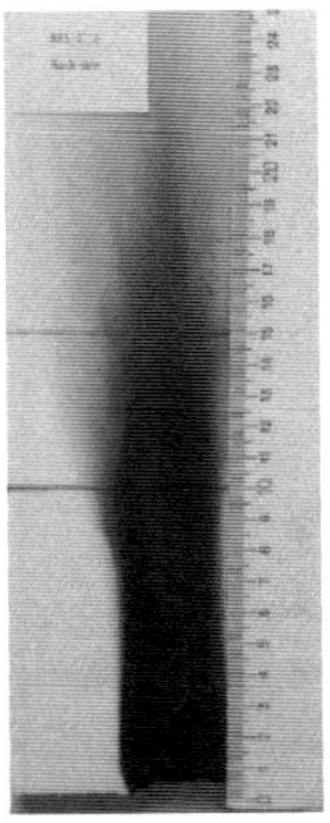

Table 3. Flammability characteristics of Cryo_PUR foams

ρ, kg/m³	Did sample ignite?	Did flame tip reach 150 mm?	Average height of the damaged area, mm
51 ± 3	Yes – 100%	Yes – 25%	130 ± 8

Figure 4. Partly burned sample of Cryo_PUR foams after flammability test

During UV-irradiation the aromatic structures are oxidized at the central methylene group, leading to highly conjugated quinone products. The accumulation of quinone products due to the chain scission of the PUR macromolecules induces formation of coloured products [15] (Figure 5a), at the same time the weaker polyether chain segments degrade under UV-irradiation. SEM images (Figure 5b) were acquired before and after exposure to UV for determination of thickness of degradation. The cell walls under the influence of UV radiation collapse first but the network of struts remains. It can be assumed that cell walls behave like thin PUR films, but struts are the thickest and most robust parts of the PUR foam. The degraded layer is clearly visible and therefore can be easily measured. Figure 5c presents the thickness of degraded layer during UV exposure. After 2 weeks the level of degradation reached the maximum, and further, deeper degradation is not observed. The thickness of degraded layer is 1.5 – 2 times thinner than for conventional foams, including foams containing renewable raw materials [16].

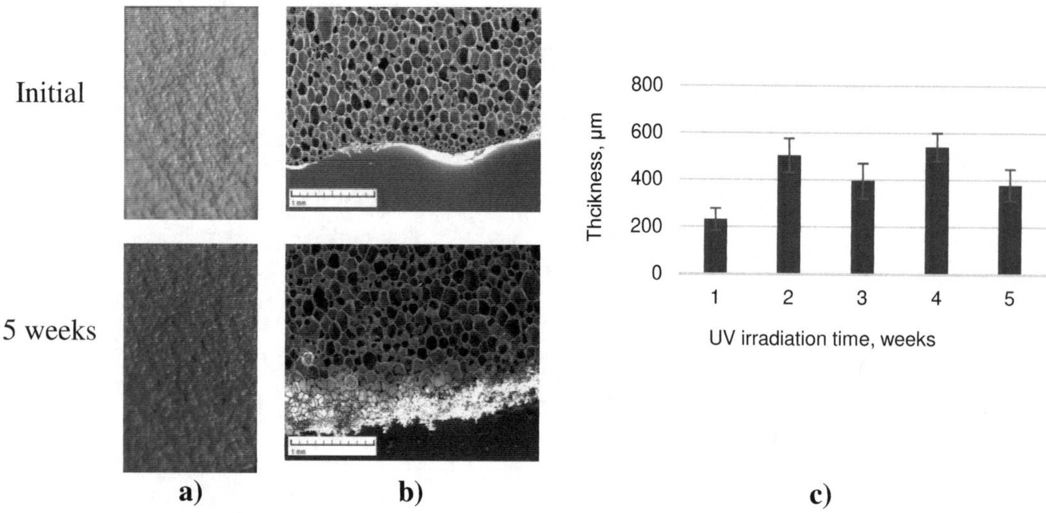

Figure 5. PUR foam samples before and after 5 weeks of artificial ageing; **a)** photographs; **b)** SEM images; **c)** thickness of degraded layer.

4. Conclusions

Rigid polyurethane foam material obtained by spraying method has been developed for the use as external insulation material for space launcher fuel tank structures. Material can be characterised by high bond strength to aluminium substrate and safety coefficient value reaching 4.7. In the same time the foam exhibits stable properties in long term storage and characterizes as self-extinguishing material. The thickness of degraded layer at accelerated UV-irradiation is less than 0.6 mm. Material is prospective not only for space applications, but also for other cryogenic fields, for example LNG transport.

Acknowledgements

The study was financed by the European Space Agency PECS project CRYOFOAMS, Contract No:4000114532/15/NL/NDe

References

[1] Graham R 2003 External Tank Thermal Protection System (NASA Facts. FS-2003-09-113-MSFC)

[2] Zhang X B, Yao L, Qiu L M, Gan Z H, Yang R P, Ma X J and Liu Z H 2012 Experimental study on cryogenic moisture uptake in polyurethane foam insulation material *Cryogenics* **52** 810–5

[3] Fischer W P P, Stirna U, Yakushin V and Cabulis U 2010 *40th Int. Conf. on Environmental Systems (Barcelona)* (Reston: American Institute of Aeronautics and Astronautics) 6295

[4] Krishnaprakas C, Narayana K B and Dutta P 2000 Heat transfer correlations for multilayer insulation systems *Cryogenics* **40** 431–5

[5] Bahadori A 2014 *Thermal Insulation Handbook for the Oil, Gas, and Petrochemical Industries. 1st ed.* (Oxford: Gulf Professional Publishing)

[6] Fesmire J E, Coffman B E, Meneghelli B J and Heckle K W 2012 Spray-on foam insulations for launch vehicle cryogenic tanks *Cryogenics* **52** 251–61

[7] Zheng J, Chen L, Cui C, Guo J, Zhu W, Zhou Y and Wang J 2018 Experimental study on composite insulation system of spray on foam insulation and variable density multilayer insulation *Appl. Therm. Eng.* **130** 161–8

[8] Lee Y, Baek K H, Choe K and Han C 2016 Development of mass production type rigid polyurethane foam for LNG carrier using ozone depletion free blowing agent *Cryogenics* **80** 44–51

[9] Cabulis U, Yakushin V and Fischer W P P 2017 Preparation of rigid polyurethane foams as inner wetted thermal insulation *Proc. of PPS-33 (Cancun)* (New York: AIP Conference Proceedings) In press

[10] Stirna U, Beverte I, Yakushin V and Cabulis U 2011 Mechanical properties of rigid polyurethane foams at room and cryogenic temperatures *J. Cell. Plast.* **47** 337–55

[11] Zhmud' N P and Yakushin V A 1986 Determination of the properties of rigid polyurethane foams in tension on ring specimens *Mekhanika Kompozitnykh Materialov* **6** 1123–7

[12] Zhang H, Fang W Z, Li Y M and Tao W Q 2017 Experimental study of thermal conductivity of polyurethane foams *Appl. Thermal Eng.* **115** 528–38

[13] Closed and open-cell spray polyurethane (PU) foam 2013 (PU Europe Factsheet No.22)

[14] Bhoyate S, Jonesku M, Radojcic D, Kahol P K, Chen J, Mishra S R and Gupta R K 2018 Highly flame-retardant bio-based polyurethanes using novel reactive polyols *J. Appl. Polym. Sci.* **135** 46027

[15] Wilhelm C, Rivaton A and Gardette J L 1998 Infrared analysis of the photochemical behaviour of segmented polyurethanes. 3. Aromatic diisocyanate based polymers *Polymer.* **39** 1223–32

[16] Paberza A, Stiebra L and Cabulis U 2015 Photodegradation of polyurethane foam obtained from renewable resource – pulp production byproducts *J. Renew. Mater.* **3** 19–27

IOP Conf. Series: Materials Science and Engineering **500** (2019) 012010 doi:10.1088/1757-899X/500/1/012010

Prediction method of electrical conductivity of nano-modified glass fibre reinforced plastics

A Aniskevich, S Stankevich and J Sevcenko

Institute for Mechanics of Materials, University of Latvia,
23 Aizkraukles str., Riga, LV-1006, Latvia

E-mail: Andrey.Aniskevich@pmi.lv, obis8053@gmail.com

Abstract. Glass fibre reinforced plastics (GFRP) is non-conductive construction material, however with carbon nanotubes (CNT) modifying it can get additional functionality due to its gained electrical conductivity. Main aim of the study is to check functionality of the prediction method of electrical conductivity of GFRP with nano-modified epoxy matrix using structural approach. GFRP composites under investigation were based on unidirectional (UD) Glass fiber and two matrixes modified with carbon nanotubes. Electrical conductivity of epoxy resin modified by CNT (concentrations < 1 %) was modelled using structural approach. Electrical conductivity of unidirectional GFRP layer was measured experimentally and modelled based on conductivity of polymer matrix and assuming non-conductive fibres. It was supposed that epoxy matrix and composite on different structural levels follow the Ohm's law. Two components of tensor of electrical conductivity for monotropic material in main axis of symmetry were calculated for composites by several different ways. Control experiment was performed on a sample of GFRP monolayer cut from the unidirectional plate under different angles. The calculated data are in a good agreement with the experimental values.

1. Introduction

Glass fibre reinforced plastics (GFRP) is a construction material, that has grown within global market for 9% over last 5 years according to the world's largest market research store "*Research and markets*" [1]. Being part of a composite, glass fibres stand for good mechanical properties of the whole system keeping it still 15-20% lighter than traditional metals. Owing to GFRP light weight, they are used as substitutes for steel and aluminium in various industries. GFRP is being originally electrical insulator, however, modifying matrix of GFRP with electrically conductive carbon nanotubes (CNT) makes the whole system improve its mechanical and electrical properties. The electrically conductive composites can be produced with high electrical conductivity at volume content of CNTs below 1%. Consequently, the carbon nanotubes are extending the usage range of GFRP with modified matrix in biomedical, electronic, automotive and aerospace industries [2]. Despite other modifications, with ability to control the electrical conductivity of GFRP rises potential of composite to detect structural damage of material (damage monitoring), screen the electromagnetic irradiation and drain electrostatic charging [3-7]. CNT-modified matrix is usually considered as an isotropic material and variety of models are applied for estimation of its electrical conductivity depending on the content of conductive fillers [8-10]. To define electric current density in composites, its anisotropic electrical conductivity and stacking sequence of the laminated plates should be taken into account. For that reason, finite element modelling (with high computational costs) could be applied to calculate electric current distribution in each subpart of the composite. Aside from this, electromechanical modelling [11] and analytical calculation of the electric function (with simplification for orthotropic materials) [6] are also used as an alternative. And yet, there is an issue with making the easiest and possibly the fastest method for GFRP electrical conductivity prediction.

From other side, structural approach is wide used in mechanics of composite (glass fibre reinforced plastics) for prediction of e.g. their elastic properties [12]. Main aim of the study is to check functionality of the prediction method of electrical conductivity of GFRP with CNT-modified epoxy matrix using structural approach.

2. Materials

Composite under investigation was GFRP based on unidirectional (UD) glass fibre (511 g/m^2) and CNT modified epoxy matrix. During this study 2 polymer matrices were used for different samples:

1. Aradur 3486 epoxy resin + Araldite LY 1564 hardener with CNT masterbatch Epocyl™ R128-02 (Nanocyl). Epoxy resin was premixed with masterbatch during 20 min, later hardener was added and mixed 10 min more. Vacuum chamber was used for matrix degassing for 40 min. CNT concentration of 0.75 wt% was selected for polymer matrix to provide optimal combination of viscosity and conductivity of the matrix based on previous research [8, 13]. Vacuum bagging was used during 30 min to remove excess of the epoxy matrix before hardening. Hardening of nanomodified matrix, composite was performed in low vacuum at 50 °C during 24 h. Final size of samples was 200×150×3 mm. Considering the reinforcement coefficient 0.8 (by mass) of final composite, its CNT concentration is 0.15 wt%.
2. Nano-Force® E200 epoxy resin (masterbatch) contains less than 0.7 wt% CNT (the exact amount of CNT was not specified by the manufacturer) + E hardener system. Epoxy resin and hardener were mixed in 100 : 30 ratio for 10 min at the room temperature of about 20 °C. After the mixing, prepared epoxy system contained less than 0.5 wt% CNT. Composite production steps were identical to previous sample type. Final size of samples was 200×150×3 mm. Considering the reinforcement coefficient 0.55 (by mass) of prepared composite, its CNT concentration is estimated less than 0.2 wt%.

UD plates consist of 8 UD layers $[0°]_8$ with fibre direction oriented along axis 1. Reinforcement coefficient of UD plates and laminates was 0.7 (Epocyl™ 128-02) and 0.47 (Nano-Force®) by volume. Samples of the composite for measurements of components of resistivity tensor in main axis of symmetry of the material were strip shape with sizes 70×10×3 mm cut along and across fibre directions (in directions 1 and 2, respectively). Resistivity measurements were performed along 70 mm side. Samples for measurement of resistivity off-axes were square plates 60×60×3 mm with fibre orientation 45° to square sides.

Cross ply symmetrical composite laminates consisted totally of 8 UD layers $[0/90]_4$. Plate sizes were 190×140×2.6 mm. Resistivity measurement were performed in both in-plane directions. Control experiment was performed on plate of 60×60×2.6 mm cut under 45° (only for Epocyl™ R128-02) to the composite main axes of symmetry.

Typically, 3 to 5 samples were tested in most of experiments and average data with standard deviation are presented in figures. Exception were two control samples with orientation 45°.

Resistance was measured by two-electrode method on direct current within linear segment of voltage-current dependence (that follows Ohm's law) up to 40 V. Silver paste contacts were created on opposite faces of samples and cover all its surfaces. DMM4020 Digital Multimeter by Tektronix, Inc. was used for voltage measurements with resolution 100 μV in range up to 20 V. Laboratory Power Supply PS 200 B by Elektro-Automatik GmbH was used for power supply that provides DC stability better than 0.02 %.

3. Modelling of composite electrical conductivity

3.1. Micro-scale
Fibre reinforced UD composite in micro-scale could be considered as a set of long parallel fibres placed in a polymer matrix. Tensor of electrical conductivity of this monotropic material in main axis of symmetry could be written as

$$s_{ij} = \begin{pmatrix} s_{11} & 0 & 0 \\ 0 & s_{22} & 0 \\ 0 & 0 & s_{33} \end{pmatrix} \quad (1)$$

If fibre direction is along axis 1, one can assume that $s_{22} = s_{33}$ and only two independent components of the tensor fully characterise the material. Components of the tensor for UD GFRP monolayer s_{11} and s_{22} could be calculated using conductivity of its structural components: matrix (superscript "m") and fibre (superscript "f"). Rule of mixture (ROM) is the most reasonable to calculate longitudinal component of conductivity. Several equations could be used for calculation of transversal component similarly as it is accepted in calculation of thermal conductivity, diffusivity, e.g. [12]

$$s_{11} = \eta s_{11}^{f} + (1-\eta)s^{m},\; s_{22} = s^{m}\left[1 + \frac{\eta}{s^{m}/(s_{22}^{f} - s^{m}) + (1-\eta)/2}\right] \quad (2)$$

3.2. Macro-scale: monolayer
Sample of UD composite could be cut off main axes. In this case coordinate axis are rotated in plane 1–2 on angle θ and the tensor components are transformed as

$$s'_{kl} = \begin{pmatrix} s'_{11} & s'_{12} & 0 \\ s'_{21} & s'_{22} & 0 \\ 0 & 0 & s'_{33} \end{pmatrix} \quad (3)$$

where $s'_{kl} = s_{ij}\cos(\alpha_{ki})\cos(\alpha_{lj})$ and $\alpha_{2'2} = \alpha_{1'1} = \theta$. Respectively for this case:

$$s'_{11} = s_{11}\cos^2\theta + s_{22}\sin^2\theta,\; s'_{22} = s_{11}\sin^2\theta + s_{22}\cos^2\theta, \quad (4)$$

$$s'_{12} = s'_{21} = (s_{22} - s_{11})\sin\theta\cos\theta \quad (5)$$

$$s'_{33} = s_{33}. \quad (6)$$

Let us consider a specific case for a sample cut with $\theta = 45°$. In-plane tensor components could be calculated using equations:

$$s'_{11} = \frac{1}{2}(s_{11} + s_{22}),\; s'_{22} = \frac{1}{2}(s_{22} + s_{11}),\; s'_{12} = \frac{1}{2}(s_{22} - s_{11}). \quad (7)$$

It follows from equation (7) that for $\theta = 45°$ in-plane components of conductivity are equal to average value of both components in main axes.

3.3. Macro-scale: laminate with symmetrical layup
A set of stacked UD layers creates a laminate and is the most interesting for practical application. Two specific cases of symmetric layup $[0/90°]_n$ and $[\pm\theta]_n$ are presented in Figure 1.

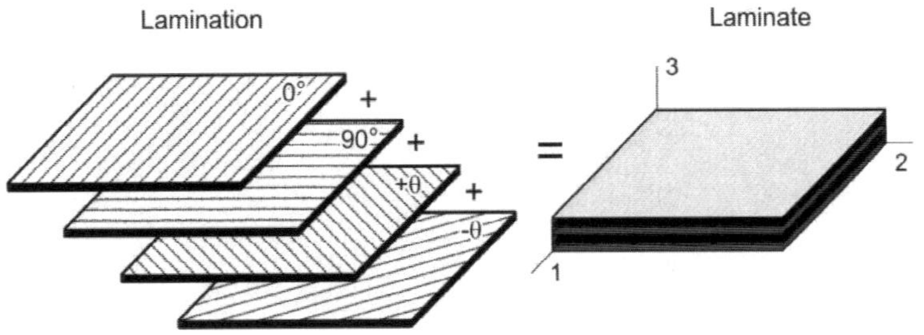

Figure 1. Macro-scale of composite with symmetrical layup with orientation of layers 0/90°, ±θ.

If a lamina of thickness H consists of N layers' conductivity may be calculated using expression

$$S'_{kl} = \frac{1}{H}\sum_{i=1}^{N} h_i s'_{kl(i)} \tag{8}$$

In-plane conductivities 11 and 22 can be written as

$$S'_{11} = \frac{1}{H}\sum_{i=1}^{N} h_i (s_{11}\cos^2\theta_{(i)} + s_{22}\sin^2\theta_{(i)}) \tag{9}$$

$$S'_{22} = \frac{1}{H}\sum_{i=1}^{N} h_i (s_{11}\sin^2\theta_{(i)} + s_{22}\cos^2\theta_{(i)}) \tag{10}$$

Let's consider two specific cases of symmetrical layup:

1. The first one, $\boldsymbol{\theta} = \mathbf{0/90}°$ layup

$$S'_{11} = \frac{1}{2}(s_{11} + s_{22})\text{ , } S'_{22} = \frac{1}{2}(s_{22} + s_{11}) \tag{11}$$

2. The second case, layup with orientation of layers $\boldsymbol{\theta} = \pm\,\mathbf{45}°$. This gives expressions

$$S'_{11} = \frac{s_{11}}{H}\sum_{i=1}^{N} h_i \cos^2\theta_{(i)} + \frac{s_{22}}{H}\sum_{i=1}^{N} h_i \sin^2\theta_{(i)} \tag{12}$$

$$S'_{22} = \frac{s_{11}}{H}\sum_{i=1}^{N} h_i \sin^2\theta_{(i)} + \frac{s_{22}}{H}\sum_{i=1}^{N} h_i \cos^2\theta_{(i)} \tag{13}$$

Simplify we will get similar to equation (11):

$$S'_{11} = \frac{1}{2}(s_{11} + s_{22}) \tag{14}$$

$$S'_{22} = \frac{1}{2}(s_{22} + s_{11}) \tag{15}$$

$$S_{33} = s_{33} = s_{22} \tag{16}$$

$$S_{12} = S_{21} = 0 \tag{17}$$

4. Experimental results and discussion

Two components of the tensor were experimentally measured and it was determined that the degree of anisotropy (ratio of the tensor components) is higher than one order of magnitude. Tensor components of resistivity s_{11} and s_{22} for non- and reinforced samples are shown in Figure 2.

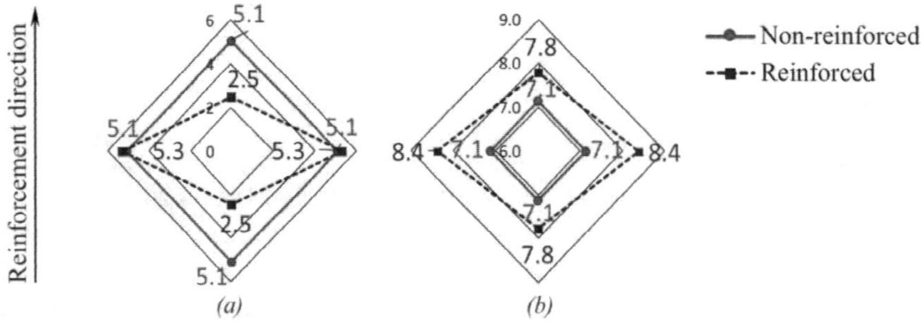

Figure 2. Polar plot of resistivity (ln ρ, Ohm·m) for non- and reinforced modified polymer matrices (*a* – Epocyl™ R128-02; *b* – Nano-Force® E200).

Due to anisotropy of reinforced modified polymer matrices, untypical data of conductivity of fibres s_{11}^{f} and s_{22}^{f} was supposed for calculations to avoid discrepancy in conductivity of the components and composite, and may be considered as some effective characteristic of the component but not of fibre material – glass itself. Possible reason of this discrepancy may be difference in bulk and micro conductivity of the components and boundary layer of matrix, but this is a point of interest for further experimental research and modelling.

Calculated and experimental data for nanomodified composites based on Epocyl™ and Nano-Force® masterbatches for all cases are presented in Figure 3.

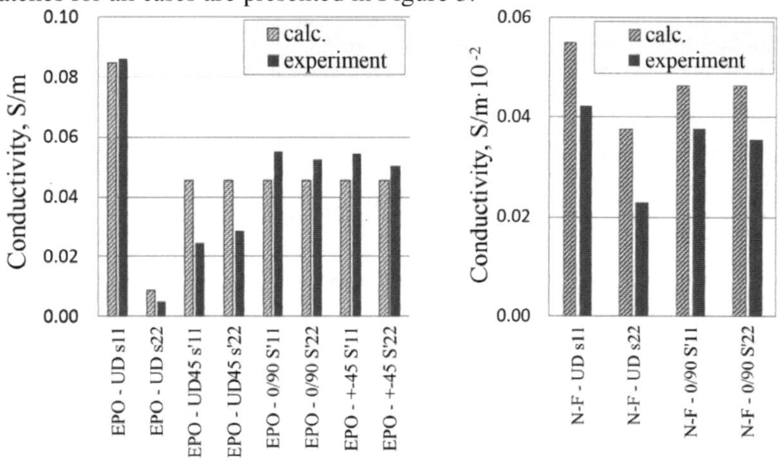

Figure 3. Experimental and calculated data of conductivity of CNT modified polymer matrices (*EPO* – Epocyl™ 128-02 [14], *N-F* – Nano-Force® E200), UD composite in main axes (UD s_{11} and UD s_{22}), UD composite cut under θ = 45° (UD45 s'_{11} and UD45 s'_{22}), cross ply [0/90°]$_4$ laminate (S'_{11} and S'_{22}), and laminate [+-45°]$_4$ (S'_{11} and S'_{22}).

Conductivity for composite based on Epocyl™ 128-02 is greatly higher than for Nano-Force® E200 one. These are different materials from two manufacturers and CNT concentrations are different. CNT concentration of polymer matrix for Epocyl™ 128-02 based composite is higher than for Nano-Force® E200 one. More detailed investigation is needed for in-depth understanding of composite reinforcement-conductivity correlation.

Despite the strong conductivity difference of tested materials, the calculated data are in a good agreement with the experimental values for both composites within small and high CNT concentration range.

5. Conclusions

The effective electrical conductivity of the matrix and composites were determined experimentally. Anisotropy due to orientation of non-conductive fibers was taken into account by introducing the corresponding anisotropic conductivity tensor for each ply. Measurements of electrical conductivity were made for unidirectional single- and multi-ply composites cut on various angles, as well as for orthotropic cross-ply GFRP laminates. The experimental and calculated data for both polymer matrices is in reasonable agreement, thus prediction method based on structural approach can be used for GFRP electrical conductivity calculating. Additional functionality of the composite could be used for monitoring of damage in the GFRP lamina with CNT-doped polymer matrix via electrical conductivity methods and their application for non-destructive in-service integrity monitoring of multifunctional composite panels in constructions.

Acknowledgments

Authors thank technician Viktor Novikov for sample preparation. The research leading to part of these results has received the funding from European Regional Development Fund project. ERDF project identification No.: 1.1.1.1/16/A/141.

References

[1] JEC Group 2018 Overview of the glass fibre composite market *JEC Composites Magazine* **123** 13-14
[2] Bandaru P 2007 Electrical properties and applications of carbon nanotube structures Journal of nanoscience and nanotechnology **7** 4-5
[3] Viets C and Schulte K 2014 Damage mapping of GFRP via electrical resistance measurements using nanocomposite epoxy matrix systems *Composites Part B: Engineering* **65** 80-8
[4] Haj-Ali R, Zemer H, El-Hajjar R and Aboudi J 2014 Piezoresistive fiber-reinforced composites: A coupled nonlinear micromechanical–microelectrical modeling approach *International Journal of Solids and Structures* **51** 491-503
[5] Liberson S, Walsh R and Roemer J 2009 Damage quantification in electrically conductive composite laminate structures *Defense Technical Information Center* p 10
[6] Yamane T and Todoroki A 2017 Analysis of electric current density in carbon fiber reinforced plastic laminated plates with angled plies *Composite Structures* **166** 268-76
[7] Naghashpour A and Van Hoa S 2014 A technique for real-time detecting, locating, and quantifying damage in large polymer composite structures made of carbon fibers and carbon nanotube networks *Structural Health Monitoring* **14** 35-45
[8] Kulakov V, Aniskevich A and Ivanov S 2017 Effective electrical conductivity of carbon nanotube–epoxy nanocomposites *Journal of Composite Materials* **51** 2979-88
[9] Georgousis G and Pandis C, et al. 2015 Strain sensing in polymer/carbon nanotube composites by electrical resistance measurement *Composites Part B: Engineering* **68** 162-9
[10] Rajinder P 2007 On the electrical conductivity of particulate composites *Journal of Composite Materials* **41** 2499-511
[11] Park B, Okabe T, Takeda N and Curtin W 2002 Electromechanical modeling of unidirectional CFRP composites under tensile loading condition *Composites Part A: Applied Science and Manufacturing* **33** 267-75
[12] Christensen M 2012 *Mechanics of Composite Materials* Dover Publications **54** 414
[13] Kulakov V, Aniskevich A and Poltimae T 2015 Effective electrical conductivity of CNT-epoxy nanocomposites *20th International Conf. on Composite Materials* (19-24 July 2015, Copenhagen, Denmark)
[14] Aniskevich A, Stankevich S and Sevcenko J 2018 Structural approach for prediction of electrical conductivity of nano-modified glass fibre reinforced plastics *18th European Conf. on Composite Materials* (24-28 June 2018, Athens, Greece) p 7

IOP Conf. Series: Materials Science and Engineering **500** (2019) 012011 doi:10.1088/1757-899X/500/1/012011

The effect of pH on the ability of different lignins to stabilize the oil-in-water emulsion

Galia Shulga, Sandra Livcha, Brigita Neiberte, Anrijs Verovkins, Sanita Vitolina, Elina Zhilinska

Laboratory of Lignin Chemistry, Latvian State Institute of Wood Chemistry, 27 Dzerbenes St., LV-1006, Riga, Latvia

E-mail: shulga@junik.lv

Abstract. The ability of aspen soda and wheat straw soda lignin to stabilize the "rapeseed oil-in-water" (O/W) emulsion, depending on the pH value and concentration of the applied water solutions as well as on the lignin chemical composition and molecular mass, and the sizes and zeta potential of lignin particles was studied. It was shown that the O/W emulsion stabilized with the alkaline water solutions of wheat straw lignin demonstrated the highest stability in alkaline medium, while the ability of aspen soda lignin to prevent the coalescence of oil droplets manifested itself in a whole range of the studied pH values. With decreasing pH values of the wheat straw lignin water solution, the drop in the separation volume of the O/W emulsion was insignificant with the exception for the lignin enhanced concentrations in acidic medium. The aggregative stability of the O/W emulsion, which was stabilized with aspen soda lignin, increased with decreasing pH values and increasing lignin content in the emulsion. The findings indicated that the stabilization of the O/W emulsion with the alkaline lignin solutions proceeded via the electrostatic stabilization mechanism, while in the neutral and acidic medium, this mechanism was more complicated, with a growing role of the ability of lignin to form interfacial polymeric films at the liquid/liquid interface. The comparison of the stability of the O/W emulsion showed that aspen soda lignin was a better stabilizer of the O/W emulsion than wheat straw lignin.

1. Introduction

The need for efficient surface active agents, including detergents, surfactants and dispersants, grows annually and has led to ongoing researches to develop the surface active agents not only with an enhanced surface activity, but also with a pronounced biodegradability. In this connection, much attention is given to the rational utilization of biomass for the replacement of the existing non-eco-friendly products with bio-based ones. One of the representatives of biomass is wood lignin that accounts for 15-30% of the total biomass. It has an amorphous polyphenolic nature arising from the enzyme-mediated dehydrogenative polymerization of three phenylpropanoid monomers – coniferyl, sinapyl and p-coumaryl alcohols – in wood and grass. Lignin content generally varies from 24 to 33% in softwood, from 19 to 28% in hardwoods and from 11 to 27% in nonwoody lignocellulosic fibres [1, 2]. The peculiarities of the structure and chemical composition of lignin, combined with its renewability, low cost, biocompatibility and biodegradability, make the design of new value added lignin-based products a challenge for fundamental and applied research. The chemical properties of lignin are highly dependent on the wood or grass species, and the pulping and recovery process. These factors affect the chemical composition, reactivity and properties of lignin. According to the literature analysis, the application of technical lignins for obtaining surface active agents such as stabilizers, emulsifiers, detergents and dispersants for treatment of asphalt, latex, soap, cement, clay, etc. is a very promising field [3-9], which takes the first place in comparison with lignin application in other areas, namely, for obtaining resins, biocomposites, concrete and chemicals.

In this work, aspen soda and wheat straw soda lignin were comparatively evaluated as a potential stabilizer of the oil-in-water emulsion.

2. Experimental

Hardwood lignin was obtained by soda delignification of pre-hydrolyzed aspen sawdust. The pre-hydrolysis was performed under laboratory conditions with the purpose to enhance the yield of lignin during wood delignification as well as to decrease the energy consumption of aspen sawdust in the process of its milling for obtaining microparticles used as a filler for wood-polymer composites [10]. The hydrolysis parameters of the sawdust were 0.1% HCl at 60°C for 3 h. The delignification was carried out with a 4.0% NaOH solution at 165°C at the hydromodulus 1:7 (sawdust/water) for 1.5 h using a laboratory autoclave at a pressure 0.45 MPa. Soda lignin was precipitated from the black liquor with a 20% sulphuric acid, with the following filtration, washing with distilled water and drying. Wheat straw soda lignin was a commercial product of the biorefinery process (GreenValue SA, Switzerland).

The chemical composition of lignin was studied by elements (Elementar Analysensysteme GmbH, Germany) and functional groups (methoxyl and aliphatic hydroxyl groups (OH_{alif}), carbonyl groups, phenolic hydroxyl groups (OH_{ph}), carboxyl groups) analysis using the Fibok-Shvappakh method, acetylation, interaction with hydroxylamine hydrochloride, and potentiometric and conductometric titration (InoLab level 3), according to Zakis [11]. The average molecular weights of lignins (Mw, Mn) were studied by SEC-MALS20 (Malvern, United Kingdom) in DMSO with lithium bromide as an eluent at 60°C. The average molecular mass (M_w) of aspen and straw lignin was 10.8 kDa and 15.7 kDa, respectively. The oil-in-water (40/60 v/v) emulsions (O/W) were prepared with a Disperser T10 (IKA) (γ = 9500 rpm) for mixing of 1 min. A commercial rapeseed oil (saturated fat acids – 61%, polyunsaturated fat acids – 32%, saturated fat acids – 7%) was used as the oil phase. The water phase of the emulsion was adjusted to various pH values by adding 0.1 M NaOH or 0.1 M HCl. The stabilizing effect of lignin was evaluated according to the final volume of the separation (H) of the emulsion after its centrifugation with a centrifuge Hettich EBA 20 at 6000 rpm for 30 min. The velocity of separation was evaluated on the basis of the volume of the released water phase during the first ten minutes after the preparation of the emulsion. Due to the good comparability of the obtained results, a number of the repeated measurements of the final separation volumes of the emulsion samples and their separation velocities, obtained by the centrifugation for each sample, was equal to three, and the final results represented the arithmetic mean values.

The size and zeta-potential (Z) of lignin particles and oil droplets were determined with a ZETASIZER NANO ZS Malvern Instrument (Malvern, United Kingdom) at 25°C. Five replicates were made for each measuring, and standard deviation for each index was calculated.

3. Results and Discussion

The chemical composition of aspen soda lignin and wheat straw soda lignin is given in Table 1. According to the obtained data, the lignins differ in the content of the functional groups. Straw lignin contains more carboxyl and carbonyl groups, but less phenolic and aliphatic hydroxyl groups in comparison with aspen lignin. The degree of oxidation calculated as the ratio O/C for straw and aspen lignin was 0.51 and 0.47, respectively. A higher ratio of the degree of oxidation for straw lignin may be associated with the higher content of carboxyl and carbonyl groups. The ability of the lignins to stabilize the rapeseed oil-in-water emulsion was examined as a function of pH and the concentration of the applied lignin solutions. The separation process of the emulsion resulted in a distinct interface between the water and oil phase. The initial emulsion samples had a white colour, but after a few

Table 1. Chemical composition of lignins

Lignin	C %	H %	O %	N %	S %	OCH_3 %	CO %	OH_{ph} %	COOH %	OH_{alif} %
Aspen	63.3	6.5	29.8	0.2	< 0.3	18.30	2.12	4.49	3.15	3.31
Straw	61.2	5.7	31.3	1.6	< 0.3	12.50	2.69	3.57	5.95	1.46

minutes, the lower part of the test-glass was coloured with a brown colour. Figure 1 shows the values of the separation volume of the rapeseed oil-in-water (O/W) emulsion depending on the pH of the water phase, containing the lignins with a concentration of 0.006%-0.1%. The O/W emulsion stabilized with the alkaline straw lignin solutions demonstrates a higher stability compared with aspen lignin. With increasing content of straw lignin in the water solution, the stability of the emulsion changes insignificantly. The emulsion prepared with the neutral solutions of straw lignin is characterized by the lowest stability that slightly increases with growing lignin content in the O/W emulsion. In acidic medium, the ability of straw lignin to prevent the separation of the O/W emulsion does not change for its low concentrations in the water solutions but grows for a concentration of 0.05-0.1%.

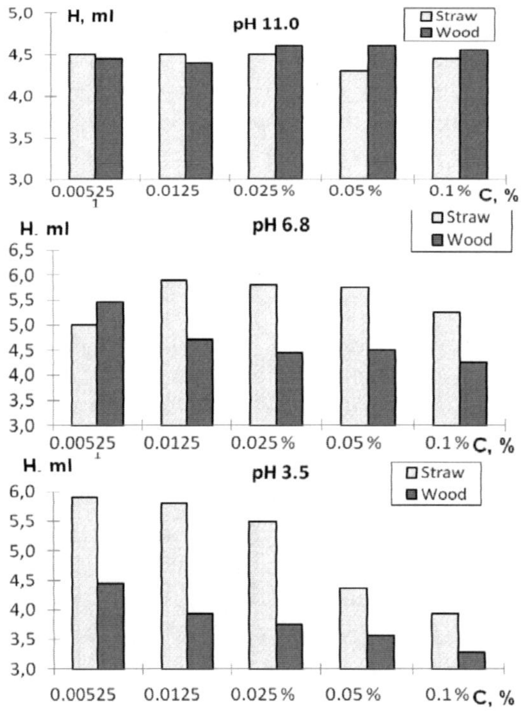

Figure 1. Separation volume (H) of the O/W emulsion depending on the lignin solution concentration at different pH values

In contrast, the highest ability of aspen lignin to stabilize the emulsion appears in neutral and acidic media. With decreasing pH value and increasing content of aspen lignin in a water solution, the stability of the emulsion essentially increases and is characterized by the remarkably lower separation volumes compared with the emulsion stabilized with straw lignin. The effect of the pH value and concentration of the lignins on the velocity of the emulsion separation for the first ten minutes after its preparation is given in Table 2. It can be seen that, for straw lignin, the lowest velocity values of 7.5-

$9.5 \cdot 10^{-2}$ ml/min are observed in alkaline medium, but the highest separation velocity, reaching 55-$59 \cdot 10^{-2}$ ml/min, takes place when the straw lignin neutral and acidic solutions are applied. Aspen lignin is characterized by the lowest velocity of separation of $3.5\text{-}7.5 \cdot 10^{-2}$ ml/min in the alkaline and acidic medium, while in the neutral medium it has the moderate values of $17.5\text{-}22.5 \cdot 10^{-2}$ ml/min.

It is known that an emulsion can be stabilized via different mechanisms, namely, the introduction of a charge at the O/W interface or the formation of a steric interfacial film at the interface. In our case, the presence of ionic groups such as carboxylic and phenolic hydroxyl groups imparts an essential negative charge for lignin macromolecules in alkaline solution.

Table 2. Velocity of emulsion separation after the first ten minutes as the emulsion was prepared

Concentration, 10^{-2}, %/ Velocity, 10^{-2}, ml/min	**Aspen lignin**			**Straw lignin**		
	pH			pH		
	11.0	6.8	3.5	11.0	6.8	3.5
0.63	9.5	22.5	19.0	10.0	59.0	55.0
1.25	9.5	50.0	13.0	9.5	56.0	54.0
2.50	9.0	18.0	8.0	8.0	54.5	48.0
5.0	8.0	18.5	5.5	7.5	27.0	30.0
10.0	7.5	17.5	3.5	6.5	20.0	17.0

The lignin macromolecules, being adsorbed at the oil droplets, stabilize the O/W emulsion by forming the inner Helmholtz layer, which leads to the increase of the zeta potential and the aggregative stability of the emulsion due to the repulsion interaction between the negatively charged emulsion droplets.

According to Table 3, wheat straw soda and aspen soda lignin have the high values of zeta potential at pH 11.0. In neutral and acidic medium, the electrostatic repulsion plays a second role in the emulsion stability, but the ability of the lignins to form steric semi-rigid polymeric films at the interface comes to fore. Taking into account the values of the degree of oxidation of both the lignins, aspen soda lignin is more hydrophobic than wheat straw lignin that positively affects its surface activity.

Table 3. Size and zeta potential of the particles of hardwood soda and wheat straw soda lignin

Lignin	S, nm			Z, mv		
	pH 11.0	pH 5.8	pH 3.5	pH 11.0	pH 5.8	pH 3.5
Aspen	164 ± 8	192 ± 15	120 ± 19	-51 ± 3	-33 ± 5	-19 ± 3
Straw	363 ± 10	288 ± 21	212 ± 28	-36 ± 5	-31 ± 4	-29 ± 4

On the other hand, straw soda lignin forms associates of a larger size in the water solutions in the whole range of the studied pH values compared with aspen soda lignin (Table 3) that may be associated with a higher value of its molecular mass compared with aspen lignin. This may hamper the adsorption and orientation of the lignin macromolecules at the oil/water interface and negatively affects the formation of the interface polymeric film. Thus, aspen soda lignin is a better stabilizer for the rapeseed-water emulsion in the studied interval of concentrations and pH values of the applied water solutions.

Another way to determine the effectiveness of the emulsion stabilization is based on the evaluation of the sizes and zeta potential values of the formed droplets in the treated emulsion. Figure 2 shows that, with increasing aspen soda lignin concentration in alkaline medium from 0.006 % to 0.1%, the size of the emulsion droplets grows from 398 nm to 1260 nm, while the separation volume of the emulsion increases.

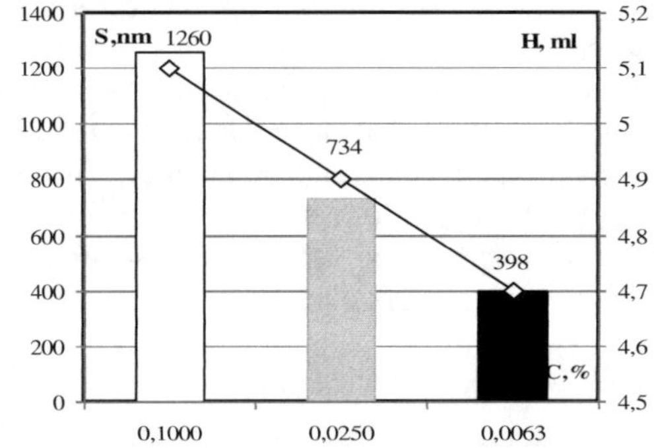

Figure 2. Size of emulsion droplets and separation volume (◊) of the emulsion depending on the applied concentration of the alkaline aspen lignin solution

Compared with the lowest applied aspen lignin concentration, the droplets' size increases almost threefold for the emulsion stabilized with the highest concentration of the aspen lignin alkaline solution. It is obvious that the higher the formed emulsion particle size, the faster the separation and the loss of the stability of the O/W emulsion.

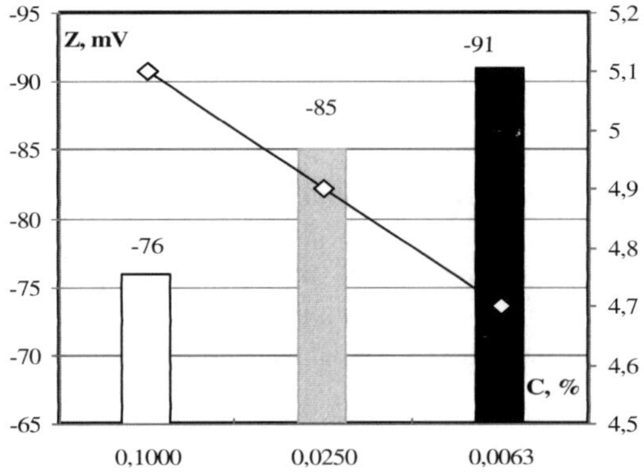

Figure 3. Zeta potential of emulsion droplets and the separation volume (◊) of the emulsion depending on the applied concentration of the alkaline aspen lignin solution

According to Figure 3, the emulsion droplets characterized by an enhanced stability have the highest negative zeta potential in a range of -76 – -91 mv that testifies the electrostatic mechanism of the enhanced stabilization of the O/W emulsion with lignin in alkaline medium.

4. Conclusions

A strong pH dependency of the surface activity of aspen soda and wheat straw soda lignin at the rapeseed oil-water interface was observed. Different mechanisms of the stabilization of the O/W emulsion with lignin particles are suggested. The findings indicate that the stabilization of the emulsion with alkaline lignin solutions proceeds via the electrostatic mechanism, while in neutral and acidic medium, with increasing lignin concentration, the stabilization mechanism is more complicated, with the growing role of the formation of steric semi-rigid polymeric films. Due to a smaller hydrophobicity and the pronounced ability to form large macromolecule associates in water solutions, which are favoured by the peculiarities of its chemical composition, wheat straw lignin is a weaker stabilizer for the O/W emulsions compared with aspen soda lignin.

IOP Conf. Series: Materials Science and Engineering **500** (2019) 012011 doi:10.1088/1757-899X/500/1/012011

5. Acknowledgements
The study was financed by the Latvian State Programme "ResProd" and the Latvian State Institute of Wood Chemistry Bioeconomy grant „AtkritBiomasKM".

6. References
[1] Fengel D and Wegener G 1984 *Wood - Chemistry, Ultrastructure, Reactions*, Walter de Gruyter Berlin and New York
[2] Forss K G and Fremer K E 2003 *The Nature and Reactions of Lignin – a New Paradigm*, OY Nord Print Ab Helsinki
[3] Chilingarian G and Voravutr P 1981 *Drilling and Drilling Fluids* Elsevier Scientific Pub. New York
[4] Aguie-Beghin V, Baumberger S, Monties B and Douillard R 2002 *Fr. Langmuir* **18** 5190
[5] Baumberger S, Aguie-Beghin V, Douillard R, Lapierre C and Monties B 1997 *Industrial Crops and Products* **6** 259
[6] Gundersen S A, Sæther Ǿ, Sjoblom J 2001 *Colloids and Surfaces A: Physicochem. Eng. Aspects* **186** 141
[7] Shulga G, Shakels V, Brovkina J, Solodovniks P and Skudra S 2008 *Proceedings of the 7th World Surfactants Congress 'CESIO 2008'* Paris France 98
[8] Shulga G, Skudra S and Shakels V 2011 *Proceedings of the 8th International Scientific and Practical Conference - Environment. Technology. Resource,* Rezekne Latvia 276
[9] Li S, Willoughby J A and Rojas O J 2016 *ChemSusChem* **9** 2460
[10] Shulga G, Neiberte B, Verovkins A, Vitolina S, Jaunslavietis J and Ozolins J 2016 *Materials Science (Medžiagotyra)* **22** 370
[11] Zakis G F *Functional Analysis of Lignins and their Derivatives*, Tappi Press, Atlanta, 1994

IOP Conf. Series: Materials Science and Engineering **500** (2019) 012012 doi:10.1088/1757-899X/500/1/012012

Ethylene-octene copolymer and carbon black composite electro-thermal properties for self-regulating heating

A Berzina[1*], I Klemenoks[1], V Tupureina[2] and M Knite[1]

[1] Institute of Technical Physics, Faculty of Materials Science and Applied Chemistry, Riga Technical University, Riga, 3/7 Paula Valdena Street, Latvia
[2] Institute of Polymer Materials, Faculty of Materials Science and Applied Chemistry, Riga Technical University, Riga, 3/7 Paula Valdena Street, Latvia

*Astrida.Berzina_1@rtu.lv

Abstract. Carbon black – ethylene-octene copolymer (EOC-CB) composite electro-thermal properties were investigated for using the composite as self-regulating heating material. The materials electrical resistance change with temperature change was investigated. Additionally, while a voltage of 30 V was applied, the sample temperature and electrical current strength was measured. The samples with 15 to 20 phr CB showed positive temperature coefficient of resistivity and could be used as self-regulating heating materials. For different concentrations of CB the achieved equilibrium temperatures were between 30°C to 60°C (with 30 V) and could be used to comfortably warm the human body.

1. Introduction

A lot of scientific effort has been put into the research of materials that exhibit positive temperature coefficient of resistivity (PTC). The electrical resistivity of materials with PTC effect increases nonlinearly with temperature [1]. This effect usually occurs with materials at their phase transition temperature – usually melting temperature – that for polymers is rather high compared to the human body's temperature [2]. Materials that have PTC can be used as self-regulating heating materials. For these materials as long as a voltage is applied to the material, the temperature increases to an equilibrium temperature and stays at that temperature.

The mechanism of PTC is still not fully understood and explained. There are many models and theories [3–6], which try to explain the effect, but all struggle with some aspect of explanation of the experimentally acquired data. This is because it is almost impossible to ascertain of the models rightness or wrongness experimentally. Nevertheless, the widely accepted opinion is that electrical resistance in polymer composites can be described by the percolation theory – the electro-conducting particles form a conductive network throughout the polymer matrix, which is responsible for conductivity [7,8]. Between close particle aggregates, a thin polymer layer prevents direct conduction, but through this layer electron tunneling can occur [8,9]. The main effect responsible for increase in resistance with increased temperature is the matrix volume expansion, either due to the difference in coefficients of linear thermal expansion between polymer and particles or due to phase transition – mostly melting of crystalline regions. Therefore, the most accepted theory is based on electron tunneling – electrons tunnel through thin dielectric polymer layers, which are between particles. The tunneling probability is highly dependent on layer thickness, so if the polymer matrix expands, the resistance increases exponentially [10,11].

Many materials have been shown to exhibit PTC behavior. The most popular being high density polyethylene as matrix material [12–15], but other types of matrix materials are epoxy [16], polypropylene [17], poly(vinylidene fluoride) [18] with different types of filler, including carbon nanotubes [14,16,17], carbon nanofiber [12,13], Ni [18] and Ag nanoparticles [19]. The biggest downside to these materials is, that the PTC temperature is above 130°C, which is way too high to use as a self-regulating heating element for the human body.

In this paper we propose a new PTC exhibiting composite material made of ethylene-octene copolymer (EOC) and carbon black (CB). Our goal is to get the PTC peak temperature comparatively low – in the range of 50-60°C, so it can be used as a self-regulating warming device for the human body. The composite's resistance vs temperature, particle distribution, and equilibrium temperatures for different phr (parts per hundred rubber) of CB have been characterized.

2. Materials and methods

The samples are prepared by mixing ethylene-octene copolymer (EOC) granules (38% octene content, Engage 8200, Dow) with carbon black (CB) powder (Printex XE-2, Evonik Industries) and dicumyl peroxide (crosslinking agent) on a heated two-roll mill (Labtech Engineering). Following that, the sample is vulcanized at 170°C under 6 kPa for 10 min. Silver paint (Spi supplies) electrodes are painted on the flat surface on opposite ends of the sample. The sample length is 95 mm (space between electrodes is 85 mm), the sample width 75 mm and thickness 0.7 mm. Samples with 10 - 30 phr of CB are prepared.

DSC measurements are made using Mettler Toledo instrument in a temperature range from 20 to 200 °C with speed 10°/min in a heating-cooling-heating regime.

The resistance dependence on temperature measurements are carried out in a SNOL 58/350 laboratory oven with an additional thermocouple attached to the sample. The temperature and current/resistance dependence on applied voltage are carried out in room temperature with a thermal insulator beneath the sample. The applied voltage is constant 30V.

For the electro conductivity measurements in microscale atomic force microscope (AFM) measurements (Smena, NT-MDT) are performed in contact mode using an electro conductive tip (CSG10/Pt, NT-MDT). A round sample of 18 mm diameter and 1 mm thickness is used. The mathematical processing to characterize the particle distribution in the composite matrix in microscale have been made as explained in [20]. In AFM electro-conducting mode the current between the AFM tip on the composite surface and the bottom electrode has been measured at an applied voltage of U = 1V. If the particles form a channel, which goes through the sample, then a current begins to flow and an increased intensity peak is observed. The electro conducting channel distribution *dIndex* is calculated by dividing the map into 9 squares and calculating the percentage of area taken up by electro conducting channels in each square. The size distribution *sIndex* is calculated by acquiring the channel size distribution, and evaluating agglomerate presence. The average of the two indexes is *cIndex* – which describes the sample overall.

3. Results and discussion

3.1. Thermal properties

Table 1. DSC measurement values for melting heat, crystallization heat and peak temperatures.

	Melting heat (Jg^{-1})	Melting peak temperature (°C)	Crystallization heat (Jg^{-1})	Crystallization peak temperature (°C)
10 phr	-11.14	59.72	8.97	37.46
15 phr	-10.28	59.39	7.37	36.78
20 phr	-11.36	58.88	6.08	39.78
25 phr	-9.48	56.74	5.03	36.59
30 phr	-7.69	54.54	4.86	39.45

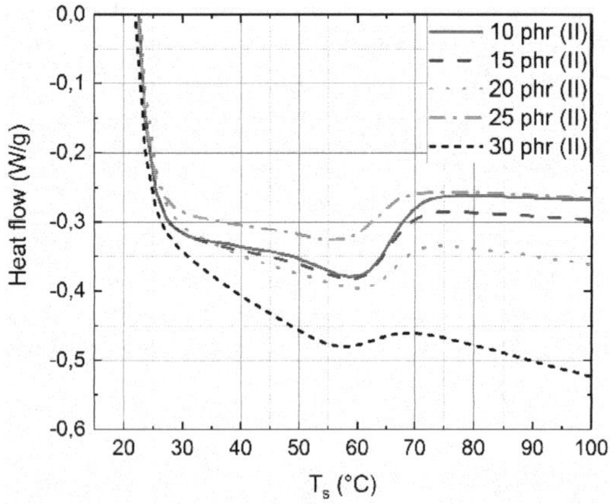

Figure 1. DSC measurements of EOC-CB composites in the second heating cycle.

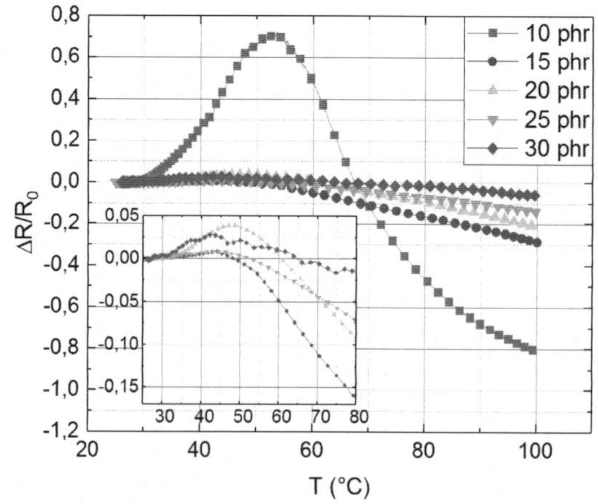

Figure 2. Relative electrical resistance change with temperature for different phr of CB samples.

DSC thermograms of EOC-CB for the second heating cycle are shown in figure 1. As can be seen, the melting occurs at approximately 55 – 60°C, which is consistent with peroxide cured EOC in different literature [21]. During the cooling cycle the matrix undergoes crystallization at approximately 36 - 40°C. The values for melting heat, crystallization heat, and peak temperature are shown in table 1. The PTC effect is considered to be until around 55 °C when the matrix undergoes the most significant expansion. This is confirmed by resistance-temperature measurements (shown in figure 2). The highest relative resistance change maxima is for 10 phr, which has the smallest CB concentration and has the most crystalline matrix (see table 1). All the other samples show an increase in resistivity until 50°C, however the increase is relatively smaller, due to the large amount of CB preventing the polymer from having more crystalline regions. Contrarily, for the material to work as a heating element for voltages around 30 V and lower, the base resistance needs to be appropriately small. Samples with 10 phr did not undergo noticeable heating if 30 V were applied, so the next best concentration for our purposes seems to be in the range of 15 phr – 20 phr.

3.2. Electrical heating properties

A 15 phr sample heating experiment in the time scale is given in figure 3. First, base electrical resistance is measured for 30 seconds, then a constant voltage of 30 V is applied and the current strength is measured. After 25 minutes an equilibrium state is reached, in which the temperature of the sample does not increase significantly after that point. Then the voltage is switched off and the resistance of the sample while cooling/relaxing is measured. By increasing the temperature, the electrical resistance increases, however the calculated power (P = IU) decreases. This shows that the sample exhibits PTC effect. After 5 minutes of heating, the sample has already reached ~65% of the maximum equilibrium temperature, which shows that the sample heating rate is moderate (0,7°C/min). In figure 4, you can see the maximum temperature reached (equilibrium) while electrically heating the samples with 30 V for each filler load.

In figure 5, the current strength vs time for 15-25 phr CB samples is shown. For 15 and 20 phr of CB the current strength decreases (resistance increases) with time and temperature increase. The electro-conducting particle amount for 15 and 20 phr of CB is low enough so that the damage to the electro-conducting network done by thermal expansion is the dominant mechanism for sample electrical resistance increase (current strength decrease). For 25 phr of CB sample the current strength first drops slightly, but afterwards increases significantly. At higher loads of CB the mechanism for resistance change is interchanging between two effects – the thermal expansion with consecutive resistance increase and the resistance decrease due to formation of new conductive channels in the region of melting when the temperature increases. If the filler load is sufficiently high, then the temperature increase results in an increased electrical current flow through the sample. In this case, the thermal fluctuation induced tunneling conduction takes place between electro-conducting particles and the electrical resistance is reduced [22].

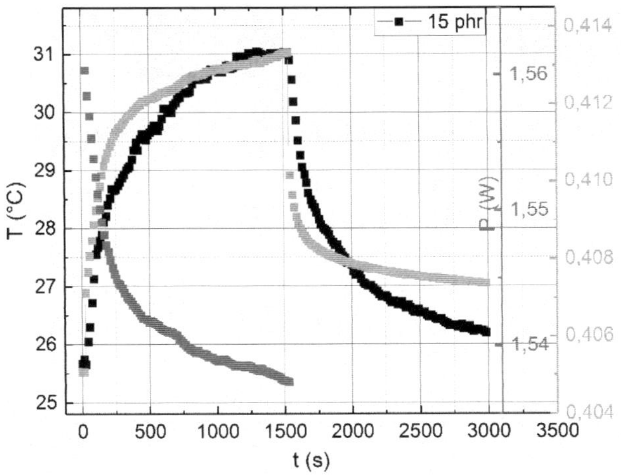

Figure 3. Temperature, resistance and power curves vs time for 15 phr CB samples. For the first 25 minutes the applied voltage is 30 V.

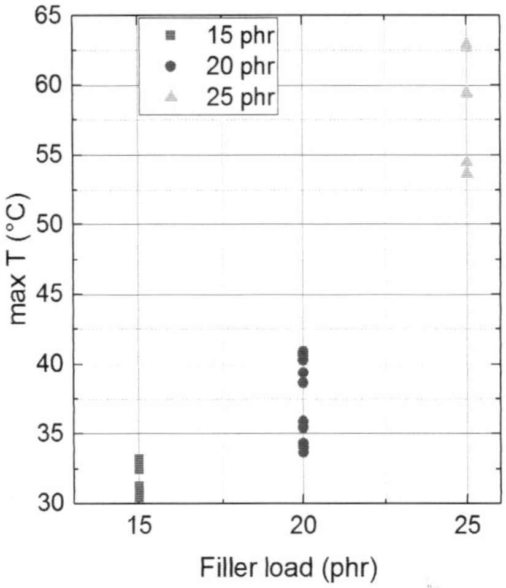

Figure 4. Maximum temperature reached for different phr of CB samples while electrically heating with 30V.

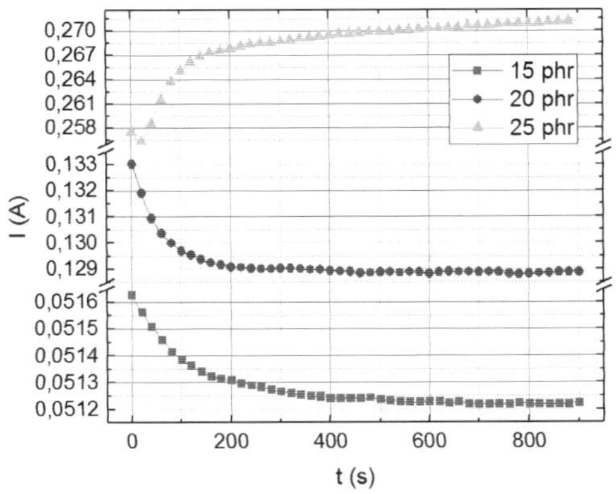

Figure 5. Current strength vs time, while electrically heating 15, 20 and 25 phr of CB samples.

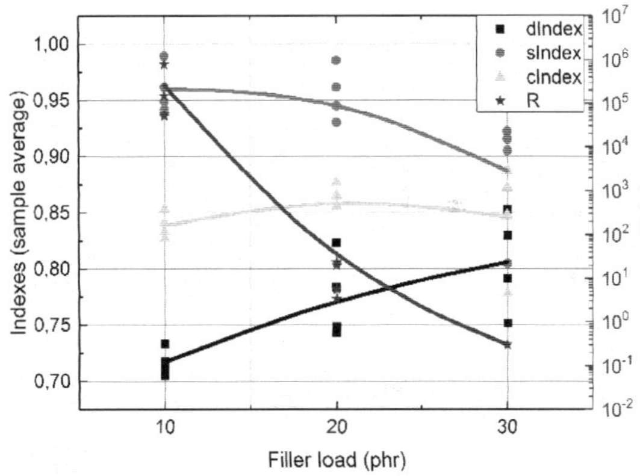

Figure 6. *sIndex*, *dIndex*, *cIndex* (left axis) and sample electrical resistance (right axis) depending on parallel samples.

3.3. Nano-scale characterization

The electro conducting channel distribution indexes for parallel samples for 20 and 30 phr , compared to 10 phr samples (figure 6), are quite largely dispersed, which can be explained by overload of CB particles. The effect is more pronounced for 30 phr samples with an electrical resistance under an ohm. These samples have a lot of filling material what makes them more electro-conducting and the channels fuse together on the AFM electro-conducting mode map (figure 7), reducing the *sIndex* values. Overall, for different phr samples the highest *cIndex* (mathematical average of the other two indexes) value is for samples with 20 phr CB, what seems to be the optimal concentration for particle distribution. This concentration showed promise also from the thermal point of view, so it seems to be

IOP Conf. Series: Materials Science and Engineering **500** (2019) 012012 doi:10.1088/1757-899X/500/1/012012

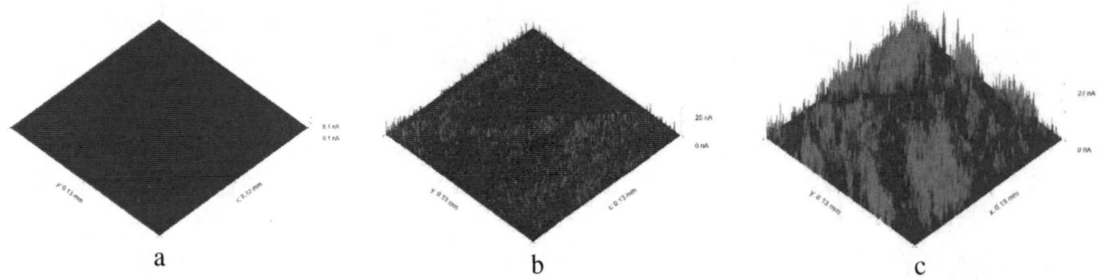

Figure 7. AFM electro-conducting mode images (130x130 μm) of 10 (a), 20 (b) and 30 (c) phr of CB. Red peaks show electro conducting channels and blue planes – dielectric regions.

the best for a self-regulating heating material, as the electrical resistance is a couple of ohms and the dispersion is quite uniform.

4. Conclusions

In this paper we propose a new PTC exhibiting composite material made of ethylene-octene copolymer and carbon black. The composite's most intense PTC effect is in the 30-50°C region that corresponds to maximal temperatures with 30 V at around 30-34°C for 20 phr samples, therefore it can be used as a self-regulating warming device for the human body. With a higher filler load, the required voltage to get an equilibrium temperature of around 38°C can be minimized, and as a result, the self-regulating device could be powered with a USB connection (5V, 0.5A).

References

[1] Voet A 1981 Temperature Effect of Electrical Resistivity of Carbon Black Filled Polymers *Rubber Chem. Technol.* **54** 42–50

[2] Narkis M, Ram A and Flashner F 1978 Electrical properties of carbon black filled polyethylene *Polym. Eng. Sci.* **18** 649–53

[3] Kohler F 1962 Resistance element, US3243753A

[4] Ohe K and Naito Y 1971 A New Resistor Having An Anomalously Large Positive Temperature Coefficient *Jpn. J. Appl. Phys.* **10** 99–108

[5] Meyer J 1973 Glass transition temperature as a guide to selection of polymers suitable for PTC materials *Polym. Eng. Sci.* **13** 462–8

[6] Gunes I S, Jimenez G A and Jana S C 2009 Carbonaceous fillers for shape memory actuation of polyurethane composites by resistive heating *Carbon* **47** 981–97

[7] Strumpler R and Glatz-Reichenbach J 1999 Conducting Polymer Composites *J. Electroceramics* **3** 329–46

[8] Balberg I 2002 A comprehensive picture of the electrical phenomena in carbon black-polymer composites *Carbon* **40** 139–43

[9] Kirkpatrick S 1973 Percolation and Conduction *Rev. Mod. Phys.* **45** 574–88

[10] Chin S J, Vempati S, Dawson P, Knite M, Linarts A, Ozols K and McNally T 2015 Electrical conduction and rheological behaviour of composites of poly(ε-caprolactone) and MWCNTs *Polymer* **58** 209–21

[11] Knite M, Ozols K, Fuith A, Aulika I and Orlovs R 2016 Photo-thermal electrical resistance response of polyisoprene/nanographite composites *Polymer* **85** 55–60

[12] Zhang X, Zheng S, Zou H, Zheng X, Liu Z, Yang W and Yang M 2017 Two-step positive temperature coefficient effect with favorable reproducibility achieved by specific "island-bridge" electrical conductive networks in HDPE/PVDF/CNF composite

Compos. Part A Appl. Sci. Manuf. **94** 21–31

[13] Zhang X, Zheng X, Ren D, Liu Z, Yang W and Yang M 2016 Unusual positive temperature coefficient effect of polyolefin/carbon fiber conductive composites *Mater. Lett.* **164** 587–90

[14] Zha J W, Wu D H, Yang Y, Wu Y H, Li R K Y and Dang Z M 2017 Enhanced positive temperature coefficient behavior of the high-density polyethylene composites with multi-dimensional carbon fillers and their use for temperature-sensing resistors *RSC Adv.* **7** 11338–44

[15] Asare E, Evans J, Newton M, Peijs T and Bilotti E 2016 Effect of particle size and shape on positive temperature coefficient (PTC) of conductive polymer composites (CPC) - a model study *Mater. Des.* **97** 459–63

[16] Cao X, Lan Y, Wei Y, Zheng G, Dai K, Liu C and Shen C 2015 Tunable resistivity-temperature characteristics of an electrically conductive multi-walled carbon nanotubes/epoxy composite *Mater. Lett.* **159** 276–9

[17] Li G, Hu C, Zhai W, Zhao S, Zheng G, Dai K, Liu C and Shen C 2016 Particle size induced tunable positive temperature coefficient characteristics in electrically conductive carbon nanotubes/polypropylene composites *Mater. Lett.* **182** 314–7

[18] Kono A, Shimizu K, Nakano H, Goto Y, Kobayashi Y, Ougizawa T and Horibe H 2012 Positive-temperature-coefficient effect of electrical resistivity below melting point of poly(vinylidene fluoride) (PVDF) in Ni particle-dispersed PVDF composites *Polymer.* **53** 1760–4

[19] Rybak A, Boiteux G, Melis F and Seytre G 2010 Conductive polymer composites based on metallic nanofiller as smart materials for current limiting devices *Compos. Sci. Technol.* **70** 410–6

[20] Berzina A, Tupureina V, Klemenoks I and Knite M 2016 A Method for Dispersion Degree Characterization Using Electro Conductive Mode of Atomic Force Microscopy *Procedia Computer Science* **104** 338–45

[21] Babu R R, Singha N K and Naskar K 2011 Effects of mixing sequence on peroxide cured polypropylene (PP)/ethylene octene copolymer (EOC) thermoplastic vulcanizates (TPVs). Part. II. Viscoelastic characteristics *J. Polym. Res.* **18** 31–9

[22] Sheng P, Sichel E K and Gittleman J I 1978 Fluctuation-induced tunneling conduction in carbon-polyvinylchloride composites *Phys. Rev. Lett.* **40** 1197–200

Acknowledgments
The financial support of RTU Doctoral Research Scholarship is greatly acknowledged.

IOP Conf. Series: Materials Science and Engineering **500** (2019) 012013 doi:10.1088/1757-899X/500/1/012013

Properties of sustainable composites based on bio-based polyamides and recycled carbon fibers under static and cyclic loads

M. Feldmann

Kassel, Germany

E-mail: Feldmann82@arcor.de

Abstract. The demand for high-performance materials increases in the last decade and sustainability of materials becomes more important as well. Different bio-based polyamides of type x.10 were compounded with 30wt% recycled carbon fibers to achieve a material with high mechanical properties with low CO_2 emissions. The micro structure was evaluated by using high-resolution tomography (voxel size under 1μm) and scanning electron microscopy (SEM). Mechanical tests were done to characterize the quasi-static and fatigue performance of the composites. The carbon fiber composites show a significant increase of Youngs' Modulus and tensile strengths compared to the neat polyamides. The fatigue strength measured by an increasing stress fatigue test is approx. 67% of the tensile strength of each studied composite. Besides the short carbon fibers (mean value about 80μm) larger impurities from the recycled carbon fibers were observed. Due to that and the fiber recycling process, the tensile strengths of these sustainable composites are lower compared to composites with virgin carbon fibers and approximately on the level of glass fiber reinforced polyamides.

1. Introduction

The technical requirements of fiber reinforced composites increased in the last few decades to compete with steels in structural applications e.g. in the aircraft industry. Thermosets resins are widely used in this field as matrix material. The amount of carbon fiber reinforced thermosets is about 50% by weight in a modern aircraft [1]. Due to this the amount of high value production waste and components with drawbacks in reuse increases, too. In addition to that, sustainability becomes more and more important in all areas of plastic industry. Recycling is an opportunity and an important issue in all areas of industries. To enhance the sustainability of parts based on carbon fiber reinforced thermosets a recycling method based on pyrolysis is commercially available and provides short carbon fibers [2][3][4]. They can be used as a reinforcement for thermoplastic materials in injection molding or extrusion processing. The recycled carbon fibers (rCF) offer higher mechanical properties compared to glass fibers and a lower price and carbon foot print compared to virgin carbon fibers [2].

Different papers about rCF and composites with rCF are already available. E.g. Feng et al [5] found that the tensile strength and the Youngs' modulus of composites based on PA 6 with rCF can be increased by a fiber treatment with DGEBA epoxy resin. A maximum tensile strength of 170 MPa and a Youngs' modulus of 20 GPa were achieved in this investigation.

Composites based on polyamides and short carbon fibers with a fiber content of 30% by weight are commercially available and structural parts like gearings and lever-arms are made of those composites [6]. Therefore, for such applications cyclic loads become more important and has to be determined.

There are just a few studies focusing on bio-based polymers using short carbon fibers, but in terms of sustainability bio-based plastics are also in the focus of interest. Up to now, the focus of bio-based plastics is mostly on packaging and single-use products. There are just a few applications using engineering bio-plastics such as bio-based polyamides. The author [7][8] studied different polyamides with glass and cellulose fibers and found approx. 7.2 GPa Youngs' modulus for PA 6.10 and PA 10.10 with 30wt% E-glass fibers and a tensile strength of about 130 MPa for PA 10.10 30GF and about 140 MPa for PA 6.10 30GF. Besides this, composites with regenerated cellulose fibers show a higher temperature sensitivity and therefore higher requirements regarding processing are necessary [8][9]. Compared to glass fibers higher impact properties can be achieved with the regenerated cellulose fibers as a reinforcement in several matrix materials and also bio-based polyamides [7]. Benefits in reduced water uptake and carbon footprint are also known for the bio-based polyamides [10].
The aim of this study is to compound bio-based polyamides with rCF and investigate the mechanical properties of the sustainable composites. The characterization of the microstructure should give an impression about the effect on the mechanical properties.

2. Materials and Methods

2.1. Materials

The bio-based polyamides 10.10 and 6.10 from the company Evonik Industry and the bio-based polyamide 4.10 from the company DSM were used as a matrix material. All bio-based polyamides are based on castor oil. Polyamide 10.10 is fully bio-based, the polyamide 6.10 approx. 62% and the polyamide 4.10 71% bio-based. From polyamide 10.10 to 6.10 to 4.10 the tensile strength, tensile modulus and heat distortion temperature increases. All investigated matrix materials show significant lower carbon footprints compared to common petrol-based polyamides with similar specifications [10].

Recycled short carbon fibers (chopped) with a treatment for polyamides from the company CarboNXT were used as fiber reinforcement.
A stabilizer package for polyamides (Polyad ®201) from the company PolyAd Services was used to protect the composite against processing and service degradation.

2.2. Composites Preparation

The materials were compounded with a co-rotating twin-screw extruder from the company Leistritz with a screw diameter of 18 mm and a L/D ratio of 40. The additive was fed together with the polymer in the main barrel and the fibers were fed directly in the melt by using a side feeder. The screw configuration was as described in [8] and the processing temperatures were set to 20°C above the melting point of each matrix material. The strand was pelletized into 3 mm granules.

After compounding the dried granules (less than 0.2% moisture) were processed with an injection moulding machine C320 Golden Edition from the company Arburg with a screw diameter of 25 mm. The mold temperature was set to 80°C and the melt temperature 20°C above the melting point.

2.3. Characterization

All composites were characterized at dry state.

The tensile test was performed according to ISO 527 and a fatigue test with increasing stress was performed to characterize the mechanical properties. The fatigue test starts with 30 MPa (upper stress level, significantly under the tensile strength or yield strength) and the stress increases every 2.5k cycles by 2.5MPa. The frequency was set to 5 Hz and the stress ratio (lower stress to upper stress) was 0.1. The and cycles to failure and the failure stress were evaluated.

Additionally, scanning electron microscopy (SEM) and high-resolution X-ray computer tomography (voxel size 0.9 µm) were performed to characterize the microstructure of the composites. The SEM images were done on the break surface after tensile test of polyamide 10.10 specimen with 30wt% rCF. The X-ray tomography was performed on a cutout at the center in longitudinal direction of a polyamide 10.10 test specimen. The half the volume of the cross-section was measured in respect to a homogeneous material behavior and to achieve a more symmetric material thickness during the measurement (in direction of rotation). The measured volume was 10 mm by 4 mm by 5 mm.

3. Results

3.1. Tensile Properties

The results of the tensile test are shown in Figure 1. As expected, the strength and Youngs' modulus of neat PA 4.10 is higher compared to PA 6.10 and PA 10.10 due to their chemical structure and number of amine-groups in the macromolecule. The composites follow the trend of the matrix materials and the rCF lead to a significant increase in strength and Young's modulus. A factor of about 20 was observed for the Youngs' modulus compared to the neat matrix material (up to 25 GPa for PA 4.10 with 30wt% rCF). This is a factor of 3 compared to composites with 30wt% glass fibers (see [7]). The tensile strength is slightly higher compared to glass fiber reinforced composites (see [7]) and approx. 30% lower compared to composites with virgin carbon fibers. The tensile strength of dry-state PA 4.10 30GF is 170 MPa, the Youngs' modulus is 9.5 GPa according to the suppliers' datasheet. A commercially PA 4.10 with 30wt% carbon fibers is not available, but comparable to the bio-based PA 4.10 is the common petro-based PA 6.6. In datasheets for PA 6.6 with 30wt% virgin carbon fibers a tensile strength up to 300 MPa and a Youngs' modulus up to 24.5 can be found.

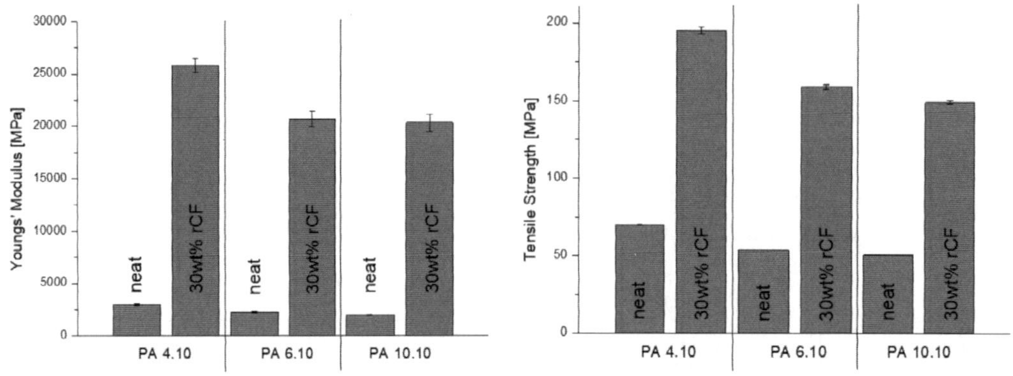

Figure 1: Youngs' Modulus (left) and tensile strength (right) of all composites and the neat matrix materials

3.2. Fatigue Properties

Figure 2 show the failure stress and the cycles to failure of all composites and the unreinforced bio-based polyamides. It can be studied that the failure stress of all composites ranges between 66% and 68% of the quasi-static tensile strength. This does not correlate to the failure stress of the unreinforced polyamides. For the polyamides the highest relative stress was observed for PA 4.10 compared to PA 6.10 and 10.10. This might be caused by the higher heat distortion temperature of PA 4.10 and therefore the resistance against the warming due to the cycle loads. Parallel to the higher failure stresses the cycles to failure increases as well. For the PA 4.10 more than 100k cycles occur before failure. The lowest cycle number was observed for the PA 10.10 composites (71k cycles). Unreinforced polyamides achieve the lowest number of cycles to failure starting from 9k cycles for PA 10.10 to 11k cycles for PA 6.10 and 28.6k cycles for PA 4.10.

IOP Conf. Series: Materials Science and Engineering **500** (2019) 012013 doi:10.1088/1757-899X/500/1/012013

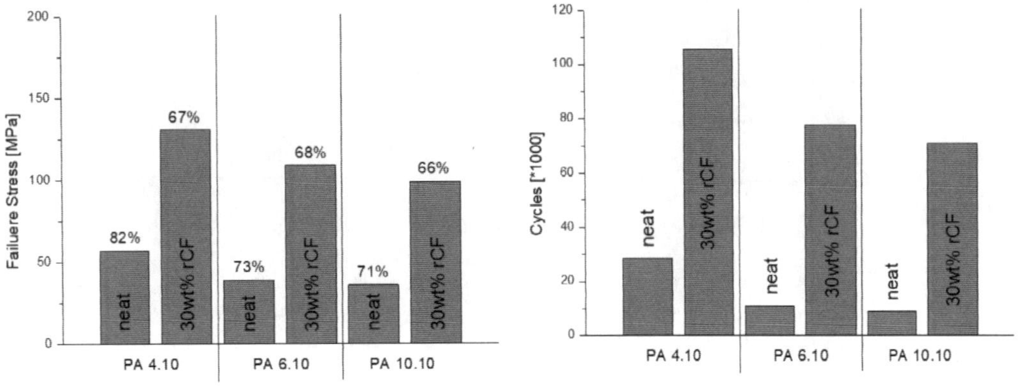

Figure 2: Fatigue properties of all composites and matrix materials – Failure stress (left) and cycles to failure (right)

3.3. Scanning Electron Microscopy (SEM)
The microstructure of a representative break surface of a PA 10.10 30rCF composite is shown on Figure 3. The fiber distribution is homogeneous and the fiber matrix adhesion is high based on visual inspection. Besides the single filaments also larger agglomerates and larger impurities can be seen on the break surface. The larger impurities cannot be allocated with fiber-like material. The fiber agglomerates could be a result of the fiber treatment for polyamides after pyrolysis. Obviously, those were not fragmented during compounding.

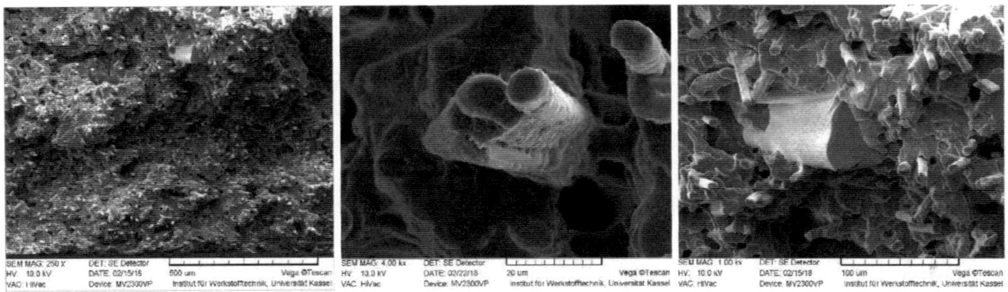

Figure 3: Microstructure of polyamide 10.10 with 30wt% rCF – Overview (left), fiber agglomeration (center) and impurity (right)

3.4. High Resolution X-ray Computer Tomography
Figure 4 shows the microstructure of a cutout of a PA 10.10 specimen with 30wt% rCF including fibers and different impurities. The impurities are significant larger compared to the fibers. The mean value of the fiber length is approx. 80 µm and the diameter is about 8 µm. The impurities have a length of approx. 200 µm and also a significant higher diameter without the typical carbon fiber shape. Unreinforced polyamides do not show these impurities. Based on the images, it can be concluded, that the impurities are mostly no fiber agglomerates and must be contained in the rCF.

Figure 4: Microstructure of polyamide 10.10 with larger impurities; Colored impurities (left), enlarged impurity (center) and sliced impurity (right)

4. Conclusions

Fully and partly bio-based polyamides were compounded with recycled carbon fibers and characterized. The composites allow to achieve outstanding properties in short fiber reinforced polyamides for injection molding applications with respect to sustainability. The significant higher Youngs' modulus and the lower density of the composites compared to glass fiber reinforces polyamides lead to benefit for light weight parts. The lower tensile strength of the composites with rCF compared to composites with virgin carbon fibers are based on the recycling process. The large impurities act as failures in the microstructure of the material. The fatigue performance of the composites measured by a fatigue test with increasing stress show a failure stress of about 67% of the tensile strength of each composite. Caused by the higher properties of PA 4.10 the mechanical properties of PA 4.10 composites are higher compared to PA 6.10 and PA 10.10 composites.

Acknowledgments

The investigations were done during my scientific work at the Institute of Materials Engineering/ Polymer Engineering at the University of Kassel. I acknowledge for the support by Mr. Fabian Kottusch, for the collaboration with Dr. A. al Mamun from the company HP Pelzer/Witten Germany and the companies Evonik Industries, CarboNXT, PolyAd and DSM for supporting the investigation with the raw materials.

References

[1] Mallick, P. K. (2007). Fiber-reinforced composites: materials, manufacturing, and design. CRC press.
[2] Pimenta, S., & Pinho, S. T. (2011). Recycling carbon fibre reinforced polymers for structural applications: technology review and market outlook. Waste management, 31(2), 378-392.
[3] PTS, P. S. (2014). Highly efficient production and characterization of CFRP made from recycled carbon fibers. SAMPE Journal, 50(3).
[4] Job, S. (2014). Recycling composites commercially. Reinforced Plastics, 58(5), 32-38.
[5] Feng, N., Wang, X., & Wu, D. (2013). Surface modification of recycled carbon fiber and its reinforcement effect on nylon 6 composites: Mechanical properties, morphology and crystallization behaviors. Current Applied Physics, 13(9), 2038-2050.
[6] https://www.luvocom.de/en/startseite/ (website, 20.09.2018)
[7] Feldmann, M., & Bledzki, A. K. (2014). Bio-based polyamides reinforced with cellulosic fibres–processing and properties. Composites Science and Technology, 100, 113-120.
[8] Feldmann, M., Heim, H. P., & Zarges, J. C. (2016). Influence of the process parameters on the mechanical properties of engineering biocomposites using a twin-screw extruder. Composites Part A: Applied Science and Manufacturing, 83, 113-119.

[9] Feldmann, M. (2016). The effects of the injection moulding temperature on the mechanical properties and morphology of polypropylene man-made cellulose fibre composites. Composites Part A: Applied Science and Manufacturing, 87, 146-152.

[10] https://www.vestamid.com/product/vestamid/en/products-services/VESTAMID-terra/ (website, 20.09.2018)

Influence of different carbon allotrope filler on the VOC sensitivity of polymer composite

Gita Sakale, Santa Stepina, Gita Revalde, Igors Klemenoks, Sandra Guzlena, Maris Knite

Institute of Technical Physics, Faculty of Materials Science and Applied Chemistry, Riga Technical University, P. Valdena 3/7, Riga, LV-1048, Latvia

Santa.Stepina@rtu.lv

Abstract. In this work, both, pristine and hybrid polyisoprene as well as ethylene vinyl acetate copolymer nanocomposites with carbon nanoparticles and carbon nanotubes were developed and studied for their application in toluene sensors. The response to toluene was determined for pristine and hybrid composite samples. Hybrid samples were compared to pristine composite samples and the composite with highest sensing effect is obtained.

1. Introduction

Electrically conductive polymers are subjects of intensive research due to great interest in different technological applications [1,2]. Our research is devoted to the development and studies of hybrid nanocomposites for their application in Volatile Organic compound (VOC) vapour sensors [3]. Detection of VOC vapour including toluene is significant for air quality monitoring. VOCs are believed to be the major contributors to poor indoor and outdoor air quality. For example, toluene is used frequently in many work operations involving paint, metal cleaners, adhesives and other products. In last 50 years, air quality sensor market is dominated by metal oxide gas sensors. Polymer based composite sensors could be a good alternative over metal oxide sensors, having several advantages like low price and functioning at room temperature. Polymer based composites also can be easily produced (dip coating, spin coating, inkjet or screen printing [4]) and the cost of one sensor unit can be comparatively low. Many properties of polymer based composites are described with percolation phenomena[5–8]. However, the results are contradictory and depend on particle dispersion technique, physical characteristics of particles and type of polymer used.

In this work both, binary and hybrid polyisoprene (Pi) or ethylene vinyl acetate copolymer (EVA) with carbon nanoparticles and carbon nanotubes were developed and studied. All elaborated composites were tested for response to toluene vapours because toluene can be found in common household products (paints, paint thinners, adhesives, synthetic fragrances and nail polish), cigarette smoke and even in gasoline at large quantities. Synergetic effect was observed and enhanced sensing properties were achieved for certain composites.

2. Materials and methods

2.1. Materials

Pi was chosen as one of the matrix because it is known as an elastomer with good both gas permeation and elastic properties. Pi was mixed with additives (sulphur, zinc oxide, stearic acid) to enable crosslinking during the curing of the raw composite mass. For second set of samples as matrix material EVA (vinyl acetate content 40%) was used. One type of filler, the extra electro-conductive, highly structured carbon black nanoparticles (CB; Printex XE2), was purchased from Degussa Co. CB has mean particle size 30 nm, dibutylphthalate (DBP) absorption - 380 ml/100 g, specific surface area - 950 m2/g. Two types of CNT (carbon nanotubes) were used as filler material as well. First type of CNT was purchased from Sigma Aldrich; the outer diameter of tubes is in range from 40 to 60 nm, inner diameter is in range from 5 to 10 nm, length 0.5-500 μm, and aspect ratio 12500. These comparatively long CNT further will be designated as CNT_l. Second type of CNT were obtained from CheapTubes; outer diameter is in range of 50-80 nm, inner diameter is from 5 to 15 nm, length 0.5-2 μm, maximal aspect ratio 40. Further these short CNT will be designated as CNT_s.

2.2. Sample preparation

Elaboration of the composite samples was performed with liquid mixing (using chloroform as solvent). At first, a solution of the liquid matrix-filler was prepared. The solution was coated on epoxy laminate substrate by dip-coating method. Dip coating procedure was followed by drying at room temperature. Composites with Pi were also cured during compression molding at both temperature of 150 °C and pressure of 30 atm. Composites containing EVA were left air dried without compression molding. More detailed instructions for preparation of the composites are given in our previous work [9]. The dimensions of the obtained samples were 10 mm x 14 mm x 0.05 mm.

2.3. Methods

Keithley picoammeter 6487 was used to measure steady state electrical resistance of composites. KinTek FlexStream™ automated permeation tube system was used for generation of different toluene vapor concentrations in a set with nitrogen as a carrier gas. Special VOCs sensing chamber was originally created for simultaneous testing of 4 composite samples. To minimize the humidity influence on the response of the sensors to definite VOCs concentration, the chamber was designed as a closed flow system connected to KinTek FlexStream™. During sample recovery the chamber was purged with pure nitrogen. Tests of toluene response determination were carried out at the room temperature (23 °C). Agilent data acquisition switch unit 34972A was applied to measure sample electrical resistance during sensing tests for toluene.

3. Results

Based on our previous researches on Pi and EVA composites as VOC sensor materials [10–14], in this work will be researched hybrid composites and its sensing abilities compared with pristine composites made with CNT and CB. If only CNT are used as filler, electro-conductive network is formed from CNT and the distance between adjacent CNT (existence of tunneling currents) determines the CNT contribution in the electro-conductive network. If both CB and CNT are used and the concentration of CNT is larger than CB, main electro-conductive network is formed by CNT, however CB aggregates contribute in electro-conductive network formation due to connection of CNT branches as described by Jan Sumfleth et al. [8]. Some of CB aggregates are settled between CNT branches, additional electro-conductive branches are joined to the electro-conductive filler network and the resistance of the composite decreases. If the CB content in CB composite is equal to or more than 5phr or CB is more than CNT in the hybrid composites, main electro-conductive network is formed by CB, however CNT contribute in electro-conductive network formation due to connection of CB branches. In this case, the electro-conductive network is built from homogenously distributed CB aggregates and can be described by percolation theory.

3.1. Percolation theory

The percolation theory is used [7,15] to characterize electro-conductive network structure in composites. The peculiarities of electro-conductive network structure formed within the polymer are essential to improve the gas sensing properties. According to the percolation theory the dependence of composite conductivity (σ) on filler concentration is represented as follows [7]:

$$\sigma = \sigma_o(f - f_c)^t \,, \qquad (1)$$

where σ_o is conductivity of electro-conductive particles (S/m), f is actual filler concentration, f_c - critical filler concentration and t - critical index.

A fit of experimental percolation curves by statistical percolation scaling law (1) was performed to obtain critical percolation parameters t and f_c. Critical percolation parameter t value for PiCB is 1.078, for $PiCNT_l$ is 4.95 and for $PiCNT_s$ is 4.54 (percolation curves in figure 1). Previously it was reported that t indicates the dimensionality of the electro-conductive network [8]. If value of the t is between 1.33 and 2.00, it means that the electro-conductive network is built in two or three-dimensional network [5,16]. According to classical percolation theory, t is called universal value if it equals 2.00 (the case without both, aggregation and tunnelling junctions between electro-conductive particles). The divergence of t value can be explained by large statistical distribution of the values of distance between adjacent particles or aggregates in the composite [5,7]. According to Balberg's [7] considerations, the non-universal high t values represent large distribution with of the inter-particle distances, while low t values corresponds to the almost-equal small inter-particle distances in composite. It is clear that those network paths, where the distance between adjacent particles is shorter and, subsequently, tunnelling currents are stronger, determine the electrical conductivity of the composite.

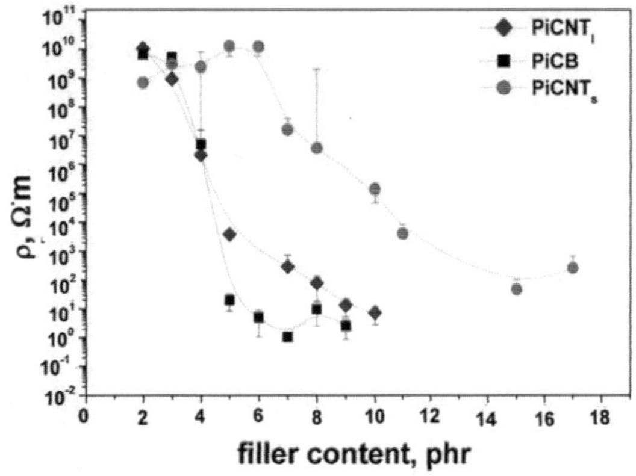

Figure 1. Electrical resistivity as function of filler concentration (parts per hundred rubber - phr) for PiCB, $PiCNT_l$ and $PiCNT_s$ composites

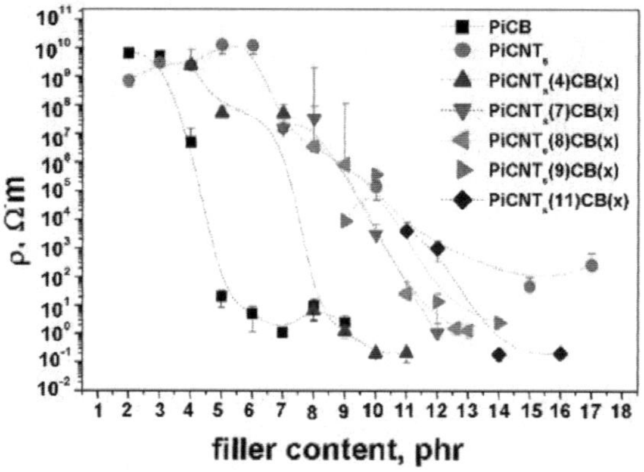

Figure 2. Electrical resistivity as function of filler concentration for PiCB, $PiCNT_s$ and $PiCNT_sCB$ composites, where CB concentration (x) in hybrid composites is varied from 1 to 7 phr.

PiCB composite (Fig.3.) has the lowest percolation critical concentration, $f_c = 2.60$, followed by $PiCNT_l$ ($f_c = 3.40$) and $PiCNT_s$ ($f_c = 6.13$). Hybrid composites were made by adding CB filler to the $PiCNT_s$ with fixed values of CNTs concentration. In Fig.2 the electrical resistivity for hybrid composite $PiCNT_sCB$ versus cumulative sum of electro-conductive filler concentration values (in phr units) is shown.

As can be seen from Fig.2, percolation curves of electrical resistivity for the hybrid composite are located between PiCB and $PiCNT_s$ curves that indicate a synergistic effect of electro-conductive fillers. By substitution certain amount of CNT with much cheaper filler – CB, electrical resistance of a composite is considerably decreased. An addition of at least 3 phr of CB is necessary to reduce electrical resistance values of the $PiCNT_s$ below 100 MΩ.

As it can be seen from the percolation data of PiCB composite (Fig.2), when CB content is of 5 phr, the composite already has considerably low electrical resistance. It means that the main electro-conductive network is formed by CB aggregates. In turn, if hybrid composite is made using CB equal to or more than 5 phr, then the electro-conductive network is generally formed from CB aggregates, where CNT contribute in electro-conductive network as connectors of CB aggregates. As a result, electrical resistance of hybrid composites is decreased even more comparing to pristine PiCB composite. Synergetic effect is expressly remarkable in prepared hybrid composites when short CNT are used. In these composites both, entanglement and folding of CNT, are negligible.

As second set of composites was compared previously investigated [12,14] EVA composites with CNT_s and CB conducting fillers. As well as hybrid composites made by using both fillers at once. For EVA CB composite f_c is 3 phr, but for EVA CNT_s composite f_c is 13 phr. It is necessary to make samples with resistivity below 100 MΩ, so the more simple measuring devices could be used. Accordingly, the best composition was EVA CB composite with 7.75 phr CB and in case of CNT_s it was 30 phr. In hybrid case such were EVA composites with CNT_s content above 10 phr.

3.2. Sensing effect

For application of developed composites in toluene sensing in air, the change of the relative electrical resistance of composites at the presence of toluene vapour has been determined. In Fig.3 the relative electrical resistance change versus toluene concentration for $PiCNT_s(7)CB(5)$, $PiCNT_s(9)CB(5)$ and PiCB(7) composites is shown. The samples were exposed to toluene vapor concentrations 90, 60 and 30ppm. Exposure was done until saturation was reached. After exposure to toluene, the samples were purged with pure carrier gas (N_2) until the initial electrical resistance was reached. No external heating was applied to sensor element during these measurements.

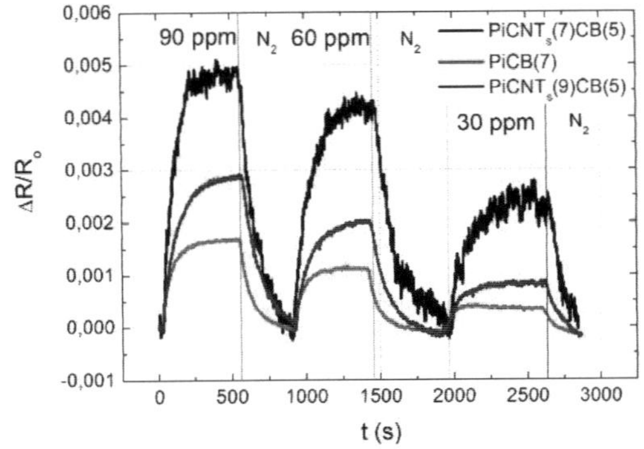

Figure 3. Relative electrical resistance change of $PiCNT_s(7)CB(5)$, $PiCNT_s(9)CB(5)$ and PiCB(7) composite versus time, when samples are exposed to toluene concentrations of 90, 60 and 30ppm.

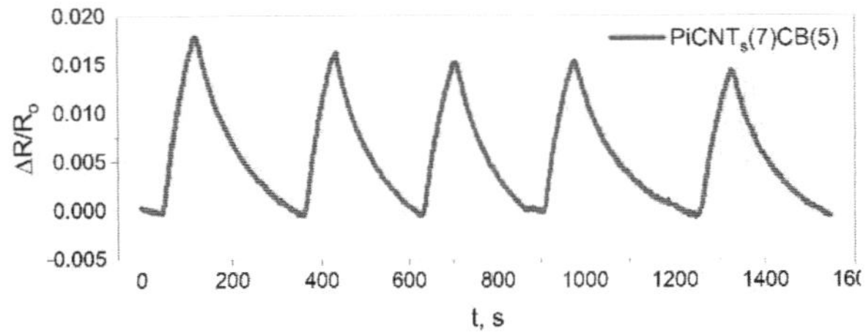

Figure 4. $PiCNT_s(7)CB(5)$ relative electrical resistance change versus time, when samples are exposed to toluene for 60s at 400ppm in five repeated cycles.

The sensing mechanism of the VOCs for this kind of composites was discussed in our previous works [13,17]where the change of the electrical resistance was explained basing on electron quantum tunneling theory:

$$\frac{R}{R_o} = \left(\frac{s}{s_o}\right) \exp[\gamma(s - s_o)], \tag{2}$$

where R_o is the initial electrical resistance, R - electrical resistance after matrix swelling, s_o – initial distance between adjacent particles, s – distance after matrix swelling, and γ is calculated from:

$$\gamma = \frac{[4\pi(2m\Phi)0.5]}{h}, \tag{3}$$

where m is mass of the electron , Φ – the height of the potential barrier between adjacent particles, h – Plank's constant. It has been proved experimentally that VOCs sensing mechanism is based on polymer swelling induced increase of distance between electro-conductive nanoparticles that causes exponential decrease of the tunnelling currents [13,17].

Hybrid samples were exposed to toluene vapour of 400 ppm concentration for time period of 60s in five repeated cycles as can be seen in Fig.4. The sensing effect is repeatable and can be used many times in a row.

Average maximal relative electrical resistance change from five repeated measurements was calculated and the standard deviation estimated (see Fig.5). The $PiCNT_s17$ composites showed the highest sensitivity among the PiCNT composites [9] and PiCB(7) showed the highest sensitivity among the

PiCB composites. Hybrid composites with electro-conductive network built of CNT and CB (Figure 5) show higher sensitivity to toluene, compared with composites where only CNT or CB as filler is used. The highest sensitivity among hybrid composites showed PiCNT$_s$(7)CB(5). Thus, one can conclude that sensitivity of CNT composites can be considerably increased by addition of the carbon black nanoparticles in case of Pi matrix. The diffusion of toluene molecules in composite matrix obviously causes greater changes of width of tunnelling junctions between different types of electro-conductive fillers. In case of EVA matrix it can be clearly seen that EVA CB 7.75 has the highest relative electrical change compared to all created samples, even in comparison with Pi hybrid composite samples. Adding CB to EVA CNT samples gave insignificant increase in relative electrical resistance change. The rise of sensing effect when concentration of short CNTs rises (PiCNT$_s$15 -> PiCNT$_s$17) we explain due to the increase of the number of comparatively easy destroyable and renewable (in case of matrix swelling -> recovering) conductive channels in composite. In contrary, in case of long CNTl (PiCNT$_l$7 -> PiCNT$_l$ 10) the additional conductive channels have a larger reinforcement effect, it means, the conductive channels can be not so easy destroyed due to matrix swelling, so the sensing effect diminishes. More detailed review of EVA composites sensing abilities can be found in our previous studies [11,12,14].

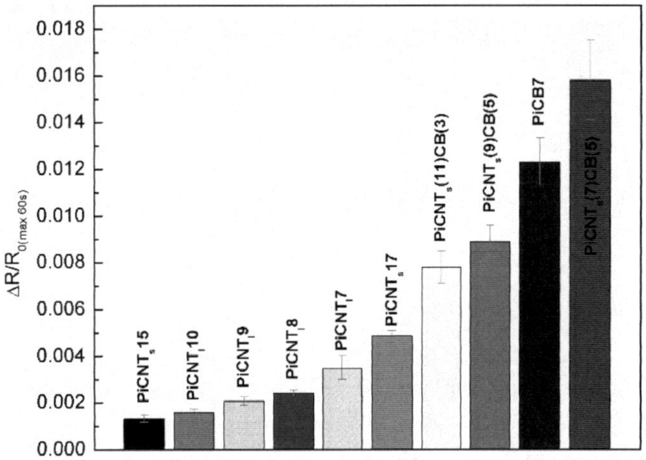

Figure 5. Average maximal relative electrical resistance change in toluene (400ppm) compared for different composite samples with Pi matrix.

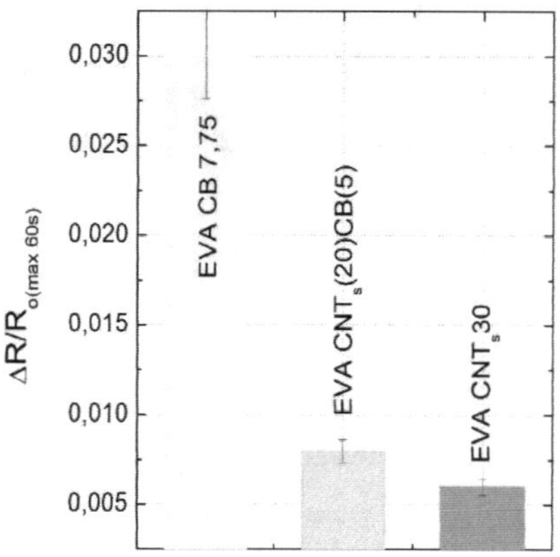

Figure 6. Average maximal relative electrical resistance change in toluene (400ppm) compared for different composite samples with EVA matrix.

4. Conclusions

In this paper hybrid composites of Pi or EVA with carbon black (CB) and carbon nanotubes (CNT) as conductive fillers for the application in VOC sensors were elaborated and studied. The electrical resistance versus the filler concentration was investigated for composites with the CB and CNT fillers separately and in the hybrid case. Electro-conductive network structure for polymer/conductive filler composites for CB and CNT was tested and analysed. Response to toluene vapour of elaborated composites was tested. Because of synergetic contribution of fillers, hybrid polymer nanocomposites show several advantages over composites where only one type of filler is used. Firstly, addition of certain amount of CB to PiCNT composites considerably decreases the composite steady state electrical resistance (that is preferable for possible sensor integration in portable VOC detection system). Secondly, the toluene sensor effect of hybrid Pi composites is greater in comparison to composites, where only CNT or CB is used as electro-conductive filler whereas EVA hybrid composite showed much lower sensing abilities than EVA CB 7.75.

It can be concluded that matrix and filler material can greatly affect composite sensing abilities and even increase relative electrical resistance change more than 10 times (EVA CB 7.75 in comparison to $PiCNT_s15$).

Acknowledgements
This work was supported by ESF Grant No.1DP/1.1.1.2.0/13/APIA/VIAA/030 and by National research program IMIS2.

References

[1] Yoo R, Kim J, Song M-J, Lee W and Noh J S 2015 Nano-composite sensors composed of single-walled carbon nanotubes and polyaniline for the detection of a nerve agent simulant gas *Sensors Actuators B Chem.* **209** 444–8
[2] Alizadeh T and Hamedsoltani L 2014 Graphene/graphite/molecularly imprinted polymer nanocomposite as the highly selective gas sensor for nitrobenzene vapor recognition *J. Environ. Chem. Eng.* **2** 1514–26
[3] Penza M and Consortium E 2014 COST Action TD1105: Overview of Sensor-systems for Air-quality Monitoring *Procedia Eng.* **87** 1370–7
[4] Crowley K, Morrin A, Hernandez A, O'Malley E, Whitten P G, Wallace G G, Smyth M R and Killard A J 2008 Fabrication of an ammonia gas sensor using inkjet-printed polyaniline nanoparticles *Talanta* **77** 710–7
[5] Dang Z-M, Shehzad K, Zha J-W, Mujahid A, Hussain T, Nie J and Shi C-Y 2011 Complementary percolation characteristics of carbon fillers based electrically percolative thermoplastic elastomer composites *Compos. Sci. Technol.* **72** 28–35
[6] Xiong Z-Y, Zhang B-Y, Wang L, Yu J and Guo Z-X 2014 Modeling the electrical percolation of mixed carbon fillers in polymer blends *Carbon N. Y.* **70** 233–40
[7] Balberg I 2002 A comprehensive picture of the electrical phenomena in carbon black–polymer composites *Carbon N. Y.* **40** 139–43
[8] Sumfleth J, Adroher X C and Schulte K 2009 Synergistic effects in network formation and electrical properties of hybrid epoxy nanocomposites containing multi-wall carbon nanotubes and carbon black *J. Mater. Sci.* **44** 3241–7
[9] Sakale G, Jakovlevs D, Aulika I and Knite M 2012 Effect of Nanotube Aspect Ratio on Chemical Vapour Sensing Properties of Polymer/MWCNT Composites *J. Nano Res.* **21** 117–23
[10] Knite M, Sakale G and Teteris V 2012 Diffusion, swelling and electrical properties of polyisoprene/multiwall carbon nanotube composites in organic solvent vapours. *J. Nanosci. Nanotechnol.* **12** 8123–8
[11] Sakale G, Knite M, Teteris V, Tupureina V, Stepina S and Liepa E 2010 The investigation of sensing mechanism of ethanol vapour in polymer-nanostructured carbon composite *Cent. Eur. J. Phys.* **9** 307–12
[12] Stepina S, Berzina A, Sakale G and Knite M 2017 BTEX detection with composites of ethylenevinyl acetate and nanostructured carbon *Beilstein J. Nanotechnol* **8** 982–8
[13] Knite M, Ozols K, Sakale G and Teteris V 2007 Polyisoprene and high structure carbon nanoparticle composite for sensing organic solvent vapours *Sensors Actuators B Chem.* **126** 209–13
[14] Stepina S, Sakale G and Knite M 2015 Ethylene vinyl acetate copolymer and nanographite hybrid composite as innovative material for chemical vapour sensing *Adv. Mater. Res.* **1117** 4
[15] Sheng P, Sichel E K and Gittleman J I 1978 Fluctuation-Induced Tunneling Conduction in Carbon-Polyvinylchloride Composites *Phys. Rev. Lett.* **40** 1197–200
[16] Bauhofer W and Kovacs J Z 2009 A review and analysis of electrical percolation in carbon nanotube polymer composites *Compos. Sci. Technol.* **69** 1486–98
[17] Sakale G, Knite M and Teteris V 2011 Polyisoprene-nanostructured carbon composite (PNCC) organic solvent vapour sensitivity and repeatability *Sensors Actuators A Phys.* **171** 19–25

Characterisation of ureaurethane elastomers with rapeseed oil-based polyol

K Mizera[1*], J Ryszkowska[1], M Kurańska[2], A Prociak[2]

[1] Faculty of Materials Science and Engineering, Warsaw University of Technology, Woloska 141, 02-507 Warsaw, Poland
[2] Faculty of Chemical Engineering and Technology, Cracow University of Technology, Warszawska 24, 31-155 Cracow, Poland

*e-mail: kamila.mizera.dokt@pw.edu.pl

Abstract. The purpose of this study was to evaluate the effect of rapeseed oil-based (RO) polyol on the thermal and mechanical properties of the tested ureaurethane elastomers (EPUUs) produced with poly(ethylene adipate) (PEA). The bio-polyol was synthesized by epoxidation and opening of oxirane rings using diethylene glycol (DG). The use of RO polyol in the amount 40 wt.% in polyol premixes changed the chemical structure of EPUUs with different hard segments (HS) content, what in a consequence indicates the change in quantity of the number of hydrogen bonds and degree of phase separation (DPS). The consequence of these changes are variation in the thermal characteristics of these materials. The DSC analysis showed that an addition of the RO polyol causes the increase in the glass transition temperature of the soft phase with an increase of HS content in EPUUs. An introduction of the RO polyol also resulted in an decrease in the elongation at break with an increase of HS content in the modified elastomers.

1. Introduction

In recent years, the introduction of plant components to polyurethane (PUR) systems meets all the ideas of sustainable development and is an important challenge for chemical companies [1]. PURs are well known and widely used because of their very good properties and applications. The production of PURs is based on the two most important components: polyols and isocyanates. Nowadays, one of the problems of PUR industry, is facing is their dependence on petroleum feedstock [2]. The increasing interest of industry of polyols from vegetable oils is observed [3]. In the PUR production the natural oil polyols can be used to replace petrochemical materials [4]. Polyols can be synthesized from natural materials, such as plants, oils or wood [5,6]. There are many publications about obtaining PURs with polyols from different vegetable oils based on soybean oil [7-9], rapeseed oil [10], sunflower oil [11], linseed oil [12], or castor oil [13]. Most of PUR materials from vegetable oil polyols are foams [14], coatings [15], films [16] and adhesives [17]. To obtain PURs from renewable vegetable oils there are some aspects that have to be considered – hydroxyl value, functionality, dangling chains and structure of polyol and their effect on properties of PUR network [16].

Polyurethane elastomers (EPUs) are the most distinguished group of polymers, because of their excellent properties such as good resistance to abrasion and to oils, grease and weather conditions

[17]. They are used for many applications: from flexible to rigid foams, to thermoplastic polyurethane and elastomers [18]. Currently, polyurethanes industry is highly dependent on oil as its two main raw materials, polyols and isocyanates, are largely petroleum. Because of global warming and the oil crisis, began to look at alternative substrates [19].

In this work, we examine bio-based ureaurethane elastomers (EPUUs) with a 40% wt. of rapeseed oil-based (RO) polyol with a different content of hard segments (HS). The bio-polyol was synthesized using epoxidation and opening of oxirane rings using diethylene glycol. The different content of rapeseed oil-based polyol was applied in order to evaluate influence on thermal stability as well as mechanical properties of the EPUUs in comparison with the reference EPUU based only on petrochemical polyol.

2. Experimental

2.1. Materials

Ureaurethane elastomers (EPUUs) were prepared by using a polyester poly(ethylene adipate) (PEA) - Polios 60/20, with molar mass of 2318 g/mol (Purinova, Bydgoszcz, Poland) and 4,4'-diphenylmethane diisocyanate (MDI) (Sigma Aldrich Co., Poznan, Poland). As chain extender the dicyandiamide (DYDI) - Omnicure 5, with molar mass of 84 g/mol (Emerald Performance Materials Company, USA) was used. During the production of EPUUs, rapeseed based polyol (RO) in an amount of 40 wt. % was added. Rapeseed polyol (RO) characterized by hydroxyl value 42.9 mgKOH/g, acid value 4 mgKOH/g, content of water 0.06 wt. % and number molecular weight 2387 g/mol was synthesized using at the Cracow University of Technology according to method described in our previous publication [20].

2.2. Synthesis of the ureaurethane elastomers

Ureaurethane elastomers (EPUUs) were synthesized by one step method, with isocyanate index I_{NCO} = 1.07. The content of hard segment (HS) of produced EPUUs was different and it is shown in table 1.

The PEA with DYDI were melted and degassed for 1.5 h at 150 ± 5°C and 6 hPa with intense stirring. Then the PEA was cooled to 80 ± 3°C before the RO polyol was added. The mixture was then stirred for 30 min. Next, the mixture was cooled to 60 ± 3°C and MDI was added and stirred for 10 min. The prepared mixture was then poured into a duraluminium mould and annealed for 18 h at 110°C. The samples were tested after 30 days of seasoning under ambient conditions.

2.3. Test methods

The Fourier Transform Infrared Spectroscopy (FTIR) was carried out using a Nicolet 7600 (ThermoElectrone Corporation, USA). Spectra were recorded as a sum of 64 scans using a spectral range from 4000 to 400 cm^{-1}, with a resolution of 4 cm^{-1}. The data analysis was performed using Omnic software. A baseline correction for CO_2 and H_2O was performed to eliminate the impact of the compound residues from the analysis.

The DSC analysis was performed using a differential scanning calorimeter Q1000 (TA Instruments, USA). Samples (5 mg) were closed in hermetic aluminum cups and heated at 10°C/min in the temperature range -90°C to 250°C.

The thermal stability was defined by thermogravimetric analysis using a Q500 analyzer (TA Instruments, USA). The tested 10 mg samples were heated from room temperature to 800°C for EPUUs with a heating rate of 10°C/min. Tests were carried out in a nitrogen atmosphere with a flow rate of 90 ml/min.

The dynamic mechanical thermal analysis (DMTA) was carried out on a TA Instruments DMA Q800 (USA). This study was conducted using cast rectangular-shaped samples with dimensions 60x10x2 mm and a two-point bending method. Samples were heated at 3°C/min in a temperature range between -100°C to 110°C, using a constant frequency of 1 Hz and an amplitude of 25 μm. The test was carried out in a nitrogen atmosphere.

The static tensile test was performed on a tensile machine (Instron 1115) according to PN-EN ISO 527. The cast paddle-shaped samples with dimensions 2,5x2 mm and measuring length 20 mm, were stretched at a speed of 500 mm/min.

Wear resistance was measured using the Schopper–Schlobach method. The cast samples had a cylindrical shape with dimensions 16 × 6mm2 and were tested according to ISO 4649.

3. Results and discussion

3.1. The chemical constitution

The FTIR analysis confirmed the occurrence of groups characteristic for EPUUs. The FTIR spectra of RO polyol and the obtained materials are compared in figure 1.

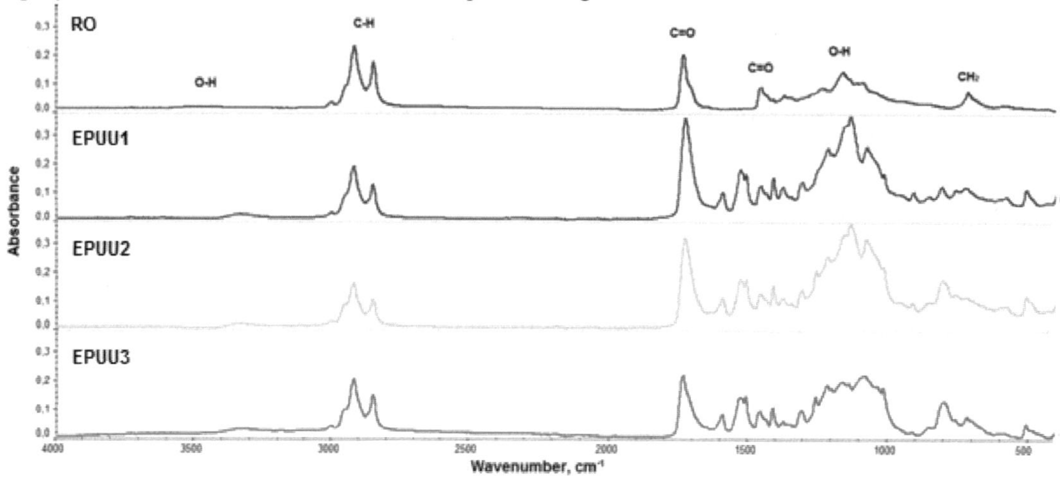

Figure 1. Fourier transform infrared spectroscopy (FTIR) spectra of RO polyol and tested EPUUs

The interpretation of the FTIR spectra of tested EPUUs shows signals in the range of 3349-3342 cm^{-1}, which are the result of symmetric and asymmetric stretching vibrations of the N-H groups present in urea and urethane derivatives . This bond is involved in the formation of hydrogen bonds. The signals at 2960-2953 cm^{-1} and 2874-2872 cm^{-1} are corresponding to symmetric and asymmetric CH_2 stretching vibrations, respectively. The multiplet signal at 1728-1727 cm^{-1} indicate the presence of C=O bond stretching vibrations of the carbonyl groups in hard segments. The signal at 1598-1597 cm^{-1} are corresponding to aromatic groups. In the range of 1534-1530 cm^{-1} the urethane N-H bending vibrations and C-N stretching vibrations are observed. The signals at 1460-1455 cm^{-1} are corresponding to scissoring vibrations of C-H bond. Signals in the range of 1212–1219 cm^{-1} are the result of C-N stretching vibrations. The signals in the range of 1260-1060 cm^{-1} are corresponding to the C-O bond, attributed to the flexible segments.

For RO polyol the signal at 3468 cm^{-1}, 3006 cm^{-1}, 2922 cm^{-1}, 2853 cm^{-1}, 1741 cm^{-1} and 722 cm^{-1} are observed and was discussed in our previous work [21].

With the increasing amount of HS content, the absorbance of bands originating from symmetric and asymmetric stretching of bonds C-H decreases, for tested materials. The absorbance of the bands originating from the stretching vibrations bonds C-O at 1159-1156 cm^{-1} also decreases with the higher HS content. The intensity of bands originating from the vibrations of the carbonyl bonds C=O, also decreased with the increase of HS content.

The data analysis was performed using Omnic software. A baseline correction for CO_2 and H_2O was performed to eliminate the impact of the compound residues from the analysis. On the basis of the intensity of the bands, hydrogen bonding index (R) was calculated by equation:

$$R = \frac{A_1 + A_2}{A_3 + A_4}$$

where: A_1 and A_2 are the areas originating from vibrations bound by the hydrogen bond of the carbonyl groups of urea 1630-1680 cm^{-1} and urethane 1705-1727 cm^{-1}; A_3 and A_4 are the areas originating from vibrations unbonded with the hydrogen bond of the carbonyl groups of urea 1690-1701 cm^{-1} and urethane 1736-1760 cm^{-1}.

The value of degree of phase separation (DPS) was determined based on the R index and calculated by equation:

$$DPS = \frac{R}{R+1}$$

Calculated values of R and DPS were calculated and are summarized in table 1.

Table 1. Hard segments (HS) content, hydrogen bond index (R) and degree of phase separation (DPS) for the tested EPUUs

Sample	HS, %	R	DPS
EPUU1	24	1.13	53.1
EPUU2	28	1.32	56.9
EPUU3	33	1.04	51.0

The DPS is the highest for EPUU2. For this material the highest formation of hydrogen bonds between rigid segments is observed. The characteristic of DPS values are related with physical cross-linking of the examined EPUUs.

3.2. Thermal properties

The Tg temperature determined from DSC analysis gives an information about structure of the soft phase in the tested materials. Values of Tg determined for EPUUs were within the range between -28.5°C and -26.2°C (table 2). The highest Tg value indicates that macroparticle fragments forming the soft phase of EPUUs have somewhat lower flexibility.

On the basis of TGA thermograms of tested EPUUs the temperature at 5% mass loss ($T_{5\%}$) as well as the mass remaining after annealing at 800°C (U_{800}) were determined (table 2).

No significant changes were observed in temperature associated with the beginning of degradation of materials ($T_{5\%}$). After degradation at 800°C (U_{800}) there remains 6.0-8.3% mass for EPUUs. The residue at 800°C decreased with an increase of HS content in EPUUs.

Table 2. Thermal and mechanical properties of tested EPUUs

Sample	Tg, °C	$T_{5\%}$, °C	U_{800}, %	Tα, °C	δ_M, MPa	ε, %	E'_{-30}, MPa	E'_{20}, MPa	E'_{60}, MPa	ΔV, mm^3
EPUU1	-27.0(0.2)	289(0.1)	6.0(0.1)	-20.1(0.2)	11.64	444	7559	39	15	184.1(0.1)
EPUU2	-28.5(0.3)	288(0.4)	8.1(0.3)	-21.8(0.1)	13.02	292	10440	124	77	139.0(0.2)
EPUU3	-26.2(0.1)	290(0.1)	8.3(0.1)	-18.7(0.1)	17.42	167	4366	180	91	169.7(0.4)

*Values between the parenthesis are standard deviations

3.3. Mechanical properties

The viscoelastic properties of the tested EPUUs were investigated using DMTA. The curves of EPUUs obtained from measurement using DMTA are shown in figure 2, storage modulus (E'), loss modulus (E") and loss tangent (tan δ), respectively. The results from the DMTA curves are listed in table 1.

Storage modulus E' at three different temperatures: -30°C, 20°C and 60°C (E'_{-30}, E'_{20} and E'_{60}) were determined (table 1). The E' in minus temperatures is highest for EPUU2. The E'_{20} (above glass transition temperature Tg) increased with the increase of RO polyol content (table 2). The storage modulus at high temperature (E'_{60}) also increased with the increase of RO polyol content. These result allows for the application of materials of high mechanical demanding.

From the analysis of the curve of loss modulus (E") (figure 2), taking into account the shape, position and height of the peak, it is possible to determine the degree of order and a free movement of the polymer chains. From this curve the α relaxation temperature, related to the glass transition temperature of soft segments (Tα) was determined as a maximum peak (table 2). The lowest temperature Tα for EPUU2 is observed. This shown that for this material, the increase in the flexibility of the soft segments of the material microdomains, is observed.

Tan δ value corresponds to the ratio of loss modulus (E") to the storage modulus (E'). On the basis of tan δ curves, the ability of a material to absorb and accumulate energy can be evaluated. Materials with high vibration-absorption and energy accumulation capacity are considered to have tan $\delta > 0.3$ across a wide temperature range [21]. For all tested EPUUs the maximum peak of tan δ is higher than 0.3.

Both DSC and DMTA analysis showed significant differences between the soft phase characteristic for tested materials. Both methods showed an increase in Tg and Tα for material EPUU3, with the highest content of HS.

Thetensile strength (δ_M) increased with an increase of HS content. Elongation at break (ε) decreased with an increase of HS content (table 2).

The usefulness of EPUUs in many applications is evaluated by wear resistance. The abrasive wear (ΔV) is lowest for EPUU2. This material have highest wear resistance.

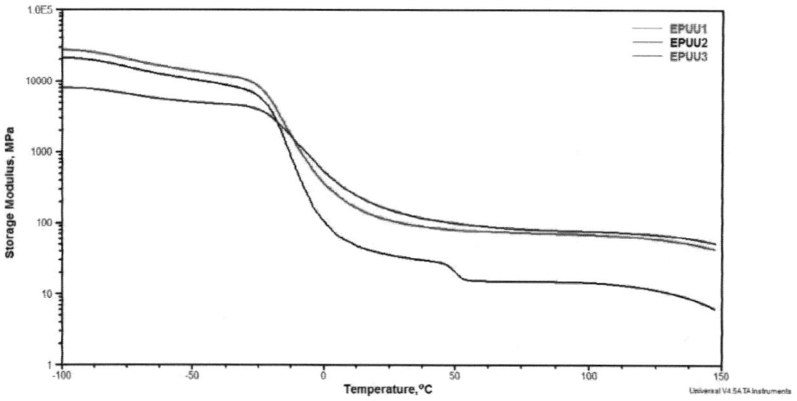

a)

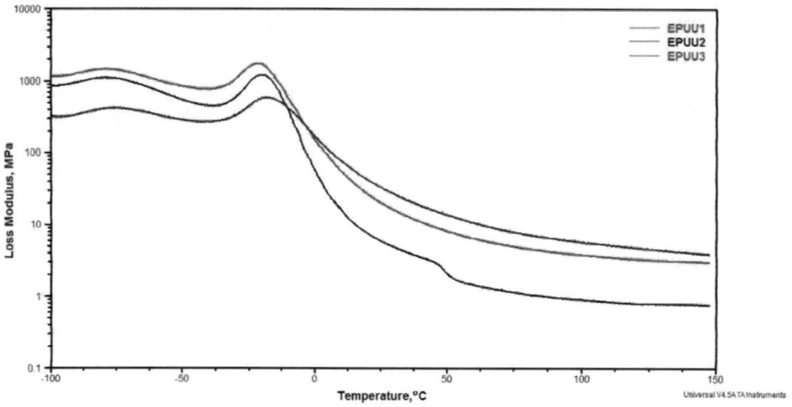

b)

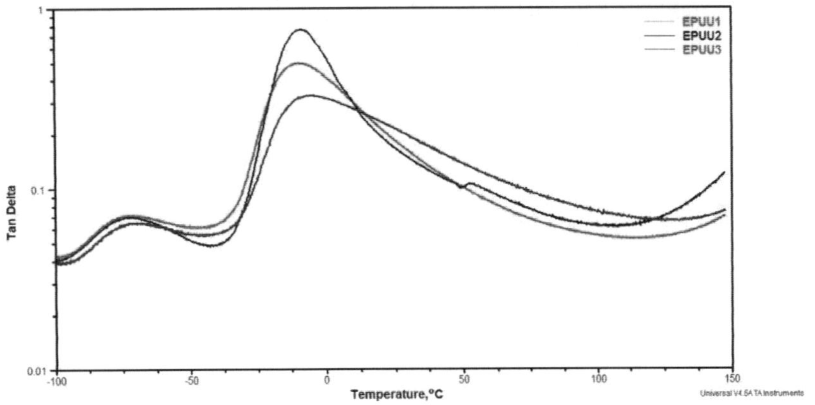

c)

Figure 2. The DMTA curves of (a) storage modulus, (b) loss modulus and (c) tan δ of tested EPUUs

4. Conclusions

According to the conducted study, the introduction of rapeseed oil-based polyol into the ureaurethane elastomers have different influence on their properties. The addition of RO polyol causes changes in the structure of soft phase in tested materials. The analysis of the DSC thermograms suggested that the increasing content of hard segments, results in higher glass transition temperature of the soft phase in EPUUs. This relationship has been confirmed by DMTA analysis. Wear resistance was highest for material EPUU2 with a 28 wt. % of HS content.

These studies showed that it is possible to control the properties of EPUUs with RO polyol by changing the hard segments content.

References

[1] Kurańska M, Prociak A, Kirpluks M and Cabulis U 2015 *Ind.Crops Prod.* **74** 849-857
[2] Datta J and Głowińska E 2014 *Ind. Crop Prod.* **61** 84–91
[3] Bhoyate S, Ionescu M, Radojcic D, Kahol PK., Chen J, Mishra SR and Gupta RK 2017 *J. Appl. Polym. Sci.* DOI: 10.1002/app.46027
[4] Kurańska M and Prociak A 2016 *Ind. Crops Prod.* **89** 182-187
[5] Kairyte A and Vejelis S 2015 *Ind. Crops Prod.* **66** 210-215
[6] Gaidukova G, Ivdre A, Fridrihsone A, Verovkins A, Cabulis U and Gaidukovs S 2017 *Ind. Crops Prod.* **102** 133–143
[7] Yang LT, Zhao CS, Dai CL, Fu LY and Lin SQ 2012 *J. Polym. Environ.* **20** 230–236
[8] Mizera K and Ryszkowska J 2016 *Polym. Degrad. Stab.* **132** 21–31
[9] Mizera K and Ryszkowska J 2018 *J. Elast. Plast.* DOI: 10.1177/0095244318772323
[10] Hu YH, Gao Y, Wang DN, Hu CP, Zu S, Vanoverloop L and Randal D 2002 *J. Appl. Polym. Sci.* **84** 591–597
[11] Das B, Konwar U, Mandal M and Karak N 2013 *Ind. Crops Prod.* **44** 396–404
[12] Yadav S, Zafar F, Hasnat A and Ahmad S 2009 *Prog. Org. Coat.* **64** 27–32
[13] Thakur S and Karak N 2013 *Prog. Org. Coat.* **76** 157–164
[14] Kairyte A, Vaitkus S, Vejelis S, Girskas G and Balciunas G 2018 *Ind. Crops Prod.* **112** 119-129
[15] Javni I, Zhang W and Petrovics Z 2004 *J. Polym. Environ.***12** 123–129
[16] Frifrihsone A, Stirna U, Lazdina B, Misane M and Vilsone D 2013 *Eur. Polym. J.* **49** 1204-1214
[17] Mizera M, Kirpluks M, Cabulis U, Leszczyńska M, Półka M and Ryszkowska J 2018 *Ind. Crops Prod.* **113** 98–110
[18] Chattopadhyay DK and Rayu KV 2007 *Prog. Polym. Sci.* **23** 352–418
[19] Prasanth KSP, Mohanan A, Bouzidi L and Narine SS 2016 *Ind. Crops Prod.* **94** 431–444
[20] Zieleniewska M, Leszczyński MK, Kurańska M, Prociak A, Szczepkowski L, Krzyżowska M and Ryszkowska J 2015 *Ind. Crops Prod.* **74** 887–897

IOP Conf. Series: Materials Science and Engineering **500** (2019) 012015 doi:10.1088/1757-899X/500/1/012015

[21] Mizera K, Ryszkowska J, Kurańska M and Prociak A 2018 *Polym. Intern.* **67** 1605–1614

IOP Conf. Series: Materials Science and Engineering **500** (2019) 012016 doi:10.1088/1757-899X/500/1/012016

Compression Properties of knitted Supports with silicone elements for scars treatment and New Approach to Compression Evaluation

Laima Muralienė, Daiva Mikučionienė, Akvilė Andziukevičiūtė-Jankūnienė, Virginija Jankauskaitė

Department of Production Engineering, Kaunas University of Technology, Studentų Str. 56, LT-51424, Kaunas, Lithuania

La.milasiute@ktu.edu

Abstract. Innovative compression products for scars treatment have to fulfil various functions and contains extra elements for different purposes. It is already known that all rigid elements, inserted into support, can change elasticity of entire product – there is a strong linear dependence between the rigid element relative area and compression generated by the knitted orthopaedic support. The aim of this research was to investigate the mechanical behaviour of different knitted compression apparel products with silicon elements for scars treatment. The unique complex mean of bandages for compression treatment was created in cooperation with JSC „Ortopedijos klinika" and JSC „Plėtra". Products were manufactured from knitted material with an elastic structure and layer of biocompatible silicon with medical plants substrate which creates antihypertrophic effect for scars treatment. Knitted compression fabrics were made by knitting two types of elastomeric yarns together (4.4 tex PU core double covered with 2.2 tex PA6.6; 33 tex PU core double covered with 2.2 tex PA6.6). Samples were stretched till the selected tensile force and hold in this position for 120s in case to measure elongation in more accurate way. According to results, recommended elongation for selected products was found.

1. Introduction

The use of compression textile products for medical purposes has increased significantly since 1970. Originally, these products were used to exert pressure along the human body for the treatment of scars resulting from burns, and treatment of post-surgical condition. Knitted medical textile garments are widely used because of their greater flexibility, higher porosity, more flexible structure and better forming technology [1]. Nowadays innovative products for scars treatment have to fulfil various functions and contains extra elements for different purposes. Compression apparel for treatment after burns (usually 3th grade burns) has to generate not less than 24 mmHg compression and has to be worn at least 12 months, 23 hours per day [2, 3]. The average pressure level applied to the treated tissues is about 25 mmHg [4].

Compression apparel is often designed from knitted fabric with additional details for different purposes. It is known that there is a high correlation between the mechanical properties of the fabric and generated pressure [5]. Orthopaedic supports often have added silicone elements such as modified biocompatible silicon components or other parts for functional application and may also comprise other components, such as straps, fasteners including disengage able two-part fastener system, such as

Velcro (the brand name of the first commercially marketed fabric hook-and-loop fastener) for engaging the support with the body [6]. All rigid elements, inserted into support, can change elasticity of entire product. It is already known that in the area of low extensions, there is a strong linear dependence between the rigid element relative area and compression generated by the knitted orthopaedic support – compression linearly increases by increasing the area of the rigid element [7, 8]. Moreover, additional rigid elements can significantly affect generated compression or can even change the compression class of the product. It was established that rigid element, that occupies ~ 8% of the total area of a support, enhances the tensile force as well as compression up to 15% even at low elongation (10%). This influence depends on the level of elongations in which orthopaedic support is used. Compression, generated by the support with 25% relative area covered by a rigid element, increases up to ~ 17% at 10% fixed elongation and up to ~ 24% at 20% fixed elongation. However, if area covered by a rigid element is up to 3% and such support is used in area of low deformations (up to 10%), it is not necessary to assess the influence of relative rigid area on compression of the knitted support [7, 8].

Compression products usually are worn for a medical reasons and are indicated for long lasting term of wearing. After securing product to the body part, relaxation process appears and compression value may differ during the wearing time [9]. It is known that relaxation process of different type and nature of the yarns appears differently [10].

The aim of this study was to investigate the mechanical behavior of different knitted compression products with silicon elements for scars treatment.

2. Materials and Methods

The unique complex mean of bandages for compression treatment was created in cooperation with JSC „Ortopedijos klinika“ and JSC „Plėtra“. Experimental samples were manufactured on a flat double needle-bed 14E gauge knitting machine CMS 340TC-L (f. STOLL, Germany). Plain knitted fabric and three different location supports (ankle, knee and wrist) were designed and manufactured. Combined "Bird eye" and rib 1x1 patterns were used for the samples. Products were manufactured from knitted material with an elastic structure and layer of biocompatible silicon with medical plants substrate which creates antihypertrophic effect for scars treatment. Samples of supports for different locations were made by knitting two types of elastomeric yarns together: 4.4 tex PU core double covered with 2.2 tex PA6.6 and 33 tex PU core double covered with 2.2 tex PA6.6. The schematic structure of the knitting pattern is presented in the figure 1.

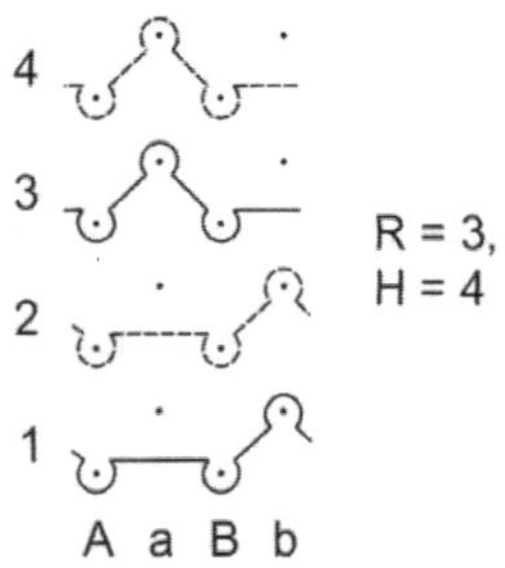

Figure1. Schematic structure of the knitting pattern (H is the height of the pattern repeat, R is the width of the pattern repeat).

The tensile behavior of the knitted fabric tested was evaluated using universal testing machine ZWICK/Z005 according to Standard LST EN ISO 13934–1:2000. According to the shape of the samples of supports special claps were designed and used for this research. The tensile speed was 100 mm/min; pretension – 2N, sensor – 5kN. According to the results of the latest research [9], samples of supports for different locations were stretched till the selected tensile force and held in this position for 120s in case to evaluate relaxation process and to measure elongation in more accurate way. Results of each point are calculated as average of five tensile tests. Compression is calculated by the Laplace formula and depends on pressure P (Pa), tensile force in the knitted sample F (N) and the area of the knitted sample S (m^2):

$$P = \frac{2\pi F}{S} \tag{1}$$

All experiments were carried out in a standard atmosphere for testing according to Standard LST EN ISO 139:2005.

Structure parameters of knitted samples were analysed according to British Standard BS 5441:1998.

3. Results and discussion

It is well known that mechanical properties of plain knitted fabric has valuable influence to the mechanical properties and behaviour of the complete knitted garment. From the previous research it is also known that additional rigid elements can significantly affect generated compression or can even change the compression class of the product [8]. Nevertheless, same purpose products, produced from the same material but appointed for different locations appears in various forms, dimensions and surface areas occupied by rigid elements. Also, in order to generate compression, perimeter of compression product and perimeter of the limb have to differ. It leads to assumption that tensile force values may be different to fulfil requirements of generated compression for such a products.

In order to investigate influence of different form, measurements and surface area occupied by the silicon element on the compression parameters, three different body locations were chosen. For further analysis supports for knee, ankle and wrist were designed and manufactured. All prepared samples were measured and structure parameters were calculated. It is recommended that compression products for scars treatment have to generate 33.2 ÷ 42.5 kPa compression and to belong to the second compression class (according to the Standard RAL-GZ 387/1:2008). According to the target compression (33.2 ÷ 42.5 kPa) the range of target tensile force was calculated by Laplace formula and average target tensile force was chosen. Obtained results are presented in the table 1.

Table 1. Measurements and structure parameters of tested knitted supports.

Location of the support	Knee	Ankle	Wrist
Surface area, m^2	0.1026	0.0550	0.0378
Surface area of the silicone element, m^2	0.0027	0.0027	0.0041
Surface area of the silicone element, %	2.6%	4.9%	10.9%
Course density, cm^{-1}	7.5	7.5	8
Wale density (right), cm^{-1}	9	9.5	9 / 14
Wale density (left), cm^{-1}	14	14	14
Calculated target tensile force, N	27-34	10-13	12-15
Chosen average target tensile force, N	30	12	14

Compression products usually are worn for a medical reasons and have to fulfil medical requirements such as appropriate value of generated compression. In order to achieve target tensile force and generate target compression, selected samples were strained till fixed tensile force and hold in this position for 120s. The period of relaxation (120s) was based on the obtained experimental result from the previous research – because the relaxation processes of inner structure the tensile force decreases significantly during this period and after this period the relaxation process is considerably slower [9]. Due to the compression garments for scars treatment are usually indicated and being worn for a long-term (e.g. 12 hours a day), the importance of initial value of generated compression is less substantial.

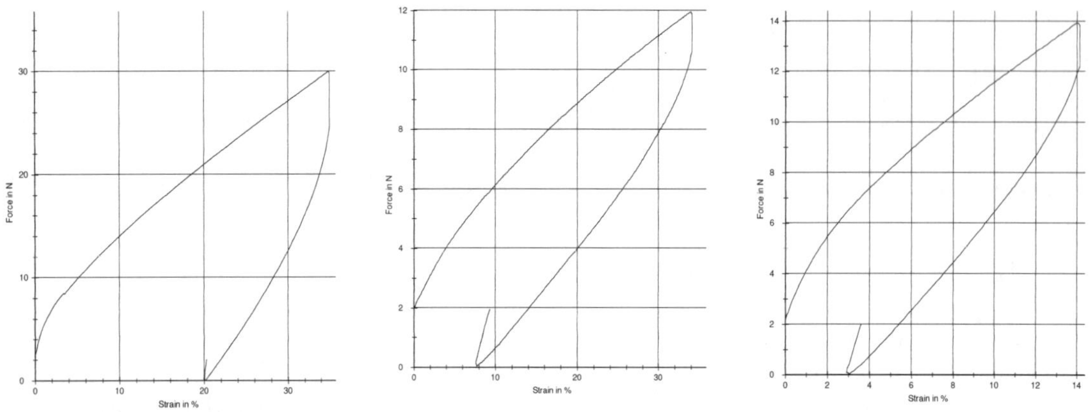

a) b) c)

Figure 2. Characteristic force-strain curves of tensile test of supports for different locations: a) knee, b) ankle, c) wrist.

Characteristic force-strain curves of tensile tests are presented in the figure 2. The variation coefficient of elongation ranges from 3.5 to 7.5. It is well known that in order to generate compression perimeter of support and the perimeter of the limb have to differ. According to the results of the tensile tests, the target elongation was estimated. It was found that to achieve target compression, tested samples of knee supports have to be stretched till 20% elongation, ankle supports – till 20% elongation and wrist supports – till 15% elongation. Results confirms the assumption that tensile force values may be different to fulfil requirements of generated compression for compression products for different locations. For this reason the perimeter of compression product and perimeter of the limb has to differ accordingly. Moreover, results of this research shown that in order to predict generated compression values not only the structure parameters of the knitted fabric but also completed product parameters have to be evaluated. More researches in this area may enable possibilities to predict generated compression in the more accurate way.

4. Conclusions

This research was based on investigation of full garment for scars treatment compression properties and recommendation for its formation. According to the results, it was found that prediction of compression properties of the same pattern knitted garment for different body locations should not be evaluated just according to the plain fabric mechanical properties. Results shown that in order to achieve target tensile force and compression values, elongation of the selected samples should be different - knee supports have to be stretched till 20% elongation, ankle supports – till 20% elongation and wrist supports – till 15% elongation. Consequently, the perimeter of compression product and perimeter of the limb has to differ accordingly. Moreover, results of this research shown that in order to predict generated compression values not only the structure parameters of the knitted fabric but also completed product parameters have to be evaluated.

References

[1] Zhang X and Ma P 2018 Application of Knitting Structure Textiles in Medical Areas *Autex R. J.* **18(2)** 181-191 DOI 10.1515/aut-2017-0019

[2] Varan N Y and Gursoy N C 2010 Textiles for burn treatments: an overview *Proc. 5th Int. Conf. ITC&DC Magic World of Textiles* (Dubrovnik, Croatia) 171-75

[3] Anzarut A et al 2009 The effectiveness of pressure garment therapy for prevention of abnormal scarring after burn injury: a meta–analysis *J. of Plastic, Reconstructive & Aesthetic Surgery* **62(1)** 77-84

[4] Legner M 1999 Medical Textile with Specific Characteristics Produced on Flat Knitting Maschines *Proc. 2nd Int. Conf.* (Bolton Institute, UK) 44-51
[5] Cheng Z, Kuzmichev V E and Adolphe D C 2017 Development of Knitted Materials Selection for Compression Underwear *Autex R. J.* **17(2)** 177-187 DOI 10.1515/aut-2016-0006
[6] Kowalski K et al 2012 Modelling and Designing Compression Garments with Unit Pressure Assumed for Body Circumferences of a Variable Curvature Radius *Fibr. & Text. in East. Europe* **20(6A)** 98-102
[7] Mikučioniene D and Ališauskiene D 2014 Prediction of Compression of Knitted Orthopaedic Supports by Inlay-Yarn Properties *Material Science* **20(3)** 311-14
[8] Mikučioniene D and Milašiūte L 2017 Influence of Knitted Orthopaedic Support Construction on Compression Generated by the Support *J. of Ind. Textile* **47(4)** 551-66
[9] Muralienė L, Mikučionienė D, Laureckienė G and Brazaitis M 2018 New Approach to Evaluation of Orthopaedic Supports Compression Properties *J. of Ind. Textile* DOI 10.1177/1528083718783312
[10] Laureckienė G and Milašius R 2017 Behaviour of Long-Lasting Stress Relaxation of Various Types of Yarns *Autex R. J.* **17(4)** 379-385 DOI 10.1515/aut-2017-0017

IOP Conf. Series: Materials Science and Engineering **500** (2019) 012017 doi:10.1088/1757-899X/500/1/012017

Influence of Side Chain Length on the Properties of Alkylated Copolymers Based on N-Methyl-N-Vinylacetamide

O A Dommes[1], A A Gosteva[1], G F Kolbina[1], O V Okatova[1], I I Gavrilova[1], E F Panarin[1, 2], G M Pavlov[1, 3]

[1] Institute of Macromolecular Compounds, Russian Academy of Sciences, Bolshoy pr. 31, St. Petersburg 199004, Russia
[2] Department of Medical Physics and Bioengineering, Peter the Great St. Petersburg Polytechnic University, Polytechnicheskaya str. 29, St. Petersburg 195251, Russia
[3] Saint Petersburg State University, Universitetskaya nab. 7/9 St.Petersburg199034, Russia

E-mail: oljaspb552@gmail.com, georges.pavlov@mail.ru

Abstract. A series of alkylated statistical copolymers of N-methyl-N-vinylacetamide and N-methyl-N-vinylamine hydrochloride containing C_6H_{13} – $C_{12}H_{25}$ side aliphatic group in the chain were studied by the methods of molecular hydrodynamics (velocity sedimentation, translational diffusion and viscous flow) in 0.1M NaCl and salt free water solutions, and by flow birefringence. The competition between electrostatic and hydrophilic/hydrophobic interactions manifests differently depending on the ionic strength and the length of the side chains. It was found that the radical length noticeably affects the characteristics of copolymer macromolecules. The hydrodynamic volume decreases and the shear stress optical coefficient increases considerably for the samples with $C_{12}H_{25}$ side chain groups. This is explained by drastic increase in the hydrophobic interactions which dominate over the electrostatic interactions.

1. Introduction

The interest in synthesis and study of new molecular carrier containers based on polymers and polymer systems for targeted transport of various biologically active substances to living organisms [1] is still ongoing. In the frame of the original "drug delivery" concept, significant efforts have been made to increase the efficiency of pharmaceutically active compounds by attempting to control the temporal as well as spatial drug distribution in vivo [2],[3].

Such biologically active substances are often hydrophobic meanwhile the efficiency of their therapeutic application is determined by the delivery to targeted organ. Accordingly, molecular systems combining hydrophobic and hydrophilic fragments are promising for use in medicine. Amphiphilic copolymers based on N-methyl-N-vinylacetamide are one of examples of such macromolecular systems. Alkylation of charged hydrophilic copolymers of N-methyl-N-vinylacetamide and N-methyl-N-vinylamine hydrochloride (MVAA-MVAH) leads to the formation of amphiphilic copolymers with a specified length of the lateral radicals. Such macromolecules have hydrophilic backbone and a number of hydrophobic side radicals and possibly may form hydrophobic internal domains in macromolecular coils.

2. Experimental part

The initial MVAA-MVAH copolymers were synthesized and then alkylated by alkyl iodides C_6H_{13}–$C_{12}H_{25}$ as described in [4] in the laboratory of hydrophilic polymers in the Institute of Macromolecular Compounds, Russian Academy of Sciences. The composition was 85(m):15(n) mol. %. The content of halide ions was determined by argentometric titration on a TVL-1 conductivity meter. The chemical structure is shown below.

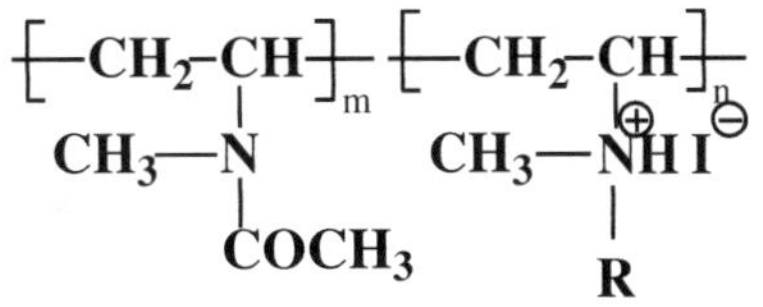

R = $-C_6H_{13}$; $-C_8H_{17}$; $-C_{10}H_{21}$; $-C_{12}H_{25}$

The nonalkylated MVAA-MVAH copolymers with similar composition and the alkylated ones with 20 mol. % of charged side aliphatic groups (from C_2H_5 to $C_{12}H_{25}$) were investigated earlier [5],[6],[7].

The following methods of molecular hydrodynamics and optics have been used: velocity sedimentation, translational diffusion, viscometry and dynamic flow birefringence (DFB). The molar masses were calculated from the data of sedimentation-diffusion analysis as described in [7]. The full complex of measurements was carried out in 0.1 M NaCl under suppression of polyelectrolyte effects. Additionally, the intrinsic viscosity [η] values at minimal ionic strength ($\approx 10^{-6}$ M, pure water) were obtained from the initial slope of ln η_r dependence on concentration *c* [8], [9].

One of the substantial properties of macromolecules is their optical anisotropy, i.e. the difference of optical properties in two orthogonal directions. The optical anisotropy of the macromolecular coil (or dispersed nanoparticle) is determined by the internal structure of the particle, by their shape and artificially induced orientation, and, finally, by the difference in refractive indices of particle and solvent [10].

Optical properties of alkylated MVAA-MVAH copolymer dilute solutions were studied in 0.1 M NaCl at 21 °C by DFB which is the most sensitive method both to the structure of the repeating unit of the macromolecule and to the conformation of its chain.

DFB was investigated on the device with a visual recording system in a dynamooptimeter [10]; the parameters of device could be found in [11].

3. Results and Discussion

Two series of samples were investigated. Each series was obtained from one MVAA-MVAH sample. The hydrodynamic data are presented in table 1 (a, b). Series 1 – samples with higher molar mass (numbers 1 – 4), series 2 – samples with lower molar mass (numbers 5 – 8).

As follows from table 1 and figure 1 (a, b), intrinsic viscosity decreases significantly when radical length is N_C=12 as in 0.1M NaCl as in H_2O due to hydrophobic interactions.

Table 1. Hydrodynamic characteristics of alkylated MVAA-MVAH copolymers in 0.1M NaCl at 25 °C

Sample no.	Radical length, N_C	Molar mass, $M \cdot 10^{-3}$, g/mol	Intrinsic viscosity [η] in 0.1M NaCl, cm^3/g	Intrinsic viscosity [η] in H_2O, cm^3/g	Diffusion coefficient $D_0 \cdot 10^7$, cm^2/s	Sedimentation coefficient $s_0 \cdot 10^{13}$, s	Mass of a unit length $M_L \cdot 10^{-9}$, g/mol·cm	Contour length $L \cdot 10^8$, cm
1	6	79	106	4500	2.4	1.9	4.18	1890
2	8	119	107	2600	2.0	2.28	4.35	2740
3	10	126	104	2700	1.91	2.32	4.52	2790
4	12	163	56	2300	2.40	3.62	4.68	3480
5	6	28	33	500	4.1	1.16	4.18	670
6	8	31	31	480	4.2	1.24	4.35	710
7	10	33	29	360	4.3	1.35	4.52	730
8	12	47	22	310	4.1	1.80	4.68	1000

Additionally the relative viscosity of solution was monitored during titration with NaCl solutions to investigate the interplay between hydrophobic/hydrophilic and electrostatic interactions. Two samples were chosen (samples 1 and 4 in table 1). The initial ionic strength of solution was $\approx 10^{-6}$M (pure water). Then 1 M and subsequently 6 M NaCl was added and the relative viscosity η_r was measured. The polymer concentration was kept virtually constant, as 0.976±0.004 g/dl. The relative viscosity η_r of sample 4 ($C_{12}H_{25}$ radical) decreased more significantly, and finally the polymer precipitated after 1.8 M NaCl in contrast to sample 1(C_6H_{13} radical) which remained soluble up to 5 M NaCl. Even dry salt addition didn't cause polymer precipitation. The results are presented in figure 1c.

At low ionic strength the electrostatic interactions are dominating in solution, and macromolecules of both copolymers have the maximal dimensions in pure water solutions. An increase in ionic strength with the gradual addition of salt leads to the progressive screening of electrostatic interactions, which in turn leads to a decrease in the coil dimensions. The latter causes an increase in hydrophobic interactions of side aliphatic chains. The hydrophobic interactions are the stronger the longer are the aliphatic side chains.

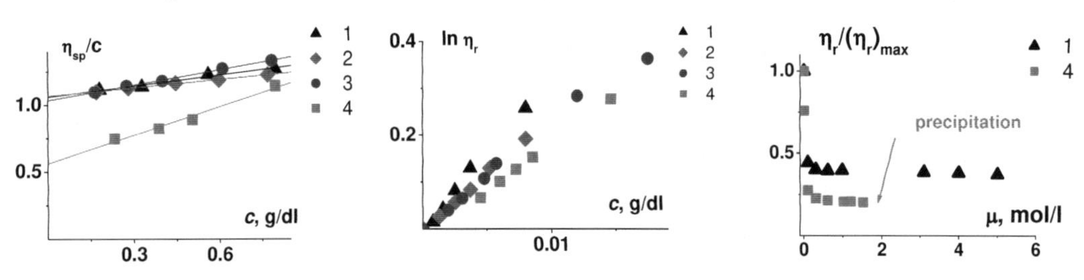

Figure 1. Viscometry data for MVAA–MVAH alkylated copolymers (samples 1 – 4 in table 1) **(a)** – Huggins [12] plot in 0.1M; **(b)** – logarithmic dependence of relative viscosity η_r on polymer concentration c in salt free H_2O; **(c)** – viscosity titration data for MVAA–MVAH alkylated copolymers (samples 1 and 4 in table 1): dependence of relative viscosity η_r normalized to its maximal value on ionic strength of solution μ (polymer concentration c = 0.976±0.004 g/dl). The initial (in H_2O) relative viscosity was η_r = 5.2 for sample 1 and η_r = 8.4 for sample 4. The measurements of sample 4 solution were finished after ionic strength μ = 1.5 mol/l because of polymer precipitation.

A powerful tool for interpreting hydrodynamic, especially viscometric, data are the Kuhn-Mark-Houwink-Sakurada plots normalized to the mass per unit length M_L [13]. The Flory-Fox relationship [14] gives a molecular meaning of the intrinsic viscosity value: $[\eta] = \Phi <h^2>^{3/2}/M$, where $<h^2>$ is the root-mean-square distance between the chain ends, Φ is the Flory viscosity parameter. The product $[\eta]M_L = \Phi<h^2>^{3/2}/L \sim V/L$ characterizes the volume V occupied by a part of macromolecule per unit of its contour length L. For linear macromolecules, the $[\eta]M_L$ = f (L) dependence (figure 2) can be considered as a distribution of linear macromolecules according to their sizes and conformations [13]. The dashed curves in figure 2 represent the dependences obtained for different linear chains and structures with different equilibrium rigidity values; the minimum values correspond to globular structures (statistical segment length A~1 nm), and the maximum values correspond to multi-stranded structures (A~200 nm). Use of this plot for new polymer systems allows to obtain qualitative estimations of their size and conformation.

It can be noted that the data for alkylated MVAA-MVAAH copolymers in 0.1M NaCl are located between flexible and rigid macromolecules, and the data in pure water solution – near extra-rigid uncharged macromolecules (figure 2).

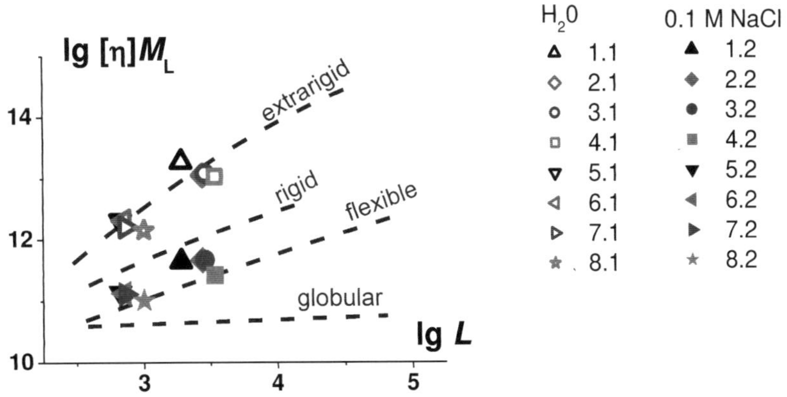

Figure 2. The double logarithmic dependence of normalized hydrodynamic volume $[\eta]M_L$ on contour length L. The curves correspond to extra rigid, rigid, flexible, and globular uncharged linear chains [13]. The data for alkylated MVAA–MVAH copolymers (samples 1–8 in table 1) in salt-free water solution are shown as open points 1.1 – 8.1 and in 0.1 M NaCl – as filled points 1.2 – 8.2. Black symbols correspond to samples containing C_6H_{13} aliphatic side chain, green symbols – C_8H_{17}, blue symbols – $C_{10}H_{21}$, red symbols - $C_{12}H_{25}$.

The most important characteristic of dissolved macromolecules in DFB is the value of shear optical coefficient $[n]/[\eta]=\Delta n/\Delta\tau$. It is related to the anisotropy of the intrinsic optical polarizability of the statistical segment ($(\alpha_1-\alpha_2)_0$) by the following relation:

$$[n]/[\eta] = \{4\pi(n_s^2+2)^2/45kTn_s\}\times\{(\alpha_1-\alpha_2)_0 + [2.60\Phi(\Delta n/\Delta c)^2/\pi^2 N_A^2]M/[\eta]\} \qquad (1),$$

here $[n]=\lim_{g,c\to 0}(\Delta n/g\eta_0 c)$, Δn is the observed DFB effect, $\Delta\tau$ is the shear stress, g is the shear rate gradient, Φ is Flory viscose parameter; n_s and η_0 are the refractive index and dynamic viscosity of the solvent; $\Delta n/\Delta c$ is the refractive index increment of the polymer-solvent system; k is Boltzmann constant; N_A is Avogadro number; T is Kelvin temperature. The second term on the right-hand side of equation (1) is the contribution of the so-called macroform effect.

Figure 3a shows the dependencies of Δn on the shear stress $\Delta\tau=g(\eta-\eta_0)$, the slope of which gives the desired value $\Delta n/\Delta\tau$ (η is viscosity of the solution). The dependence of shear stress $\Delta n/\Delta\tau$ on side radical length N_C is shown in figure 3b. The sign of the DFB value is positive and increases significantly for samples with long side chains ($N_C > 10$). This effect was observed for both molecular masses series. The large DFB value can be explained both by an increase in the order of anisotropic elements in the direction of a larger coil axis, and by an increase in the macroform effect due to the greater asymmetry of the macromolecule. The separation of these contributions requires further study in a series of polymer homologues with a fixed length of side chains and their composition.

The study of viscous flow and DFB effect in solutions of alkylated MVAA–MVAH copolymers shows that both in salt-free and in 0.1M NaCl solutions experimental characteristics ([η] и $\Delta n/\Delta\tau$) are noticeably different for samples with long side chains ($N_C = 12$) as compared with those with shorter side groups. This can be interpreted as a phenomenon of self-organization of macromolecules, which occurs due to the prevalence of hydrophobic interactions. Macromolecules of alkylated MVAA–MVAH copolymers have a hydrophilic base, hydrophobic side groups and charges on the chain, and there is a competition of all types of interactions, the contributions of which are determined by the ionic strength of the solutions.

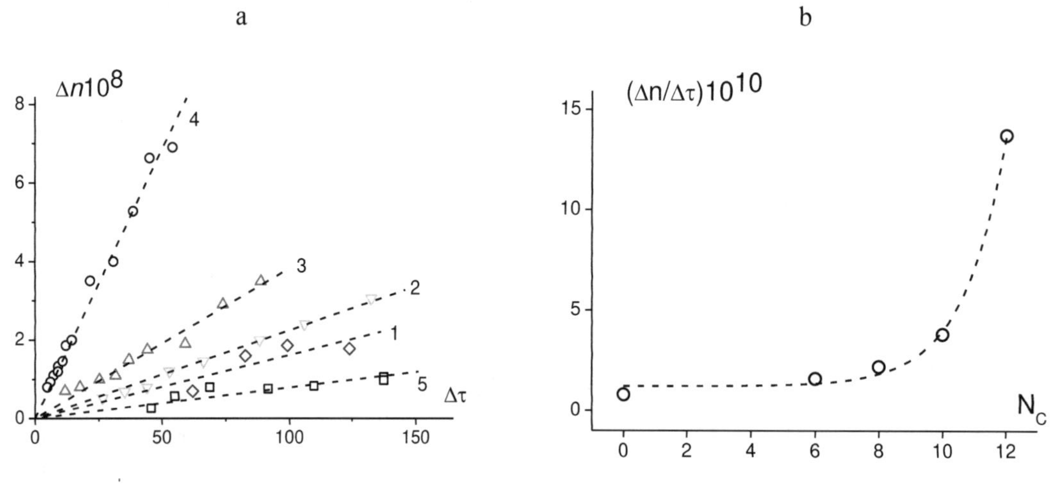

Figure 3. **(a)** – dependence of birefringence Δn on shear rate tension $\Delta\tau=g(\eta-\eta_0)$ for solutions of amphiphilic alkylated MVAA-MVAH copolymers (samples 1–4 in table 1) and 5 – unalkylated MVAA-MVAH copolymer ($N_C = 0$) in 0.1 M NaCl; **(b)** – dependence of shear stress optical coefficient or reduced birefringence $\Delta n/\Delta\tau$ on the number of carbon atoms in the side chains N_C for the same samples.

4. Conclusion

The complex study of hydrodynamic and optical properties of alkylated MVAA-MVAH copolymers containing hydrophilic backbone and hydrophobic side chain (from C_6H_{13} to $C_{12}H_{25}$) allowed the reliable analysis of their properties. The virtually monotonic change in the intrinsic viscosity and shear optical coefficient up to $N_C = 10$ followed by the sharp growth of these values for $N_C = 12$ was observed. At low ionic strength and short aliphatic side chains ($N_C < 10$) the electrostatic interactions prevail, and at medium ionic strength and long aliphatic side chains ($N_C > 10$) dominate the hydrophobic interactions. This statement is supported by comparative titration of samples with $N_C = 6$

and N_C = 12 by NaCl solutions. When the electrostatic interactions are practically "switched off" (0.1M), the self-organization of the copolymer coils is governed by hydrophobic interactions.

Acknowledgements
The reported study was funded by RFBR according to the research project № 18-33-00432.

References

[1] Vilar G, Tulla-Puche J and Albericio F 2012 *Current Drug Delivery* **9(4)** 367–394
[2] Qiu L, Bae Y 2006 *Pharm. Res.* **23** 1–30
[3] Duncan R 2003 *Nat. Rev. Drug Discovery* **2** 347–360
[4] Panarin E F and Gavrilova I I 1977 *Vysokomol. Soedin.,Ser. B* **19** 251–254
[5] Dommes O A, Okatova O V, Kostina A A, Gavrilova I I, Panarin E F and Pavlov G M 2017 *Polymer Science, Series C* **59(1)**, 125–132.
[6] Pavlov G M, Dommes O A, Okatova O V, Gavrilova I I and Panarin E F 2018 *Phys. Chem. Chem. Phys.* **20** 9975–9983
[7] Pavlov G M, Kolbina G F, Okatova O V, Gavrilova I I, Panarin E F 2015 *Doklady Chemistry* **463(1)**, 181–184
[8] Pavlov G M, Okatova O V, Gubarev A S, Gavrilova I I and Panarin E F 2014 *Macromolecules* **47(8)** 2748–2758
[9] Pavlov G M, Gubarev A S, Zaitseva I I and Sibileva M A 2006 *J. Appl. Chem.* **79(9)** 1423–1428
[10] Tsvetkov V 1989 *Rigid-chain Polymers* (New York: Plenum Press) p 490
[11] Pavlov G M, Kolbina G F, Shtennikova I N, Michailova N A, Korneeva E V 2001 *Eur. Polym. J.* **37** 1219-1225
[12] Huggins M L 1942 *J. Am. Chem. Soc.* **64(11)** 2716–2718
[13] Pavlov G M *Eur. Phys. J. E:* 2007 *Soft Matter Biol. Phys.* **22** 171–180
[14] Flory P J 1953 *Principles of Polymer Chemistry* (Ithaca, New York: Cornell University Press) p 672

PVA Hydrogel Electrolyte and Porous Polyisoprene Carbon Nanostructure Composite Based Pressure Sensitive Supercapacitor

L Lapčinskis[1*], I Cīrulis[1], P Lesničenoks[2], A Ābele[3], A Šutka[4] and M Knite[1]

[1]Institute of Technical Physics, Riga Technical University, Riga, Latvia
[2]Institute of Solid State Physics, University of Latvia, Riga, Latvia
[3]Institute of Polymer Materials, Riga Technical University, Riga, Latvia
[4]Research Laboratory of Functional Materials Technologies, Riga Technical University, Latvia

*linards.lapcinskis@rtu.lv

Abstract. Narrow operational voltage window can be considered as one of the greatest limitations for aqueous polymer electrolytes. Using freeze-thaw hydrogel preparation method we have synthesized wider potential window electrolytic polyvinyl alcohol (PVA)/Na_2SO_4 and PVA/K_2SO_4 electrolytes. Supercapacitors (SC) have been assembled using novel porous polyisoprene and carbon black composite electrodes. Our SC exhibit pressure sensitive properties therefore this effect is deeper explored here, giving explanation for capacitance increase during pressure application. It is found that up to 2 MPa the capacitance increases due to greater interface between electrode and electrolyte.

1. Introduction

Supercapacitors (SC) are short term energy storage devices which are used to give fast energy boosts when needed. Modern SC are expected to bridge the gap between batteries and conventional capacitors by providing both high power and energy. Flexible solid-state SC have extra advantages like possibility to use them in flexible electronics as well as the absence of liquid electrolyte [1-3]. However, the majority of flexible solid-state SCs can operate in a relatively narrow voltage window therefore limiting the practical use of these devices. Recently it has been stated that wider operational voltage window can be achieved using alkali metal salts as ion sources, e.g. Na_2SO_4 [4, 5].

One of the characteristics determined for SC devices is the ability to sustain stable capacitance during mechanical deformation. We have previously demonstrated that instability of capacitance under mechanical load can be successfully used for pressure sensing [6]. Mechanical load can decrease distance between electrodes, in case of piezoresistive material it can increase or decrease resistance of electrode, change the dielectric permittivity of layer between electrodes or facilitate greater interface between electrode and electrolyte.

Here we present a compressible solid-state SC with an operational voltage window of 4.4 V. PVA/Na_2SO_4 and PVA/K_2SO_4 were chosen to obtain such wide operational window in these poly(vinyl alcohol) cryogel electrolytes. Cryogel electrolytes containing H_3PO_4 and KOH, reported in literature [7-12], were also prepared to establish a comparison with pH neutral salts (Na_2SO_4 and K_2SO_4). Electrochemical characterization gives us information about limitations and energy storage properties of our device. We have investigated how separate parts of the whole supercapacitor device influence capacitance changes when it is subjected to mechanical load, therefore increasing our knowledge about interaction of SC components under pressure.

2. Results and discussion

Main components contributing to successful solid-state double-layer SC are carbon-based electrodes with current collector and ion conducting electrolyte. For electrodes flexible electro-conductive porous polyisoprene rubber was prepared to allow deformation to take place in our device. As in regard to electrolyte we chose to favour polymer hydrogel electrolyte, specifically cryogels, which are prepared by freeze-thaw method.

2.1. Porous electrode materials

Electro-conductive composite material was created by mixing polyisoprene rubber (PI) solution in chloroform with conductive carbon particles and blowing agent ChKhZ-5 particles dispersed in chloroform. The solvent was evaporated, and composite material was vulcanized under pressure. Main steps of composite material preparation are shown in figure 1.a, where last step shows crosslinked porous rubber samples ready to be used as electrodes. Special care was taken when crosslinking electrode material. First, composite material was molded by pressure in 140 °C temperature for 1.5 minutes. Then this pressure was removed, and crosslinking was continued at higher rate at 160 °C temperature. ChKhZ-5 thermally decomposes at temperature range 150-170 °C so pores are formed simultaneously with crosslinks as sample is not subjected to pressure in this stage.

We started by preparing three different electro-conductive polymer composites – PICB25, PIAC70 and PIGr65. Sample name shows that polyisoprene (PI) has been used as binder and carbon black (CB, PRINTEX-XE2), activated carbon (AC, Sigma Aldrich) and graphene (Gr, CheapTubes) as active electrode material where number shows phr (parts per hundred rubber). Phr values were chosen to be close to maximum content achievable. Electro-conductivity of these materials was measured in temperature range from 12 °C to 58 °C (figure 1.b). It is clearly seen that PICB25 sample shows superior electro-conductivity in comparison to other two materials reaching 7.4 S m^{-1} conductivity at room temperature (25 °C) while for PIAC70 it is 0.07 S m^{-1} and 0.03 S m^{-1} for PIGr65. CB was chosen for further use in samples because it possesses the highest electro-conductivity.

One of the factors that influence capacitance of SC is electrode surface area. We added pore forming agent ChKhZ-5 to increase surface area of electrode material. We expected that conductivity of material could decrease after formation of pores and to solve this we increased CB phr to maximum value we could achieve (28) before the material shows decrease of integrity. Porous electrode showed electro-conductivity of 17 S m^{-1}, which is by 11 S m^{-1} lower than non-porous material (figure 1.c), but it is still a high value for porous rubber composites.

The goal of this research was to prepare pressure sensitive supercapacitors so we tested how the conductivity of electrodes respond to mechanical load. Figure 1.c. demonstrates that for both cases of electrodes (porous and non-porous) small piezoresistive effect was observed – uniaxial pressure causes conductivity to drop. For SC this means that applied mechanical load slows down the propagation of charge to current collectors, because the resistance of electrodes increases. During fast charge-discharge cycles this effect could play a negative role.

By observing the surface of electrode material in figure 1.d we can see that pores are not uniform. Uniformity in this case is strongly dependent on content of filler particles, lower content could yield more uniform pores and higher porosity, but it would mean sacrificing electro-conductive properties.

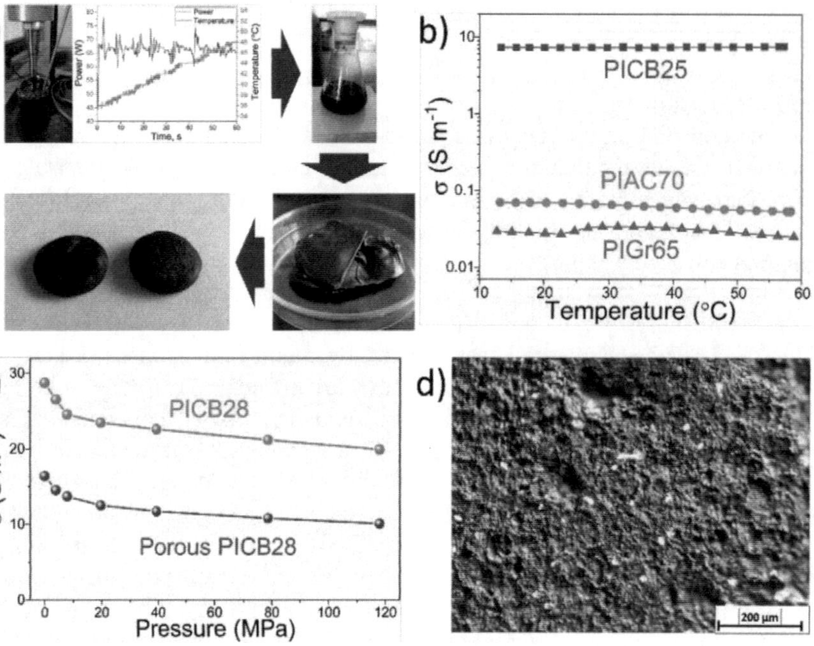

Figure 1. (a) Preparation technology of electro-conductive rubber starting with sonication of carbon particles mixed with $CHCl_3$, proceeded by mixing with PI/$CHCl_3$ solution and evaporation of solvent and lastly vulcanization of electrodes. (b) Electro-conductivity measurements of three carbon nanostructure composites. (c) Electro-conductivity of porous and non-porous PICB28 electrodes under various pressure values. (d) Optical microscopy image of porous PICB28 electrode surface, scale 200 µm.

2.2. Hydrogel electrolytes – preparation and characterization

The hydrogel electrolyte was synthesized using "Freeze-thaw" method (figure 2.a). PVA (Sigma Aldrich, Mw 130000, 99+% hydrolysed) was dissolved in distilled water at 100 °C temperature while being vigorously stirred. After complete dissolution it was cooled to room temperature and when clear, viscous solution was obtained, 20 wt% of 1 M Na_2SO_4, K_2SO_4, KOH or 20 wt% conc. H_3PO_4 was slowly added. This solution was casted in petri dishes and subjected to 3 freeze-thaw cycles in order to prepare white, opaque cryogel as shown in figure 2.a.

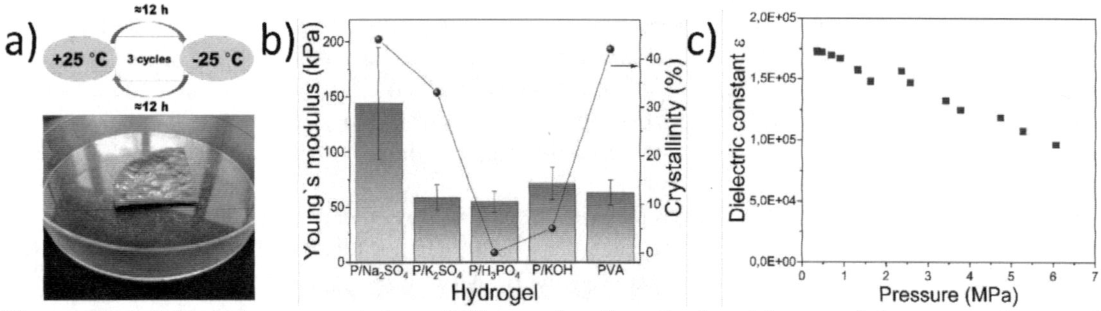

Figure 2. a) Schematic representation of "Freeze-thaw" method and image of the prepared cryogel electrolyte b) Graph showing crystallinity of prepared hydrogels calculated from DSC and corresponding Young's modulus c) Dielectric permittivity measurements when uniaxial pressure is used on PVA/K_2SO_4 hydrogel with smooth ITO electrodes.

PVA hydrogel without any ion conducting additives was also prepared using "freeze-thaw" method to determine how Na_2SO_4 and K_2SO_4 influence the structure of hydrogel. Differential Scanning

IOP Conf. Series: Materials Science and Engineering **500** (2019) 012018 doi:10.1088/1757-899X/500/1/012018

Calorimetry (DSC) of hydrogels show that addition of Na_2SO_4 hasn't made a great influence on the content of crystalline phase and for hydrogels PVA and PVA/Na_2SO_4 it is respectively 42 % and 44 % as evidenced in figure 2.b. In case of PVA/K_2SO_4 the crystallinity has slightly decreased and is around 33%. Interestingly, when H_3PO_4 and KOH is used as ion source in hydrogel the crystallinity dramatically decreases (0 % and 5 %) allowing much higher ion mobility in gel electrolyte.

Previously the influence of pressure on electrode conductivity was explored. Here we discuss influence of mechanical load on dielectric permittivity of hydrogel electrolyte. For this purpose, PVA/K_2SO_4 electrolyte was used. PVA/Na_2SO_4 showed higher Young's modulus as seen in figure 2.b and could be a better potential candidate for pressure testing, but unfortunately Young's modulus measurements of this hydrogel showed great deviations. Figure 2.c reveals that uniaxial pressure decreases the dielectric permittivity when hydrogel electrolyte is used in combination with smooth ITO electrodes. The changes in the structure of hydrogel caused by pressure reduces the free volume and amorphous phase, hence it decreases the mobility of ions and diminishes their contribution to total dielectric permittivity of hydrogel electrolyte layer.

2.3. Electrochemical characterization.

For electrochemical characterization one side of the crosslinked electrode was painted with silver paint to create current collector. SC was assembled by sandwiching the PVA/Na_2SO_4 hydrogel electrolyte between two identical electrodes. The operational voltage window was determined using cyclic voltammetry (CV). Results show that widest potential windows can be obtained for PVA hydrogels using Na_2SO_4 and K_2SO_4 (figure 3.a). These same electrolytes show the highest specific capacitance (figure 3.b) which is calculated from CV curves.

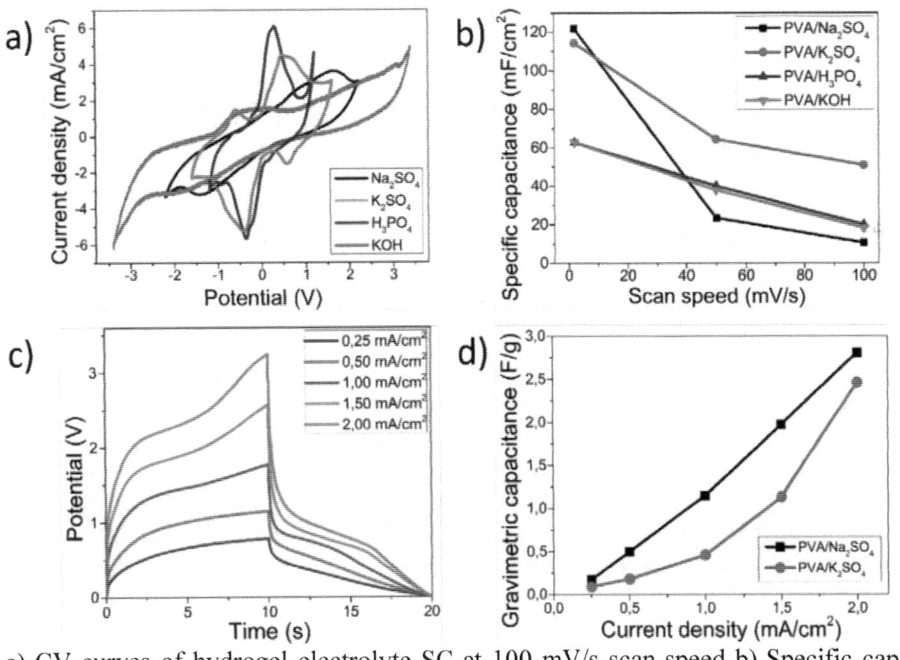

Figure 3. a) CV curves of hydrogel electrolyte SC at 100 mV/s scan speed b) Specific capacitance values obtained from CV measurements c) CP results for SC that is charged for 10 s and discharged for 10 s using various current densities (0.25-2.00 mA cm^{-2}) d) Gravimetric capacitance values calculated from charge-discharge curves.

Supercapacitors using PVA/N_2SO_4 and PVA/K_2SO_4 were chosen for further testing – for chronopotentiometry (CP) measurements. CP results show that when device is charged using various current densities for 10 s the maximum voltage increase (V = 3.25 V) is observed when 2 mA cm^{-2} is used (figure 3.c). The calculated gravimetric capacitance is lower than reported in literature, but it can be explained by the shape of electrodes. To allow compression, they have been prepared relatively thick which needs a higher amount of active electrode material. The highest gravimetric capacitance was obtained for SC using PVA/Na_2SO_4 and it was 2.8 F g^{-1} (figure 3.d).

2.4. Mechanical load impact on capacitance.

As it was established previously in subsections 2.1. and 2.2., the uniaxial pressure lowers the conductivity of electrode materials, therefore limiting charge distribution rate in electrode material, as well as decreases the dielectric permittivity of hydrogel electrolytes. Lower dielectric permittivity means that mechanical load should have a diminishing impact on capacitance of supercapacitor.

Figure 4.a demonstrates that capacitance of SC increases accordingly to applied pressure. As we have ruled out impact of increased dielectric permittivity and the electrode area doesn't increase, then we must conclude that capacitance increases due to smaller distance between electrodes and increased interfacial contact between electrode and electrolyte when uniaxial pressure is applied.

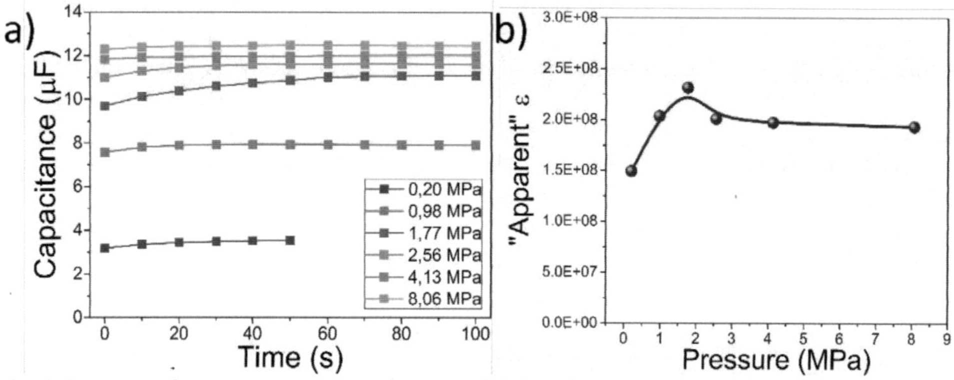

Figure 4. a) Impact of pressure on capacitance of SC using porous PICB28 and PVA/K_2SO_4 b) "Apparent" dielectric permittivity of SC device depending on the applied pressure.

To determine whether this capacitance increase is mainly due to closer distance between electrodes or better interface between electrode and electrolyte we calculated the "apparent" dielectric permittivity. Term "apparent" dielectric permittivity was used because it was calculated using the dimensions of device as the area of electrode. But in reality the rough and porous surface area of the specific electrode should be taken into account when dielectric permittivity is calculated. Therefore, the calculated dielectric permittivity is much higher (by 3 orders) than the one obtained from measurements that used ITO electrodes (figure 4.b) which is smooth and for which dimensions of device can indeed be used as the electrode area. The most interesting about this is the increase of permittivity when pressure up to 2 MPa is applied. "Apparent" permittivity slightly increases but as we have determined previously with measurements using ITO electrodes it should decrease. Hence, we can conclude that better interfacial contact is created when pressure up to 2 MPa is applied, which causes higher capacitance. When pressure is higher than 2 MPa capacitance increase can be mainly attributed to decrease of distance between electrodes because dielectric permittivity doesn't change as evidenced in figure 4. b.

3. Devices and methods

Electro-conductivity of electrode was measured with Agilent 34970A data acquisition/switch unit. Scanning electron spectroscopy images were obtained with Phenom Pro desktop SEM. Electrochemical measurements: cyclic voltammetry (CV) and chronopotentiometry (CP) were conducted with VoltLab PGZ301 potentiostat at room temperature (22 °C) using Swagelok type cell. Dielectric properties and capacitance for mechanical load dependent experiments were measured with Agilent E4980A LCR meter while mechanical load was provided by universal material testing machine Zwick/Roell Z2.5.

IOP Conf. Series: Materials Science and Engineering **500** (2019) 012018 doi:10.1088/1757-899X/500/1/012018

4. Conclusions

Use of cryogel electrolyte containing K and Na sulphates allows wider potential window of assembled devices, resulting in devices better suitable for integration in circuits. Applied uniaxial pressure decreases the conductivity of electrode material limiting charge distribution in electrode material when pressure is applied, which is not favourable for pressure sensors. Dielectric constant of electrolyte decreases under mechanical load, therefore, pressure should have a diminishing effect on capacitance. As can be seen from results effect is opposite – pressure increases the capacitance. We have concluded that weaker pressure (up to 2 MPa) increases capacitance mainly due to improvement of interfacial conditions. When pressure is higher than 2 MPa capacitance increases mainly due to smaller distance between electrodes.

Acknowledgements
Special gratitude is in order to Raimonds Orlovs for temperature dependent resistance measurements. Financial support of project 1.1.1.1/16/A/013, "Hybrid energy harvesting systems" is greatly appreciated.

References

[1] X. Long, Z. Zeng, E. Guo, X. Shi, H. Zhou, and X. Wang, "Facile fabrication of all-solid-state flexible interdigitated MnO2 supercapacitor via in-situ catalytic solution route," J. Power Sources, 325, pp. 264–272, 2016.
[2] T. Chen and L. Dai, "Flexible supercapacitors based on carbon nanomaterials," J. Mater. Chem. A, 2, 28, pp. 10756–10775, 2014.
[3] P. Yang and W. Mai, "Flexible solid-state electrochemical supercapacitors," Nano Energy, 8, pp. 274–290, 2014.
[4] Batisse N., Raymundo-Piñero E. A self-standing hydrogel neutral electrolyte for high voltage and safe flexible supercapacitors// J. Power Sources – 2017. – Vol. 348 – P. 168-174.
[5] Materials for Sustainable Energy Applications: Conversion, Storage, Transmission, and Consumption / Ed. by David Munoz-Rojas, Xavier Moya, CRC Press, 2017. – 425 p.
[6] L. Lapčinskis, A. Linarts, M. Knite, I. Gorņevs, J. Blūms, A. Šutka, "Solid-state supercapacitor application for pressure sensing", Appl. Surf. Sci., In Press, Accepted Manuscript, doi: 10.1016/j.apsusc.2018.05.036
[7] L. Sa'adu, M. a. Hashim, M. Bin Baharuddin, Conductivity Studies and Characterizations of PVA-Orthophosphoric Electrolytes, J. Mater. Sci. Res. 3 (2014) 48–58. doi:10.5539/jmsr.v3n3p48.
[8] H. Yu, J. Wu, L. Fan, K. Xu, X. Zhong, Y. Lin, J. Lin, Improvement of the performance for quasi-solid-state supercapacitor by using PVA-KOH-KI polymer gel electrolyte, Electrochim. Acta. 56 (2011) 6881–6886. doi:10.1016/j.electacta.2011.06.039.
[9] G. Ma, J. Li, K. Sun, H. Peng, J. Mu, Z. Lei, High performance solid-state supercapacitor with PVA-KOH-K 3[Fe(CN)6] gel polymer as electrolyte and separator, J. Power Sources. 256 (2014) 281–287. doi:10.1016/j.jpowsour.2014.01.062.
[10] Y.F. Huang, P.F. Wu, M.Q. Zhang, W.H. Ruan, E.P. Giannelis, Boron cross-linked graphene oxide/polyvinyl alcohol nanocomposite gel electrolyte for flexible solid-state electric double layer capacitor with high performance, Electrochim. Acta. 132 (2014) 103–111. doi:10.1016/j.electacta.2014.03.151.
[11] S. Akbulut, M. Yilmaz, S. Raina, S.-H. Hsu, W.P. Kang, Solid-state supercapacitor cell based on 3D nanostructured MnO 2 /CNT microelectrode array on graphite and H 3 PO 4 /PVA electrolyte, Diam. Relat. Mater. 74 (2017) 222–228. doi:10.1016/j.diamond.2017.03.016.
[12] K.P. Singh, R.P. Singh, P.N. Gupta, Polymer based solid state electrochromic display device using PVA complex electrolytes, Solid State Ionics. 78 (1995) 223–229. doi:10.1016/0167-2738(95)00100-K.

Impact of aging and sweat action on properties of parchment as a biopolymer of animal origin of aging

T.Kolesnyk[1], O. Andreyeva[2], A. Nikonova[3] and O. Savchuk[4]
[1]Graduate student, Kyiv National University of Technologies and Design, Ukraine
[2]Professor, Kyiv National University of Technologies and Design, Ukraine
[3]Senior Lecturer, Kyiv National University of Technologies and Design, Ukraine
[4]Professor, Educational and Scientific Center «Institute of Biology and Medicine» (Kyiv National Taras Shevchenko University), Ukraine

annavnikonova@gmail.com

Abstract. The processes of accelerated artificial aging and sweat action of leather parchment manufactured in different methods from sheep's skin were investigated. By means of standard physical and mechanical methods of analysis and method of gel electrophoresis, the effect of high temperature and sweat was analysed on the behaviour of this biopolymer over time. The multidirectional of these influences on the most important physical and mechanical properties and the structure of parchment have been founds, which can be explained by the different degree of separation of the structural elements during the liming and the presence of plasticizing components that can prevent the influence of external factors.

1. Introduction

Parchment as a biopolymer of animal origin is one of the oldest types of natural leather, whose uniqueness lies in the production specifics (the tanning process is excluded from the technological cycle), as well as in significant strength and durability due to the peculiarities of structure and properties of collagen which is its main component. This allows being to parchment one of the best materials for writing, restoration of historical manuscripts and book editions, manufacturing of decorative arts and crafts [1-2]. However, as well as many other materials, the disadvantage of parchment is instability to various types of influence (mechanical, thermal, chemical and biologically active environment, etc.) that being to change its properties, direct to the deterioration and even destruction [3]. In this case, one of the most important factors that determine the resistance of parchment material to damage is the method of treatment of hides and the type of used reagents. Therefore, in order to improve the producing technique, to expand the possibilities of using and preserving this specific material, it is important to be aware of the change in its properties over time under different conditions.

In view of the above, processes of artificial aging and sweat action of leather parchment, obtained by different technologies was investigate in many previous researches [4-7]. As determinative factors in this research, the effect of accelerated aging by high temperature and action of artificial sweat were selected which sufficiently reflect the influence that parchment can undergo during storage and exploitation.

2. Materials and methods

As investigated materials, samples of leather parchment (hereinafter simply parchment), obtained from sheep's skin by several known methods were used: group 1 (parchment for writing) according to the modern method, which involves two-stage liming with a lower (26 g/l) consumption of calcium hydroxide [4]; group 2 (parchment for restoration works) by the modern method on the basis of two-stage liming with a higher (35 g/l) consumption of calcium hydroxide [5]; group 3 (transparent parchment) according to the ancient sulphide-lime method of liming [6]; group 4 (parchment for general purpose) by accelerated method of oxidative liming [7]. Common to all methods was the presence of two consistently performed technological blocks: the first which consisted of beam house processes and operations (washing, soaking, liming, flashing), and the second which consisted of finishing processes and operations in the form of drying, grinding, plasticization; in addition, in groups 2 and 3, bleaching was carried out, and in group 3 the filling with potassium alum salt [8].

In order to studying the behaviour of parchment in time-accelerated methods of artificial aging and determination of sweat action were used. The method of artificial aging consisted of keeping the samples in climate test camber (70°C, 0% rH, and 72 h). Sweat resistance test was carried out by keeping the samples in a solution of artificial sweat which included: sodium chloride, ammonium carbonate, disodium phosphate, carbamide, lactic acid (25°C, 72 h) and after, samples dried under norm-climate condition (23±1°C, 60% rH) [10].

To establish the biochemical changes in collagen structure, the methods of gelatine yield and disc electrophoresis in polyacrylamide gel with the adding of sodium dodecyl sulphate were used [11-13]. For the yield of gelatine, the crushed parchment weighing about 0.5 g (in recalculation of absolute dry substance) was placed into a flask (100 ml volume) and 50 ml of distilled water was added. The flask was incubated in a thermostat for 2 hour at 90°C. After cooling, the contents of the flask were filtered and washed with distilled water. A Biuret reagent was added to the measuring flask and diluted with distilled water to the mark on a 100 ml volumetric flask. After holding the flask in a dark place for 20 minutes the optical density of gelatine solution was determined on a spectrophotometer (U-LAB102) at absorption mode 520 nm. The concentration of gelatine (%) was determined according to standard calibration curve [11].

For the analysis of proteins in solutions, in which samples of parchment were incubated disc electrophoresis method was used in 10% polyacrylamide gel (PAAG) in the presence of sodium dodecyl sulphate (SDS) in the Lemley system beneath denaturing conditions. Electrophoresis was carried out in a device for vertical preparative disc electrophoresis (BioRad, USA) in glass plates 1 mm thick with a current of 19 mA for concentrating gel and 35 mA for separating gel. The gels were stained with a 0.125% (w/v) Coomassie G-250 solution (dissolved in 25% (w/v) propan-2-ol and 10% (w/v) acetic acid), excessive colour washed with 0.8% (w/v) acetic acid. At staining the ratio of volumes of liquid and gel was 3:1, at washing was 5: 1 with two changes of solvent. A marker mixture of proteins with a molecular weight of 94 kDa (phosphorylase B), 67 kDa (albumin), 43 kDa (ovalbumin), 30 kDa (anhydrase), 20.1 kDa (soybean trypsin inhibitor) and 14.4 kDa (lactalbumin) were used (Amersham Biosciences) [12].

The concentration of protein in the solution was determined using Bradford method. To this end, 10 μl of 10% sodium hydroxide, 70 μl of distilled water and 2 ml of the working solution were added to 20 μl of sample. The working solution was prepared by mixing of 6 ml a stock solution (containing 10 ml of 95% ethanol, 20 ml of 85% phosphoric acid, 35 mg of Coomassie blue), 3 ml of 95% ethanol, 6 ml of 85% phosphoric acid diluting with distilled water to the mark on a 100ml volumetric flask. The concentration of the solution was measured on a Smartspec spectrophotometer (BioRad) at absorption mode 595 nm compared to control sample with distilled water. The amount of protein in solution (mg/ml) was determined according to standard calibration curve [13].

To obtain an apparent understanding of changing the properties of parchment as a result of accelerated aging (high temperature and action of artificial sweat) such important physical and mechanical indices of leather as hydrothermal stability, strength, percentage elongation, stiffness and thickness had been analysed which were evaluated by standard methods of testing leather materials on standard equipment's [14-17]. During the production of parchment to exclude the influence of topographical areas, which are different, the groups of samples have been complete by the asymmetric method [11]. The elastic-plastic properties of parchment were determined using indices of percentage elongation. The uniformity of leather properties on the area, presented as coefficient uniformity of tensile strength C_{unif}, which is the ratio of average values relatively index of tensile strength cross-vertebral direction (σ_{cv}) to the average values index of tensile strength the longitudinal direction (σ_l), calculated according to formula [11]:

$$C_{unif} = \sigma_{cv} / \sigma_l$$

Reliability of obtained results was assessed by traditional methods of mathematical statistics, using the mean square deviation and arithmetic mean [18].

3. Results and discussion

Test results of parchment samples both after accelerated aging and after resistance, testing was given in Table 1 space required from which it follows that after the heat treatment and action of sweat the yield of gelatine decreases for specimens of groups 1, 2, 4 (and especially for group 4 by 28.2 and 53.1% respectively), and increases in group 3 by 69.8 and 8.7% respectively. The least (3.6-4.0%) changes in the indicator are observed in group 1 after action of sweat and group 2 after artificial aging. Particularities of structure of investigated parchment samples, caused by a greater degree of opened op of structure, broken or weakening of the intramolecular and intermolecular bonds, are found because of deeper yield of gelatine in groups 2, 3 (19.1 and 24.9%) after artificial aging, as well as in groups 1, 2 (21.1 and 17.5%) after influence of sweat.

Table 1. Results of physical and mechanical tests of parchment

Parchment after stage	Group 1	Group 2	Group 3	Group 4
1	2	3	4	5
		Yield of gelatine (%)		
Initial sample	21.9	19.9	14.3	22.2
After aging	13.9	19.1	24.3	15.9
After sweat action	21.1	17.5	15.5	10.4
		Shrinkage temperature (°C)		
Initial sample	94.0	87.0	93.0	70.0
After aging	65.5	64.0	68.0	68.5
After sweat action	65.0	63.0	66.0	67.0
		Tensile strength σ (N/mm^2)		
Initial sample	39.4	65.8	24.8	49.3
After aging	25.9	32.6	34.4	59.5
After sweat action	29.8	24.5	30.6	39.5
		Coefficient of uniformity of tensile strength C_{unif}		
Initial sample	0.65	0.74	0.71	0.77
After aging	0.45	0.67	0.62	0.65
After sweat action	0.29	0.50	0.56	0.53
		Percentage elongation at the strain N/mm^2 (%)		
Initial sample	16.0	15.0	18.0	17.3
After aging	11.6	10.0	13.5	11.0
After sweat action	12.5	11.5	13.0	10.5
		Percentage elongation at tear load (%)		
Initial sample	41.0	40.5	46.5	33.5
After aging	19.3	21.7	28.8	24.2
After sweat action	23.0	23.5	29.0	29.0
		Thickness (mm)		
Initial sample	0.44	0.50	0.48	0.40
After aging	0.37	0.42	0.45	0.30
After sweat action	0.48	0.48	0.61	0.39

Hydrothermal stability of parchment was determined by shrinkage temperature, which decreases in all cases: for groups 1-3, significantly almost a threefold (26.4%-30.9%), for group 4 only by few percent (2.1-4.3%). It should be noted that initially for group 4 this indicator was the lowest (≤ 70°C) of all analysed groups for which it was at level of 87-94°C.

As with any other types of leather, one of the important functional properties of parchment is its strength, determined by the method of biaxial stretching. As a result of the test, the decrease in this index by 24.4-62.8% was found for group 1 and 2 after high temperature and action of artificial sweat and for group 4 only after action of sweat (by 19.9%). The strength of transparent parchment (group 3), on the opposite, increased by 23.4-38.7% for test specimens subjected to sweat aging and dry aging, respectively. Strength of the parchment made by the accelerated method of liming (group 4) changed ambiguously after the manipulations: increased by 20.7% after artificial aging and decreased by 19.9% after sweat action.

The least value index of the coefficient of uniformity of tensile strength indifferent directions of leather is observed in samples of group 1 (two-stage liming with lower consumption of calcium hydroxide) at all stages of the study; most of all in samples of group 2 (two-stage liming with higher consumption of calcium hydroxide) before and after artificial aging. In all cases, the influence of artificial aging and action of sweat is observed in reducing of the coefficient of distribution uniformity index of parchment within 9.5-30.8 and 21.1-55.4% in this order.

A decrease of percentage elongation at the strain after artificial aging by 25.0-36.4% and after action of sweat by 21.9-39.3% was experimentally established. The greatest decrease of this index 36.4-39.3% is observed for group 4 (accelerated method of liming), the most for group 3 (transparent parchment). After artificial aging and action of sweat, the percentage elongation at tensile strength of samples of all leather decreased: most for group 1 (53.0 and 43.9%), the least for group 4 (27.8 and 13.4%). By the way, the initial samples of parchment were the least flexible in group 4, and conversely in group 3, were from the beginning it had been more flexible and retained this ability after additional manipulations.

The thickness of initial parchment was at a level of 0.40-0.50 mm. The influence of artificial aging has been founds in decreasing the thickness of samples of all groups: least amount (6.3%) for group 3, most (25.0%) for group 4 and almost the same amount (15.9-16.0%) for groups 1 and 2. The influence of action of sweat on the thickness of parchment was completely different. Thus, for samples of groups 2 and 4 there is only a slight decrease (2.5-4.0%) of this index, while for group 1 and especially for group 3 samples there is a more significant increase by 9.1% and 27.1% respectively.

To assess the change in structure of parchment as a result of aging procedure the process of releasing proteins and their fragments into surrounding solution, in which the samples of this biopolymer were incubated, was investigated by determining the amount of total protein and qualitative composition of proteins-protein fragments in the solution using disc electrophoresis method. As can be seen from table 2, artificial aging and action of sweat have significant impact on native structure of parchment, resulting in a certain release of protein macromolecules and their fragments into surrounding solution. At the same time, the minimal manipulative influence on this process was found in group 2, maximal in group 1.

Table2. Total protein concentration (mg/ml) in incubated parchment solution

Stage of research	Group 1	Group 2	Group 3	Group 4
Initial sample	5.50	3.75	8.00	7.00
After aging	13.25	1.62	7.37	9.37
After sweat action	10.37	4.50	6.25	3.62

Based on the analysis of samples electrophoregram obtained by using the method of disc electrophoresis in PAAG, it is established (Table 3) that aging procedure used has different influence on the qualitative composition of proteins that are released from the parchment structure into solution. Thus, the complete absence of proteins and their fragments with a molecular weight <35 kDa and the most content of heterogeneous molecular weight group of proteins in the range of 150-100 kDa was found. Perhaps this is due to the fact that the parchment consists mainly of collagen polymer structures and the manipulations carried out lead to a partial destruction of collagen and release it into solution. This in turn, can be a reason for changing the stability and integrity of parchment. Also, a certain amount of protein with a molecular weight in the range of 100-67 and 67-35 kDa was detected. The last group, most probable, mainly consists of fragments of those fibrillar protein complexes that are part of the structure of the biopolymer.

Table3. Results of electrophoresis analysis of incubated parchment

Stage of research	Group 1	Group 2	Group 3	Group 4
1	2	3	4	5
		Molecular weight >150 kDa		
Initial sample	163	164,155,151	156	151
After aging	161	159	152	152
After sweat action	161	158	150	–
		Molecular weight 150-100 kDa		
Initial sample	143, 118, 112	133, 119, 114, 104	141, 115, 107	122, 115, 107
After aging	146, 125, 116	146, 123, 111, 103	141, 117, 107	128, 117, 110
After sweat action	143, 121, 111, 104	141, 115, 108, 100	120, 110	124, 115, 104
		Molecular weight 100-67 kDa		
Initial sample	79	90, 82, 73	79	–
After aging	–	84	89, 78	95
After sweat action	78	94, 75	92, 84	94, 84
		Molecular weight 67-35 kDa		
Initial sample	65, 45	63, 58, 52, 49, 46	63	–
After aging	–	66	62	–
After sweat action	65	62, 44, 42	62, 51, 46	64, 51, 46
		Molecular weight < 35 kDa		
Initial sample	–	–	–	–
After aging	–	–	–	–
After sweat action	–	–	–	–

4. Conclusions

The processes of artificial aging by high dry temperature and sweat of the samples of parchment, obtained from sheep's skin by different methods were investigate.

To obtain an apparent understanding of changing the properties of parchment after carried out aging such important physical and mechanical indices as hydrothermal resistance stability, strength and percentage elongation were analysed. Methods of gelatine yield and gel electrophoresis in a polyacrylamide gel were used to determine the biochemical changes of the collagen structure, as the main component of the parchment.

Experimentally it was found that in respect to the indices of hydrothermal stability and percentage elongation samples of parchment produced by accelerated method of liming showed highest resistance to the influences of high temperature and sweat while in respects of strength for transparent-parchment. The obtained data may be due to different degrees of separation of structural elements during the liming process, the presence of plasticizing component that can prevent the influence of external factors. By evaluating the process of release of parchment proteins in the surrounding solution, considerable influence of aging on the structure of parchment was revealed that could predict the probability of losing the integrity of parchment and, consequently, its resistance to wear.

References

[1] Betty M Haines 1999*Parchment: The Physical and Chemical Characteristics of Parchment and the Materials Used in Its Conservation* (Leather Conservation Centre)

[2] Della Gatta G, Badea E*et all*2005 Assessment of damage in old parchments by DSC and SEM *Journal of Thermal Analysis and Calorimetric* vol 82 pp 637–649

[3] Abdel-Maksoud G, Marcinkowska E2000 Changes in some properties of aged and historical parchment. *Restaurator* vol 21 pp 138-157

[4] Pat no 260074 SSSR, Ignatov A M, Shlyakova L A, Sarycheva V T, Dorofeev M M 1980*Method for obtaining writing parchment* app no 2471699/28-12

[5] Dzendzeliuk L S, Loda L M, StempitskaYu S 2006 Making of parchment *The theory and practice of material and artistic culture VIII electronic scientific conference Kharkiv* (KhDADM, 20 of December 2006 Ukraine) pp 17–19

[6] Kol Luy 1932 *Succulent and suede tanning* (Moscow: ONTY)

[7] Pat no 4310328 US, Monsheimer R, Pfleiderer E 1982 *Process for liming pelts of animal hides and skins* app no 12204

[8] Adakina N I, Kolesnyk T O, Andreyeva O A 2018 Technological features of different methods of manufacturing leather parchment *Herald of Khmelnytskyi national university* no1(257) pp 187–192

[9] ISO 17227:2002 *Leather Physical and mechanical tests* Determination of dry heat resistance of leather

[10] Patent 2314525 RU, Chursin V I, L'vova A N, Safronov D A 2008 *Method for determining sweat resistance of skin* app No 206117369/12

[11] Golovteeva A A *et all* 1971 *Laboratory practice at the rate of chemistry and technology of leather and fur* (Moscow: Legprom)

[12] Hockfield S, Carlson S, Evans C *et all* 1993 *Selected methods for antibody and nucleic acid probes* (Cold Spring Harbor Laboratory Press)

[13] Bradford M M 1976 A rapid and sensitive method for the quantization of microgram quantities of protein utilizing the principle of protein-dye binding *Analyt Biochem* vol 72 pp 248-254.

[14] Visotskite V K, Dobrusina S A 1995 Influence of chemical treatments on physico-mechanical properties of parchment in the process of aging *TPS St Petersburg* issue 17 pp 34-41

[15] ISO 3380:2002 *Leather Physical and mechanical tests* Measurement of shrinkage temperature up to 100 °C

[16] ISO 3376:2011 *Leather Physical and mechanical tests* Measurement of tensile strength and percentage elongation

[17] ISO 2589:2002*Leather Physical and mechanical tests* Measurement of thickness

[18] Jerome L Myers, Arnold D Well, Robert F Lorch Jr 2013 *Research Design and Statistical Analysis* (New York: Routeledge)

IOP Conf. Series: Materials Science and Engineering **500** (2019) 012020 doi:10.1088/1757-899X/500/1/012020

Use of waste from the leather industry for the production of biotechnological products based on collagen

O M Savchuk[1, 2], P A Rebrikova[1], T B Vovk[2], N G Raksha[2], L I Ostapchenko[2] and O R Mokrousova[1]

[1] Department of Biotechnology, Leather and Fur, Kiev National University of Technology and Design, 2, Nemirovicha-Danchenko str., Kyiv, 01011, Ukraine
[2] Department of Biochemistry, Educational and Scientific Center «Institute of Biology and Medicine» Taras Shevchenko National University of Kyiv, 64/13, Volodymyrska str., Kyiv, 01601, Ukraine
olenamokrousova@gmail.com

Abstract Leather industry generates vast quantities of by-products which could be a cheap and available source of collagen and collagen peptides. These peptides have been widely used in cosmetic, biomedical, and pharmaceutical industries. The main objective of present study was to obtain a biotechnological product based on collagen and to study its properties.
Lime fleshings from hides were fragmented into small pieces with size $\leq$ 0.5 cm^2. Collagen was extracted from trimming waste using 0.5 M acetic acid in the presence of 5 mM EDTA. The purity of extracted collagen was checked by sodium dodecyl sulphate polyacrylamide gel electrophoresis (SDS-PAGE). Collagen was enzymatically hydrolyzed by pepsin. To stop the reaction, the mixture was heated immediately at 100°C and then centrifuged at 4000 g for 30 min. The supernatant was filtered through ceramic membrane for separation of collagen peptides fraction (26≤kDa). Collagen peptides were prepared from the obtained collagen. The generated collagen peptides was analyzed by SDS–PAGE using 18 % resolving gel.
The influence of collagen peptides on antioxidant enzymes and lipid peroxidation were explored. Collagen peptides supplementation also lowered serum level of pro-inflammatory cytokines such as IL-1β, IL-12, while level of anti-inflammatory cytokine IL-4 was within the control value.

1. Introduction

The problem of processing and rational use of waste products of industries in recent years became particularly relevant. Wastes by leather industries have raised serious concerns on account of their environmental impacts [1]. This waste should be recycled to overcome the pollution issues which would ultimately cause health problems for human and other living beings. Therefore it is important to develop a simple and available method for the optimal utilization of this waste to provide a practically feasible and economically viable solution. The leather industry waste has a huge amount of protein-rich by-products, which after extraction could be used for different purposes. The trimmings contain high amount of fibrous protein [2], particularly collagen, which is actually a valuable bioresource due to its biological safety, excellent biocompatibility, low antigenicity, high biodegradability, and cell growth potential. Generally, collagen has been applied in different fields, among the best known are the food, cosmetic, biomedical, pharmaceutical, and film industries [3].

Nowadays, special attention is given to research into collagen peptides (CP) due to a broad spectrum of their bioactivities. CP has been found to have a wide range of functional and biological properties, including antimicrobial, anti-inflammatory, anti-ulcer, lipid-lowering, wound-healing and anti-skin-aging activities [4,5]. These bioactivities and great health-enhancing potentials have led to the use them as ingredients of functional foods or pharmaceuticals. Majority of the studies on collagen peptides focused on their anti-oxidative activity which is very important taking into account that many human diseases are known to be caused by generation of free radicals and development of oxidative stress. Collagen peptides can reduce the peroxidation of lipids or fatty acids, scavenge free radicals and chelate transition metal ions [6], which has led to the hypothesis that these peptides could be suitable supplement agents for several types of illness associated with oxidative stress.

The main objective of this study was to isolate the collagen fraction from the trimming waste of leather industry, prepared collagen peptides and then evaluated the effect of CP administration on oxidative stress markers and cytokine profile in obese rats.

2. Materials and methods

2.1. Extraction of collagen

Collagen was extracted according to the method described in [7]. Extraction of collagen from raw material of leather industry was done in two steps. At first, lime flashings from hides were washed in cold water, poured with 20 % NaCl and left for a day to precipitate non-collagen proteins, then washed with distilled H_2O to neutral pH and dried. Extraction of the collagen from the dried material was carried out using 0.5 M of acetic acid containing 5 mM EDTA for 24 h. The extract was centrifuged at 10000 g for 30 min and the residues were re-extracted with the same solution for 24 h with further centrifugation. The supernatant was combined and salted out by adding NaCl to a final concentration of 0.9 M. The precipitated collagen was separated by centrifugation at 10000 g for 30 min, redissolved in 0.5 M of acetic acid and precipitated again with NaCl. The obtained precipitate was dialyzed against distilled water and lyophilized.

2.2. Preparation of collagen peptides

The procedure of CP preparation was performed as described [8]. The obtained collagen was enzymatically hydrolyzed by pepsin (3000 U g protein^{-1}) at 37°C, pH 2.0 for 8 h. To stop the reaction the mixture was heated to boiling for 5 min and then centrifuged at 4000 g for 30 min. The supernatant was filtered through a ceramic membrane (200 μm) to separate the collagen peptides fraction with molecular weight below 26≤kDa.

2.3. Estimation of extracted collagen and collagen peptides by sodium dodecyl sulfate polyacrylamide gel electrophoresis (SDS-PAGE)

SDS-PAGE was done as described by the [9] using 4 % (w/v) stacking gel, 6 % (w/v) separating gel for estimation of purity of extracted collagen and 18 % (w/v) separating gel for collagen peptides. SDS-PAGE was performed using Mini-Protean Tetra System (Bio Rad, USA) at 19 mA for stacking and 36 mA for separating gels. After electrophoresis, the gels were stained with 2.5 % coomassie brilliant blue R-250 in 10 % (v/v) ethanol, 10 % (v/v) acetic acid, 15 % (v/v) isopropanol and the background of the gel was destained with 7 % (v/v) acetic acid for 30 min.

2.4. Animals and experimental design

A total of 30 adult male Wistar rats weighing 170±5 g were used in present study. The study protocol was approved by the Ethical Committee of ESC "Institute of Biology and Medicine" Taras Shevchenko

National University of Kyiv. All animals were housed under controlled conditions with temperature of 25±2°C, relative humidity of 60±10 %, room air changes 12–18 times/h, and a 12-h light/dark cycle. Animals were allowed free access to food and water. The potential antioxidant and anti-inflammatory effects of collagen peptides were investigated on obesity model induced by high-calorie diet [10].

2.5. Sample collection
Blood samples were collected in standard biochemical test tubes. Serum for the determination of biochemical parameters was prepared by centrifugation at 1000 g of previously incubated blood samples for 30 min at 37°C. The serum was separated and kept at -20°C until analysis. Protein concentration was determined according to the method of Bradford [11], using crystalline bovine serum albumin as a standard.

2.6. Measurement of lipid peroxidation
The lipid peroxidation (LPO) was characterized by the spectrophotometric assessment of thiobarbituric acid reactive substances (TBARS). An aliquot of 0.4 mL of blood serum was added to 1.6 mL an aqueous solution of 25 mM Tris-HCl and 175 mM KCl (pH 7.4). Total protein fraction was separated from the mixture by precipitation with 20% trichloroacetic acid and further centrifugation for 15 min at 5000 g. After addition of 1 mL of 0.8 % aqueous solution of thiobarbituric acid to 2 mL of obtained supernatant, the samples were heated for 30 min in a boiling water bath. After cooling, the optical density of the samples was determined with a spectrophotometer (Smart Spec^{TM}Plus, BioRad, USA) at 532 nm. The amount of LPO products was calculated using the molar extinction coefficient $\varepsilon_{532}=1.56\times10^5$ $M^{-1}cm^{-1}$. The TBARS level was expressed as nmol·(mg protein)$^{-1}$ [12].

2.7. Antioxidant enzyme activity assays
Superoxide dismutase (SOD) activity was assayed by the method [13] based on the capability of SOD to inhibit the autooxidation of adrenaline. Catalase (CAT) activity was measured by the method [14].

2.8. Cytokine assay
Cytokine measurements in liver tissue were done by enzyme-linked immunosorbent assay according to the standard instructions [15].

2.9. Statistical analysis
Data entry and analysis were performed using StatSoft Statistica ver. 7.0 for Windows. After testing for normality (by Shapiro-Wilk), one-way analysis of variance (ANOVA) was used to compare the means among different groups. Differences were considered to be statistically significant when $p<0.05$. Data were reported as means±standard deviation (SD).

3. Results and discussion

The purity of obtained collagen was estimated by SDS-PAGE. It's known that collagens type I are constituted of two or more different chains (heterotrimers): typically two alpha 1, one alpha 2 chains of similar molecular weight about 100-110 kDa and a beta component structure [16]. Figure 1 showed that extracted collagen consisted of both α1- and α2-chains, and the band of α1-chain was about twice as intense as that of α2-chain. The presence of band at the region of 200 kDa could be result of dimmerization of α-chains. Thus, on the basis of subunit composition and SDS-PAGE pattern, it could be speculated that collagen extracted from waste of leather industry belongs to collagens type I.

Many researches have demonstrated that collagen peptides exhibited favorable functional properties and biological activities, which had become a topic of great interest for health food and processing/preservation industries. Enzymatic hydrolysis is one of the widely used approaches for the effective release of bioactive peptides from protein sources. During hydrolysis, a wide variety of smaller peptides are generated, depending on enzyme specificity and the hydrolysis time. Hydrolysis by pepsin was used as the most suitable approach for producing of collagen peptides. Figure 2 represented the composition of the obtained collagen peptides estimated by SDS-PAGE.

It has been widely accepted that oxidative stress plays a critical role in initiating and driving the cascade of events that results in development of serious pathologies [17]. Obesity is one of the diseases which appear to be strongly associated with increased generation of free radicals and LPO reactions [18].

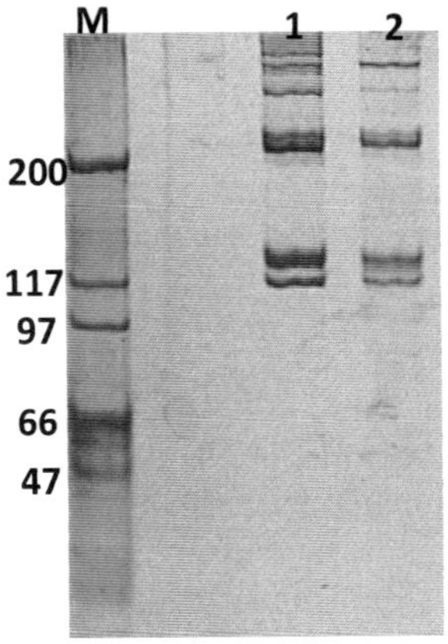

Figure 1. SDS-PAGE pattern of collagen extracted from waste of leather industry on 6 % separating gel. Lane M - high molecular weight (MW, kDa) protein markers; lane 1-2 - extracted collagen

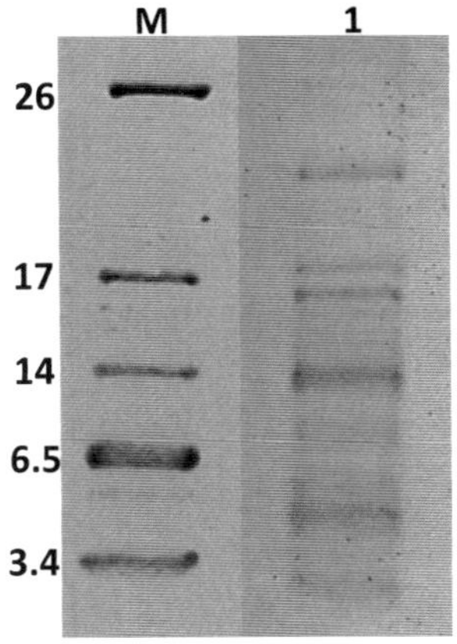

Figure 2. SDS-PAGE pattern of collagen peptides on 18 % separation gel. Lane M - low molecular weight (MW, kDa) protein markers; lane 1 - fraction of collagen peptides

The main parameter that allows to assess the intensity of oxidative processes and the stage of the oxidative stress is the amount of lipid oxidation products, in particular tiobarbituric acid reactive substances (TBARS). Normally, endogenous antioxidant enzymes are able to scavenge the excessive ROS to protect tissues from oxidative injuries. Superoxide dismutase (SOD, EC 1.15.1.1.) and catalase (CAT, EC 1.11.1.6.) are two important antioxidant enzymes that inactivate superoxide anions and hydrogen peroxide, respectively. However, when oxidative stress persists for a long time, the antioxidant resources are depleted due to a decrease in the activity of antioxidant enzymes, as well as inhibition of their synthesis.

According to our results the systemic oxidative stress manifested by the accumulation of TBARS and decrease activity of main antioxidant enzymes SOD and CAT was observed in obese animals (table 1).

We found that long-term administration of the collagen peptides to animals which were on a high-calorie diet resulted in a significant decrease in TBARS content and partly normalization activities of both SOD and CAT. This may be due to the ability of collagen peptides influence the passing of free radical processes. The exact mechanism through which peptides display antioxidant activity is not fully understood. However, several explanations have been proposed to elucidate their antioxidant properties. The antioxidant activity of bioactive peptides is related to their amino acid composition, structure, and hydrophobicity. The presence of hydrophobic amino acids also increases the affinity of peptides to the lipid system, favoring their distribution at the water-lipid interface and enhancing the radical-scavenging activity at the lipid phase [19]. It also has been demonstrated [20] that peptides could be involved in the induction of the expression of genes for nonenzymatic antioxidant components.

Table 1. The activity of antioxidant enzymes (superoxide dismutase SOD, catalase CAT, and the content of thiobarbituric acid reactive substance (TBARS) in blood serum of control rats and rats with diet-induced obesity (DIO) with or without collagen peptides administration

	The experimental groups			
	Control	Control with collagen peptides administration	DIO	DIO with collagen peptides administration
SOD activity, $U{\cdot}(mg\ protein)^{-1}{\cdot}min^{-1}$	48.2±3.3	-	23.5±4.1*	33.6±4.2*#
CAT activity, $mmol\ H_2O_2{\cdot}(mg\ protein)^{-1}{\cdot}min^{-1}$	71.6±5.6	-	13.8±3.5*	37.2±6.4*#
TBARS, $nmol{\cdot}(mg\ protein)^{-1}$	93.5±11.5	-	487.4±18.6*	356.0±18.5*#

Values are expressed as mean±SD (n=10); *p<0.05 significantly different from the control; #p<0.05 significantly different from the DIO

Substantial evidences indicate that obesity is linked to a state of chronic low-grade inflammation [21]. Adipocytes have been identified as sources of pro-inflammatory cytokines, which are involved in the triggering of chronic local inflammation in adipose tissue as well as systemic inflammation [22]. In addition, oxidative stress is associated with an irregular production of adipokines which, in turn, are potent stimulators for the production of reactive oxygen and nitrogen species by macrophages and monocytes, generating oxidative stress and, a major, irregular production of other adipokines [23]. These changes may interact among themselves and amplify, producing, in this manner, the set of metabolic and vascular alterations. Considering the relationship between oxidative stress and cytokine profile, as well as the positive effect of the CF treatment on oxidative stress biomarkers, we tested whether CP could improve the cytokine balance in obese rats.

According to the results (table 2) the increase in the pro-inflammatory cytokines IL-1β, IL-6 concentration as well as a decrease in the IL-4 concentration in blood serum of animals with obesity was observed. A slight increase in the IL-10 concentration can be considered as a compensatory reaction of the organism in response to the changes of metabolic status.

Table 2. The concentration of cytokines in blood serum of control rats and rats with diet-induced obesity (DIO) with or without collagen peptides administration

	The experimental groups			
	Control	Control with collagen peptides administration	DIO	DIO with collagen peptides administration
		Pro-inflammatory cytokines, $rel.units{\cdot}mL^{-1}$		
IL-1β	19.22±2.34	20.23±2.54	24.21±2.71	14.55±3.28#
IL-6	21.91±3.10	21.21±2.10	26.82±3.54	17.67±4.12#
		Anti-inflammatory cytokines, $rel.units{\cdot}mL^{-1}$		
IL-4	63.77±7.63	60.14±8.54	38.64±5.27*	52.34±5.17#
IL-10	47.44±4.96	49.45±5.46	49.76±4.12	52.63±6.51

Values are expressed as mean±SD (n=10); *p<0.05 significantly different from the control; #p<0.05 significantly different from the DIO

When collagen peptides were administrated to obese animals we noticed that pro-inflammatory cytokine (IL-1β, IL-6) levels were significantly decreased while concentration of anti-inflammatory cytokines (IL-4, IL-10) were increased. These results could indicate a potential application of collagen peptides against inflammatory conditions.

Conclusions

Waste of leather industry could be a profitable resource for collagen extraction and subsequent obtaining of collagen peptides. Our findings on the rat model of diet-induced obesity provide preliminary indication that CP administration can influence metabolic status under pathologies associated with oxidative stress development. This positive effect is most likely mediated by collagen peptide-induced up-regulation of antioxidant capability particularly enhancing the activities of antioxidant enzymes and reducing lipid peroxidation process that, in turn, help prevent oxidative stress and inflammation development. Taking into account results on anti-oxidant and anti-inflammatory activities of collagen peptides extracted from waste of leather industry, these peptides can be used for production of pharmacological products against diseases associated with oxidative stress.

References

[1] Kanagaraj J, Velappan KC, Chandrababu NK and Sudulla S 2006 *J. Sci. Ind. Res.* **65** 541-548.
[2] John Sundar V, Gnanamani A, Muralidharan C, Chandrababu NK and Mandal AB 2011 *Rev. Environ. Sci. Biotechnol.* **10** 151-63.
[3] Lee CH, Singla A and Lee Y 2001 *Int. J. Pharm.* **221** 1-22.
[4] Aneiros A and Garateix A 2004 *J Chromatogr B Analyt Technol Biomed Life Sci* **803(1)** 41-53.
[5] Sun XD, Li YQ and Zhao Y 2009 *China Journal of Chinese Materia Medica*, **34(23)** 3135–38.
[6] Wang H, Fu Z and Han C. 2014 *American J Marine Sci.* **2(1)** 1-8.
[7] Savchuk O, Ostapchenko L, Mokrousova O, Andreyeva O and Raksha N 2017 *Solid State Phenomena* **267** 172-176.
[8] Huang Q, Li S, Teng H, Jin Y, Ma M and Song H 2015 *Food Sci Biotechnol* **24(6)** 2193-98.
[9] Laemmli UK, 1970 *Nature* **227** 680-685.
[10] Raksha NG, Potalitsyn PY, Yurchenko AV, Halenova TI, Savchuk OM and Ostapchenko LI 2018 *Arch Biol Sci* **70(1)** 77-86.
[11] Bradford MM. 1976 *Anal Biochem* **86** 193-200.
[12] Nedzvetsky V, Andrievsky G, Chachibaia T and Tykhomyrov A 2012 *J Diabetes Metab* **3(8)** 1-9.
[13] Sirota TV. 2011 *Biochem (Moscow) Suppl Series B: Biomed Chem* **5(3)** 253-59.
[14] Korolyuk MA, Ivanov LI, Mayorova IG and Tokarev VE 1988 *Lab Delo* **1** 16-9.
[15] Halenova TI, Vareniuk IM, Roslova NM, Dzerzhynsky ME, Savchuk OM, Ostapchenko LI, Prylutskyy YuI, Ritter U and Scharff P 2016 *RSC Adv* **6** 100046-55.
[16]Kimura S, Ohno Y and Miyauchi Y 1987 *Comparative Biochemistry and Physiology* **88B(1)** 27-34.
[17] Toyokuni S 2016 *Pathol Int* **66(5)** 245-59.
[18] Furukawa S, Takuya Fujita T, Shimabukuro M, Iwaki M, Yamada Y, Nakajima Y, Nakayama O, Makishima M, Morihiro M and Shimomura I 2004 *J Clin Invest* **114** 1752-61.
[19] Qian ZJ, Jung WK, Byun HG and Kim SK 2008 *Biores Technol* **99** 3365-71.
[20] Grosser N, Oberle S, Berndt G, Erdmann K, Hemmerle A and Schröder H 2004 *Biochem Biophys Res Commun* **314(2)** 351-5.
[21] Stienstra R, Tack CJ, Kanneganti TD, Joosten LA and Netea MG 2012 *Cell Metab* **15** 10-8.
[22] Fonseca-Alaniz MH, Takada J, Alonso-Vale MI and Lima FB 2007 *J Pediatr* **83** 192-203.

[23] Fernandez-Sanchez A, Madrigal-Santillan E, Bautista M, Esquivel-Soto J, Morales-Gonzalez A, Esquivel-Chirino C, Durante-Montiel I, Sanchez-Rivera G, Valadez-Vega C and Morales-Gonzalez JA 2011 *Int. J. Mol. Sci* **12** 3117–32.

IOP Conf. Series: Materials Science and Engineering **500** (2019) 012021 doi:10.1088/1757-899X/500/1/012021

Poly(ε-caprolactone)-based biocomposites reinforced with nanocrystalline cellulose grafted with poly(L-lactic acid)

Mariia Stepanova[1], Ilia Averianov[1], Iosif Gofman[1], Olga Solomakha[1], Yuliya Nashchekina[2], Viktor Korzhikov-Vlakh[1,3], Evgenia Korzhikova-Vlakh[1]

[1]Institute of Macromolecular Compounds, Russian Academy of Sciences, Bolshoy pr. 31, 199004, St. Petersburg, Russia
[2]Institute of Cytology, Russian Academy of Sciences, St. Petersburg, Russia
[3]Institute of Chemistry, Saint-Petersburg State University, St. Petersburg, Russia

Corresponding author's e-mail: vlakh@mail.ru

Abstract. The modification of nanocrystalline cellulose with poly(L-lactic acid) was carried out using the "grafting from" approach based on the ring-opening polymerization of L-lactide initiated with cellulose hydroxyls. The modification was proved by FTIR spectroscopy, thermogravimetric analysis and dynamic light scattering. Samples of native and modified cellulose were applied as fillers for poly(ε-caprolactone) films that were studied for their mechanical properties. A significant increase of Young's modulus was observed for composite material based on PCL containing nanocrystalline cellulose modified with poly(L-lactic acid).

1. Introduction

Currently, the polyester-based materials are widely used for many applications such as food industry, surgery, drug delivery and regenerative medicine [1]. Poly(lactic acid) (PLA), poly(glycolic acid) (PGA), poly(ε-caprolactone) (PCL), poly(ω-pentadecalactone) (PDL) and their copolymers are biodegradable and one of the most commercially competitive polyesters. The major drawback of polyesters in biomedicine is their hydrophobicity, which results in uncontrollable fouling of proteins on the surface of materials upon contact with biological fluids and tissues [2]. In turn, it leads to uncontrollable cell-material interactions effects [3]. Thus the material hydrophilization is needed to improve its biological properties. These could be tuned by surface modification [4] or by addition of the special fillers [5,6], which change also the bulk properties of the materials, such as mechanical properties. To improve the mechanical properties, the polyester materials are usually reinforced with synthetic or mineral nanofillers [6,7].

In last decade, the natural polymers are received much attention due to their availability from renewable resources, which is in accordance with the Principles of Green Chemistry. Currently, the micro- and nanocrystalline cellulose (MCC or NCC) are widely considered as fillers in polymer matrices to develop fully renewable and biocompatible materials [8]. However, the application of pure cellulose micro/nanocrystals as fillers to polyesters is accompanied with the interfacial incompatibility between the hydrophobic polymer matrix and the hydrophilic cellulose [9]. This drawback could be surmounted *via* cellulose nanocrystals surface chemical modification. In particular, the adsorption of poly(ethylene glycol) (PEG) on NCC surface [10] and covalent modification of NCC surface with PLA [9] has been recently reported. In this work, we focused on preparation of the composite material based on PCL filled with NCC bearing poly(L-lactic acid) (PLLA) and study of PLLA attachment effect on the properties of formed composite material.

2. Results and Discussion

2.1. Grafting of PLLA from NCC via ring-opening polymerization of L-lactide

The modification of NCC surface by PLLA was carried out using the grafting-from strategy. In this case, the hydroxylic groups of NCC acted as initiators for ring-opening polymerization of L-lactide (Fig. 1). The characteristics of NCC are given in Tables 1 and 2. Grafting procedure was carried out both in suspension and in bulk. NCC and monomer were dried in vacuum before polymerization. The NCC/L-lactide ratio was 2 (wt/wt), but *[L-lactide]/[Sn(Oct)$_2$]* ratio was equal to 14. Other experimental conditions can be found in Fig. 1. The yield of grafted polymer was equal to 20% for in bulk polymerization and 26% for in suspension process.

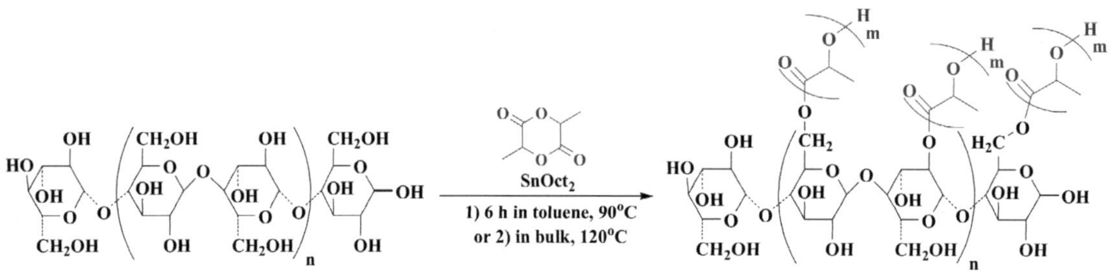

Figure 1. Scheme of PLLA grafting from NCC.

The successful modification of NCC with PLLA was proved by solid state FTIR spectroscopy (Fig. 2). For comparison, pure PLLA sample was synthesized by ring-opening polymerization. The synthesis was carried out in bulk at 130 °C for 24 h using the ratios: *[L-lactide]:[methanol]* = 10000, *[methanol]:[Sn(Oct)$_2$]* = 2. The polymer yield was equal to 83%. The obtained PLLA had following characteristics (size-exclusion chromatography in THF at 40 °C): M_w = 99000, $Đ$ = 1.26. FTIR spectra were registered for both in suspension and in bulk obtained modification products. In the spectra of NCC modified by PLLA a band at 1755 cm^{-1} corresponding to the C=O bond stretching in PLLA was detected. This band was absent in the spectra of unmodified NCC.

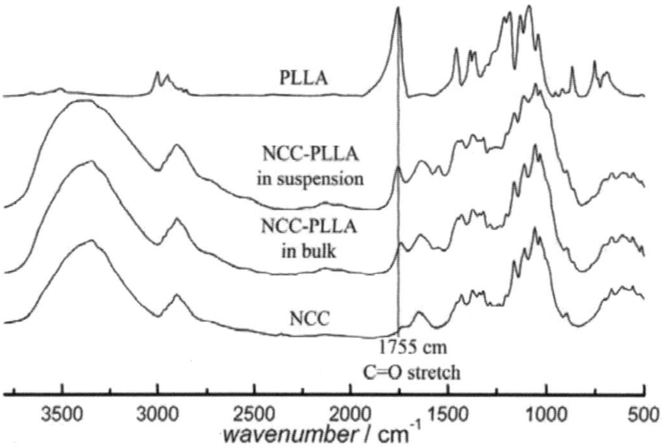

Figure 2. FTIR spectra of registered for pure NCC and PLLA as well as for NCC modified with PLLA.

Taking into account that chemical modification of polymers changes their thermal stability, the modified NCC was subjected to thermogravimetric analysis (TGA). The analysis was performed in air atmosphere at temperatures from 25 to 600 °C and heating rate 5 °C/min. The obtained TGA and differential thermal gravimetry (DTG) curves are presented in Fig. 3. Pure PLLA demonstrated the sharp loss of weigh due to destruction, while NCC and NCC-PLLA samples demonstrated an initial mass loss (about 3.5%) from 40 °C to 120 °C. At this step, the loss of the hydrated and coordinated water molecules from the cellulose environment takes place. After this the smooth weight loss occurs until 400 °C.

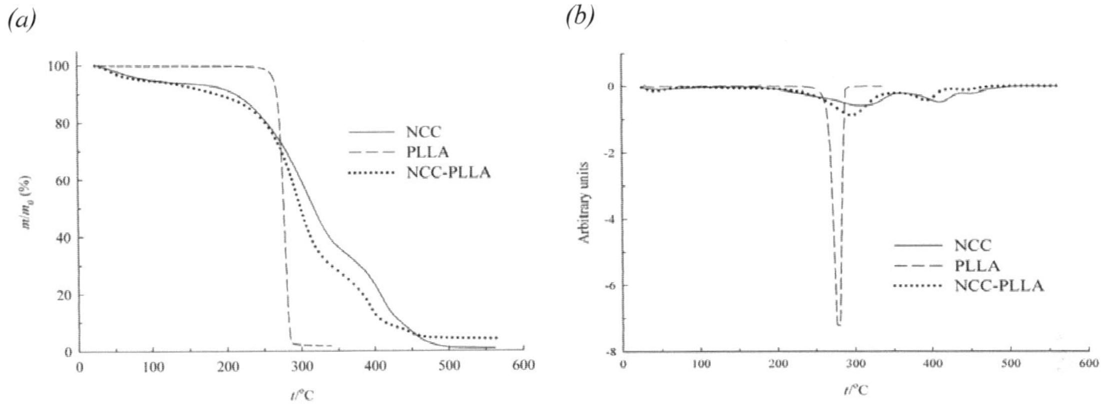

Figure 3. TGA *(a)* and DTG *(b)* curves for pure NCC and PLLA as well as NCC modified with PLLA.

Table 1 summarizes the temperatures of polymer decomposition for 5 (τ_5) and 10% (τ_{10}) mass loss. The highest thermal stability was detected for pure PLLA, whereas the lowest one was observed for pure NCC. As it was expected, the thermal stability of NCC-PLLA sample was between those demonstrated by parental polymers.

Table 1. Thermal characteristics for modified and native polymers.

Samples	τ_5/°C	τ_{10}/°C
NCC	220	241
PLLA	265	269
NCC-PLLA*	228	249

*Sample prepared in solution

Additionally, the samples of unmodified and modified NCC were analyzed by dynamic light scattering to monitor the changes in characteristics such as hydrodynamic size (D_H), polydispersity index (*PDI*) and ξ-potential of different NCC samples (Table 2). As it can be seen from the data presented, the lyophilization and following redispersion of commercially available NCC led to the increase of D_H and PDI of cellulose nanoparticles. In its turn, the NCC surface grafting with water insoluble PLLA chains resulted in additional increase of hydrodynamic diameter. Such effect was a result of hydrophobic interactions between PLLA chains in water that caused the partial NCC aggregation. As to ξ-potential of NCC, the absolute value was decreased after modification with PLLA. That also indirectly testifies the changes in surface chemistry.

Table 2. Hydrodynamic diameter *(D_H)*, polydispersity index *(PDI)* and ξ-potential of pure and modified NCC (DLS in water).

Samples		D_H, nm	*PDI*	ξ-potential, mV
NCC native˙	commercial suspension	128	0.28	– 23.0 ± 1.7
NCC native	liophylized + redispersed	264	0.58	– 25.0 ± 0.5
NCC-PLLA	liophylized + redispersed	460	0.23	– 19.6 ± 0.5

2.2. Preparation of PCL-NCC-PLLA composite films

The obtained NCC-PLLA was used as filler for PCL matrix. With aim to prepare the PCL films the matrix polymer was synthesized by ring-opening polymerization in bulk at 130 °C for 24 h. The polymerization mixture was composed in the following manner: *[ε-caprolactone]:[methanol]* = 2500, *[methanol]:[$Sn(Oct)_2$]* = 2. PCL yield was equal to 81%. The molecular weight characteristics of PCL were determined by size-exclusion chromatography in THF at 40 °C. The calibration curve built for polystyrene standards was used for this purpose. M_w and Đ were found to be 155000 and 1.53, respectively. The PCL films were prepared *via* casting of the polymer solution with or without NCC onto the flexible cellophane substrate and followed with solvent evaporation [11].

The prepared films made of native PCL and PCL filled with 5 wt% of NCC and NCC-PLLA were analyzed by polarized light microscopy (Fig. 4). The film without fillers presents the phase relief due to existence of crystalline and amorphous regions. The addition of both NCC and NCC-PLLA affects the organization of macromolecules in the material – less crystalline areas of PCL were observed by this method. The NCC particles tend to form aggregates (Fig. 4b and c). The average sizes of aggregates was estimated using Fiji ImageJ free software and were equal to 1.91 and 1.01 μm for unmodified NCC and NCC-PLLA, correspondingly.

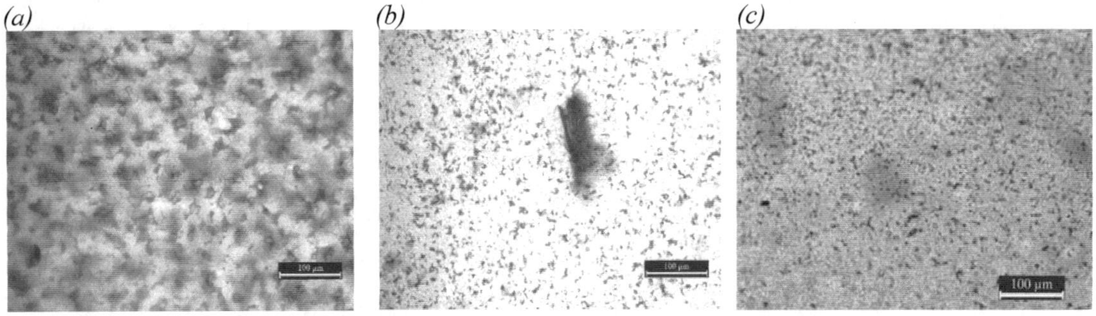

Figure 4. Images obtained by polarized light microscopy for films based on native PCL *(a)* and PCL filled with NCC *(b)* and NCC-PLLA *(c)*.

2.3. Mechanical properties of biocomposites

Mechanical characteristics of the films were determined under conditions of uniaxial extension using band-like samples of 2 mm× 20 mm. Experiments were carried out at the extension speed of 10 mm/min. The Young's modules *(E)*, tensile strengths *(σ_b)* and elongation at break *(ε_b)* were determined for three kinds of films: pure PCL, PCL filled with unmodified NCC and PCL filled with NCC-PLLA (Table 3). The characteristics of the initial PCL film were in agreement with data published elsewhere [12]. An introduction of 5 wt% pure NCC into PCL film did not influence the value of Young's modulus but led to the drastic decrease of elongation at break and tensile strength. The filling of PCL with NCC-PLLA provides the better compatibility of the nanofiller with hydrophibic polymer matrix. Comparatively to the material filled with unmodified NCC, the modification of NCC allowed for the increase of Young's modulus, σ_b and ε_b by approximately 1.5, 1.7 and 1.3 times.

Table 3. Mechanical properties of pure PCL and composite films.

Film	E, MPa	σ_b, MPa	ε_b, %
PCL	324±16	25.0±3.2	830± 58
PCL+5 wt% NCC	352±28	7.9 ±0.6	19±5
PCL+5 wt% NCC-PLLA	512 ±11	13.2±0.8	25±5

2.4. Biological evaluation of biocomposites

The effect of NCC and NCC-PLLA application on cells adhesion on the surface of PCL-based films was evaluated using bone marrow mesenchymal stem cells (MSC). The films pieces were placed into 96 well plate and MSC (5×10^3 cells per well) were seeded. After 24 hours of incubation the non-attached dead cells were washed out and the amount of attached cells was evaluated using MTT-test. Fig. 5 illustrates the dependence of formazan optical density on material composition. The optical density is proportional to the number of cells adhered to the surface of the films. In all cases, the measured parameter was at the same level. That indicates the comparable cell adhesion. Thus, the introduction of NCC into PCL as well as modification of NCC did not make materials toxic for the cells. It allows the suggestion of further surface modification of NCC with cells instructive peptides to permit the direct cell-material interactions.

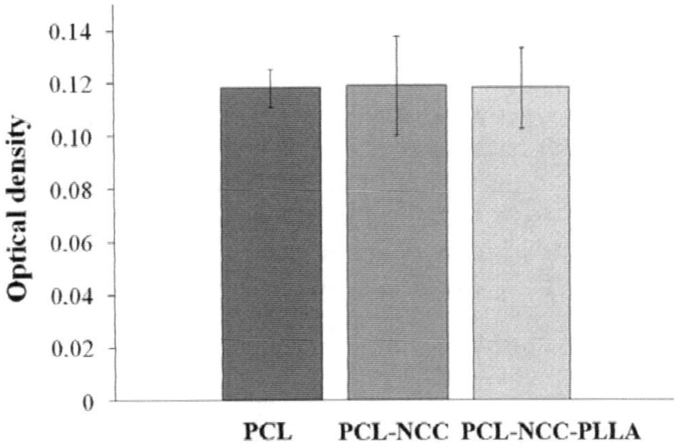

Figure 5. Adhesion of the cells (MSC) on the surface of native and composite films.

3. Conclusions

In order to enhance the compatibility of biodegradable PCL with nanocrystalline cellulose and mechanical properties of composite material, PLLA was covalently grafted onto the NCC surface *via* "grafting from" approach. The obtained composite PCL-NCC-PLLA films demonstrated significant improvement of Young's modulus comparatively to initial PCL and PCL-NCC materials. Taking into account the biocompatibility of the prepared PCL-NCC-PLLA films, the composite materials are seemed to be perspective in the field of nanobiomaterials.

Acknowledgements

This study was supported by the Russian Ministry of Education and Science (state contract no. 14.W03.31.0014, MegaGrant). The Research Center for Chemical Analysis and Materials Research of St. Petersburg State University is acknowledged for FTIR analysis.

IOP Conf. Series: Materials Science and Engineering **500** (2019) 012021 doi:10.1088/1757-899X/500/1/012021

References

[1] Manavitehrani I, Fathi A, Badr H, Daly S, Shirazi A N and Dehghani F 2016 Biomedical Applications of Biodegradable Polymers *Polymers* **8** 1

[2] Matsumoto T and Mooney D J 2006 Cell instructive polymers *Advances in Biochemical Engineering/Biotechnology* **102** 113

[3] Sheikh Z, Najeeb S, Khurshid Z, Verma V, Rashid H and Glogauer M 2015 Biodegradable materials for bone repair and tissue engineering applications *Materials* **8** 5744

[4] Haddad T, Noel S, Liberelle B, El Ayoubi R, Ajji A and De Crescenzo G 2016 Fabrication and surface modification of poly lactic acid (PLA) scaffolds with epidermal growth factor for neural tissue engineering *Biomatter* **6** e1231276

[5] Marycz K, Maredziak M, Grzesiak J, Lis A and Śmieszek A 2016 Biphasic polyurethane/polylactide sponges doped with nano-hydroxyapatite (nHAp) combined with human adipose-derived mesenchymal stromal stem cells for regenerative medicine applications *Polymers* **8** 339

[6] Tajbakhsh S and Hajiali F 2017 A comprehensive study on the fabrication and properties of biocomposites of poly(lactic acid)/ceramics for bone tissue engineering *Materials Science and Engineering C* **70** 897

[7] Xiao L, Mai Y, He F, Yu L, Zhang L, Tang H and Yang G 2012 Bio-based green composites with high performance from poly(lactic acid) and surface-modified microcrystalline cellulose *Journal of Materials Chemistry* **22** 15732

[8] Abdul Khalil H P S, Bhat A H and Ireana Yusra A F 2012 Green composites from sustainable cellulose nanofibrils: A review *Carbohydrate Polymers* **87** 963

[9] Goffin A, Raquez J, Duquesne E, Siqueira G, Habibi Y, Dufresne A and Dubois P 2011 From Interfacial Ring-Opening Polymerization to Melt Processing of Cellulose Nanowhisker-Filled Polylactide-Based Nanocomposites *Macromolecules* **12** 2456

[10] Cheng D, Wen Y, Wang L, An X, Zhu X and Ni Y 2015 Adsorption of polyethylene glycol (PEG) onto cellulose nano-crystals to improve its dispersity *Carbohydrate Polymers* **123** 157

[11] Korzhikov-Vlakh V, Krylova M, Sinitsyna E, Ivankova E, Averianov I and Tennikova T B 2016 Hydrogel layers on the surface of polyester-based materials for improvement of their biointeractions and controlled release of proteins *Polymers* **8** 418

[12] Kang D-O, Yoon H and Kim G-H 2011 Improvement of Mechanical Properties of Polycaprolactone Scaffolds by Applying Piezoelectric Vibration System Based on a Dispensing Process: A New Concept *Polymer Engineering and Science* **51** 311

IOP Conf. Series: Materials Science and Engineering **500** (2019) 012022 doi:10.1088/1757-899X/500/1/012022

Preparation of fibrous electrospun membranes with activated carbon filler

Illia Krasonu[1], Elvira Tarassova[1], Siret Malmberg[2], Victoria Vassiljeva[1], Andres Krumme[1]
[1] Department of Materials and Environmental Technology, Tallinn University of Technology, Tallinn, Estonia.
[2] Skeleton Technologies OÜ.

E-mail: illia.krasnou@ttu.ee

Abstract
Immiscible solvents blend suspensions applied to produce fibrous polymer membranes heavy filled with activated carbon (AC) particles by electrospinning method. The continuous phase of the suspension is a solution of polyacrylonitrile (PAN) in polar solvent (dimethylsulfoxide (DMSO)), the dispersed phase is suspended in non-polar solvent (hexane) AC particles. An ionic liquid 1-ethyl-3-methylimidazolium acetate (EmimAc) added to both phases to improve the electrospinning process and enhance the electrical conductivity of the electrospun mat. These types of membranes developed as electrode materials for energy storage devices.
In this study, we obtain electrospun membranes with up to 60 wt. % of activated carbon content, AC particles are connected between each other by PAN nanofibers, and covered with PAN nanofibrous mesh.

Introduction

Conductive carbonous materials intensively used for production of polymer composites. Recently electrospinning of polymer composites functionalized by different fillers are actively studied [1]. Most often fillers are dispersed in polymer material before processing [2], or material coated [3] or impregnated with fillers [4]. The first method has good homogeneity of properties through the final material, but filler particles are covered by the polymer what decreases functionality of the filler (electrical conductivity, absorption properties, etc. [5]. The second approach keeps the filler particles uncovered but the uniform distribution of particles through the volume of the product demands sophisticated impregnation techniques [6].
What concerns electrospinning of carbon-filled polymers, heavy load of carbon significantly increases the viscosity of solution that makes electrospinning process too complicated [7]. Another problem is absorption of solvent by carbon particles and thus increasing local concentration of polymer chains near the particles that leads to deposition of polymer onto the surface of particles. Electrospinning of such fluids is complicated, has low productivity and final product has not uniform morphology.
The different approach for electrospinning fluids preparation to avoid complications mentioned above could be used [8]. To keep the good fibre generation from the polymer solution, it should be spun from the proper solvent, and simultaneously, the electrospinning fluid should contain carbon particles separated from polymer chains. This problem could be solved by the preparation of the immiscible solvents blend electrospinning suspension.
We propose a new method for the electrospinning fluid preparation. The electrospinning of this type of fluids will combine both good fiber generation and high load of carbon filler with uniform particle distribution through the electrospun material. This method allows achieving electrospun mat of fine nanofibers with high load of carbonous material. The electrospinning fluid is a suspension with two phases: continuous phase is a solution of polyacrylonitrile in DMSO dispersed phase is a suspension of activated carbon particles in hexane. DMSO and hexane are not miscible, PAN soluble in DMSO and insoluble in hexane.

IOP Conf. Series: Materials Science and Engineering **500** (2019) 012022 doi:10.1088/1757-899X/500/1/012022

Materials and methods

Polyacrylonitrile 150 kDa molecular weight purchased from Polysciences Inc. DMSO purchased from Sigma Aldrich. Porous coconut shell activated carbon YP-80F, purchased from Kuraray Co. Ltd. 1-buthyl-3-methylimidazolium acetate ionic liquid synthesized in the laboratory by the methods described elsewhere [9] [10].

The sonication of suspensions is done by Bandelin Sonopuls sonification device with 3 mm MS-73 probe at frequency 20 kHz, amplitude 50% and 9 pulsed cycle, 70W generator.

The rheology studied by means of Anton Paar Physica MCR501 rheometer using cone-plate measuring geometry (cone angle = 2°). All measurements carried out in nitrogen atmosphere at temperature 21 °C. The gap = 0.051 mm was used. The oscillation tests were performed at angular frequencies $\omega = 1 \div 600$ rad/s with strain = 5% determined from the linear viscoelasticity region in the amplitude sweep test at a frequency of 1 Hz.

The homogeneity of suspensions studied by means of optical microscope Carl Zeiss Axioskop 2 in transition mode with polarized light. The morphology of fibrous mats studied by the scanning electron microscopy at Zeiss EVO MA15 in secondary electron and reflected beam regimes. To improve the image quality samples were sputtered with gold.

Electrospinning done at self-made setup, it consists of a needle pump, a high voltage supply and a grounded rotating drum collector covered with an aluminum foil. A voltage 17 kV, a distance between needle and the collector 15 cm, a pumping rate 0.7 ml/h, a needle 0.6 mm diam.

Suspension preparation

9 wt. % PAN solutions in DMSO were made by mechanical mixing at 45°C.

9 wt. % AC suspensions in hexane first mechanically mixed, then sonicated for 1 hour. A water bath used to keep suspensions cold and prevent fast evaporation of hexane. The sonication done in three intervals of 20 minutes, after each interval, the suspension weighted and the amount of evaporated hexane added to keep the concentration constant. Both liquids mechanically mixed in needed ratio and sonicated for 1 hour by the procedure described above. The stability of suspensions examined by the optical microscopy and by rheology tests.

For the stability control of PAN in DMSO + AC in Hexane suspension, the viscosity measured right after the preparation, in 12 and 24 hours after the preparation. Visually the suspension is stay homogeneous. It was found the viscoelastic behavior of the suspension at day 1 is almost identical to one on day 2. At day 3 the aggregation is observed, thus we suppose the suspension once prepared has to be electrospun within 24 hours.

Results and discussions

For comparison, we prepared roll-mill casted electrode, electrospun one solvent DMSO suspension and DMSO-hexane immiscible solvents blend suspensions.

On Figure 1 SEM, images of roll-mill casted electrode shown. The electrode consists of AC particles (big gray particles image (a)) conductive carbon black (small round shape agglomerated particles image (b)) blended with polytetrafluoroethylene (PTFE) binder. The AC load is 94 wt. %.

Figure 1. SEM image of PTFE-AC (ratio 6:94) roll-mill casted electrode

On Figure 2 the electrospun mat produced from PAN+AC suspension in DMSO is shown: AC particles are covered by PAN; thin fibers contain AC particles inclusions.

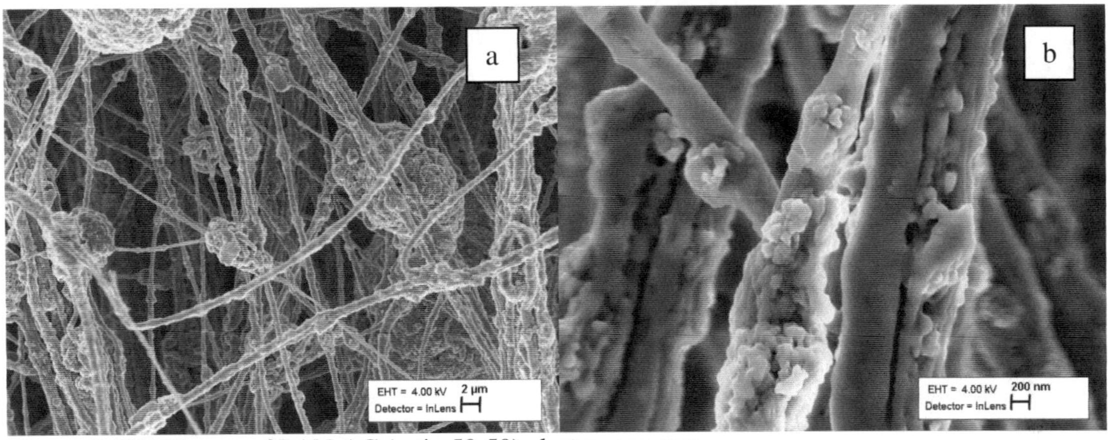

Figure 2. SEM image of PAN-AC (ratio 50:50) electrospun mat

In Figures 3 and 4 SEM images of mat electrospun from immiscible solvents, blend suspensions shown. The mats consist of separate uniform fibers, but PAN covers AC particles, one could notice that polymer coating is not uniform, it is porous and porosity decreases with AC content increase. Probably the evaporation of hexane absorbed inside AC particles prevents formation of uniform PAN coating.

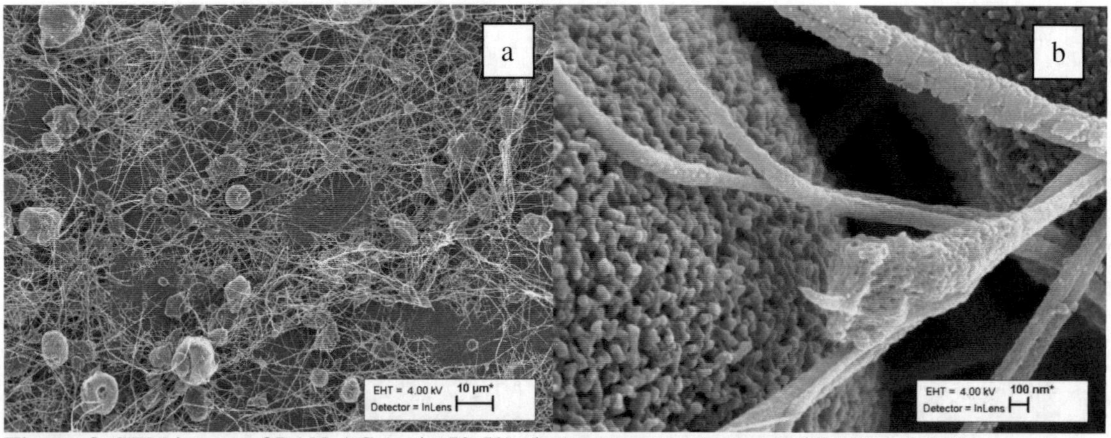

Figure 3. SEM image of PAN-AC (ratio 50:50) electrospun mat (separated PAN and AC)

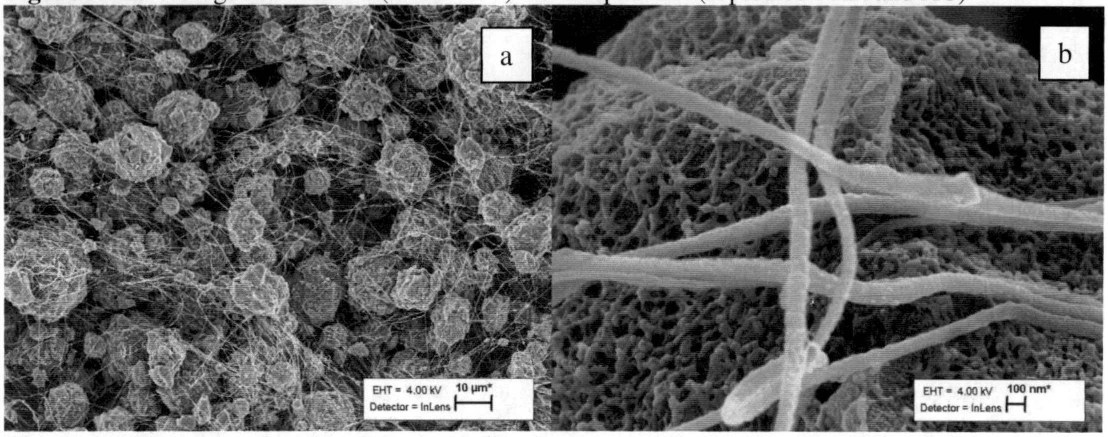

Figure 4. SEM image of PAN-AC (ratio 40:60) electrospun mat (separated PAN and AC)

Conclusion

We conclude that polymer chains cannot migrate from DMSO phase into hexane, but AC particles obviously can. As proven by SEM images, some AC particles migrate into DMSO solution of PAN and absorbs polymer on their surfaces.

The electrospinning of suspensions in which the polymer and the filler are separately dissolved or dispersed in non-mixable solvents relatively effective. Electrospun mats contain fine thin fibers and evenly distributed AC particles covered by porous layer of polymer.

Acknowledgements

The European Space Agency, project VA17066 "Fully electrospun durable electrode and electrochemical double-layer capacitor for high frequency applications".

References

1. Mrinal Bhattacharya 2016 Polymer Nanocomposites – A Comparison between Carbon Nanotubes, Graphene, and Clay as Nanofillers, Materials **9** 262
2. Anett Király, Ferenc Ronkay 1999 Effect of processing technology on the morphological, mechanical and electrical properties of conductive polymer composites, Journal of Polymer Engineering, **33**, 8, 691–699

3. Yiming Liu, Jing Wen, Yang Gao, Tianyang Li, Huifang Wang, Hong Yan, Baolong Niu, Ruijie Guo 2018 Antibacterial graphene oxide coatings on polymer substrate, Applied Surface Science **436** 624-630
4. Miyauchi, M., Miao, J., Simmons, T. J., Lee, J.-W., Doherty, T. V., Dordick, J. S., & Linhardt, R. J. 2010 Conductive cable fibers with insulating surface prepared by co-axial electrospinning of multi-walled nanotubes and cellulose, Biomacromolecules **11**(9) 2440–2445
5. Christopher Drew , Xianyan Wang , Lynne A. Samuelson & Jayant Kumar 2003 The Effect of Viscosity and Filler on Electrospun Fiber Morphology, Journal of Macromolecular Science, Part A: Pure and Applied Chemistry **40**, 12
6. Nicolas Lavielle, Anne Hébraud, Guy Schlatter, Linda Thöny-Meyer, René M. Rossi, and Ana-Maria Popa 2013 Simultaneous Electrospinning and Electrospraying: A Straightforward Approach for Fabricating Hierarchically Structured Composite Membranes, ACS Appl. Mater. Interfaces **5** (20), 10090–10097
7. Jeremy Kong Yoong Lee, Nuan Chen, Shengjie Peng, Linlin Li, Lingling Tian, NitishThakor, Seeram Ramakrishna 2018 Polymer-based composites by electrospinning: Preparation & functionalization with nanocarbons, Progress in Polymer Science **86** 40-84
8. Seema Agarwala and Andreas Greiner 2011 On the way to clean and safe electrospinning-green electrospinning: emulsion and suspension electrospinning, Polym. Adv. Technol. **22** 372–378
9. S.G. Cull, J.D. Holbrey, V. Vargas-Mora, K.R. Seddon, G.J. Lye 2000 Room-temperature ionic liquids as replacements for organic solvents in multiphase bioprocess operations, Biotechnol. Bioeng. **69** 227-233
10. X.-C. Zhang, B. Liu 2012 Two Zn and Hg bromide salts based on 1-Ethyl-3-Methylimidazolium ionic liquid: ionothermal synthesis, structures and supramolecular organization, Chem. Soc. Ethiop. **26** 407-414

IOP Conf. Series: Materials Science and Engineering **500** (2019) 012023 doi:10.1088/1757-899X/500/1/012023

The rationale behind using natural polymeric materials in shoe production

N Pervaia[1], O Andreyeva[2] and L Maistrenko[3]
1 Associate Professors, Kyiv National University of Technologies and Design, Ukraine
2 Professor, Kyiv National University of Technologies and Design, Ukraine
3 Associate Professors, Kyiv National University of Technologies and Design, Ukraine

maystrenko.la@knutd.com.ua

Abstract. In this paper, we have studied the modern natural polymeric materials which are used for producing component parts and the uppers of shoes. Having taken into account functional, economical, production and customer requirements for purpose-made shoes, we have identified a range of characteristics of leather qualities. They make it possible to provide rationalization for technological solutions, aimed at decreasing the risk of raw materials culling at all stages of a manufacturing procedure, decreasing resources and energy consumption when making products of a desired quality: tensile strength, tensile strength of surface, extension at tensile strength, water resistance of flexible leather, hygroscopicity.

1. Introduction

Under the conditions of raising quality standards of shoes with given qualities, possibilities of their certification and taking into account an increasing competition on the market, a conscious approach to shoe production becomes especially important.

It is a complex task to make sure that shoes are ergonomic. This task is solved at all production stages. It starts with constructing or choosing lasts for shoes and choosing materials for component parts of the uppers and the bottom halves of shoes [1].

Nowadays, shoe production, including purpose-made shoes, is connected with using the polymeric materials of natural, artificial and synthetic origin, as well as their combinations [2-8].

Natural leather, made of animal skins, has traditionally been considered one of the best materials for shoe making. Its properties and peculiarities of its structure determine the high quality of the finished product. Chrome-tanned leather of various kinds is the most frequently used in shoe production. It can be characterized by excellent hygienic qualities, beautiful visual appearance, an endurance of its form, softness, durability and resistance to the weather [2]. For instance, the leather for the uppers can endure 4-5 million creases without being damaged. Such a resistance to creasing is a result of the way how collagen fibres are arranged within the leather structure. Collagen is a polymer of biological origin. The spaces which exist between individual structural elements of the dermis, so-called pores, provide high air and water vapour permeability, excellent thermophysical qualities, and this also contributes to the hygienic qualities of shoes. Natural leather is easily moulded due to its resilience and plasticity. As a result, there is a good contact between the leather and shoe lasts there are not any creases and wrinkles.

It is possible to increase the quality of shoes by improving the endurance of their form, their hygienic and aesthetic properties. To achieve this, the leather or the uppers made of leather are exposed to various stimuli, e.g. high-frequency discharge plasma sources of low pressure [9]. Influence on the formation of certain properties of leather materials is also possible in their production due to the use of modern chemical materials that provide the receipt of material with predicted properties [10-15].

Thus, the data analysis revealed a large variety of natural polymeric raw materials and the urgent need for an informed approach to the selection of these materials in the production of shoes. This will improve the quality of finished shoes for different purposes.

2. Materials and methods

2.1. Materials

The materials under study have been the materials of natural origin, which have been widely used when producing the uppers of purpose-made shoes. The products assortment of natural leather is represented by the leather with natural surface (raw material – cow hide):

– **1** the hydrophobic leather «Vodogray» (technical specifications of Ukraine 19.300307856-007:2005) of chrome tanning; it is highly hydrophobic (resists to humidity for at least 3 hours); it is used for producing purpose-made military shoes, as well as sports and commercial shoes, shoes for hunters, fishers, tourists; it provides a reliable protection and comfort in any extreme conditions [16];

– **2** the Russian leather (yuft) of combined tanning methods for the uppers of shoes; used for making the shoes for military and work purposes [17];

– **3** the chrome tanned heatproof Russian leather for the uppers of shoes; used for making military and professional shoes [18];

– **4** the chrome-tanned leather for the uppers of shoes; used for producing the shoes for military servants [19];

– **5** the chrome-tanned leather for the uppers of shoes for military servants [20], which is used in the non-aggressive environment.

2.2. Methods

The quality parameters of natural leather have been analysed using standard methods of shoe material testing and standard equipment [21-24].

For analysis of leathers mechanical properties a method [21] for determining the tensile strength, elongation at 10 MPa and tensile strength of leather surface (the main indices of the mechanical properties of natural leather) has been chosen. This method is applicable to all types of leather. The testing parameters: speed of extension – 100±20 mm/min, maximum tensile stress – 980 H, load cell – 0-100 kg, labor part of sample - 5×50 mm, parallel experiments – not less than 4).

The rationale behind using certain materials for shoe making has been provided using the technique of multi-criteria compromise optimization [25]. This method is used in the form of the following generalized objective function when searching for the optimal technological parameter value or values in the case when there are more than two quality parameters for compromise optimization:

$$Y_{gen,r} = \sqrt{\sum_{j=1}^{m} \left(1 - D_{jr}\right)^2 * W_j^{\ 2}}, \qquad (1)$$

where $Y_{gen,r}$ – is the generalized objective function value for the r-th parameter, and in case of searching for optimum it approaches ($Ygen,\ r \to 0$) and allows us to measure the proximity of this point to the hypothetical optimal value in the form of the code, which equals 1; Djr – brought to the interval 0-1 value of the j-th feedback (quality parameter) in the r-th test of this experiment, depending on the aim chosen for a given quality parameter this value is calculated using different formulae; Wj – the value of the j-th quality parameter (feedback) virtually equals $1/j$; m – the number of criteria of feedbacks quality.

In addition, the graphical-numeral technique (the method of «polygon») has also been used to make a complex assessment of the shoe materials qualities. This technique is based on the combination of the graphical and quantitative methods. The distinctive feature of this technique is the choice of the material-reference standard, which has the best or high enough qualities among the materials being compared. To ensure that the quality assessment of the material under test is objective, the

IOP Conf. Series: Materials Science and Engineering **500** (2019) 012023 doi:10.1088/1757-899X/500/1/012023

comparative analysis of the range (in our case six) of parameters, which characterise the most important qualities, is applied. To do this, the centre of the polygon must be found and the shape must be divided into the equal parts with the help of triangles. Each cathetus of the triangle corresponds to the biggest value of a quality parameter of the material- reference standard. The rating of objects under comparison is assigned by measuring the correlation between the area of the shape obtained for each object and the shape area for the object-reference standard.

3. Results and discussion

We have established the nomenclature of natural materials quality parameters in order to provide the rationale behind using them in shoe production. We have taken into account the standard requirements on the correspondence between a material and a reference document or the specified requirements. We have also considered the material compliance with consumer requirements. The order of the parameters choice for establishing the nomenclature of the quality parameters was the following: to identify the obligatory and recommended standardized parameters for the product given and the materials for its production → to analyse if the standardized parameters characterize the correspondence between the material qualities and the consumer and production requirements in an adequate way → the standardized parameters nomenclature completion with the additional parameters, which contribute to the probable assessment of the compliance of a given material with the consumer requirements.

The choice of the quality parameters has been made taking into consideration the purpose and the conditions of use of a product, the consumer requirements analysis, the aims of providing the quality control and imposing the requirements on the products quality parameters.

Considering the functional, consumer, production and economic requirements for the purpose-made shoes, we have identified the range of leather properties parameters for the uppers of shoes, which allow us to provide the objective rationalization for the technological solutions, which decrease both the risks of raw materials culling at all stages of a manufacturing procedure and the consumption of resources and energy when producing a product of a quality needed. In our opinion, the following parameters of reliability and longevity can be classified as such: tensile strength, tensile strength of surface, percentage extension at 10 MPa.

The ergonomic requirements for the uppers of shoes material provide the comfort of a consumer during exploitation of a product. These requirements are realized through such quality parameters (properties) of materials as water vapour permeability, water resistance of flexible leather, hygroscopicity.

When comparing the properties parameters of natural leather for the uppers of shoes (Table 1), using the technique of multi-criteria compromise optimization.

Table 1. Natural leather parameters for the uppers of shoes.

Position (type of leather)	Tensile strength, MPa **TS**	Tensile strength of surface, MPa, **SF**	Percentage elongation at 10 MPa, % **L10**	Water vapour permeability, mg/sm^2· hour **VP**	Water resistance of flexible leather, min **WR**	Hygroscopicity, % **H**
1	19.3	18.2	33	4.3	180	18
2	15.9	14.4	19	2.5	37	12
3	17.7	16.3	20	4.1	35	13
4	15.0	14.0	27	5.0	30	11
5	19.0	17.9	30	5.3	30	20

We have identified the smallest value of the generalized objective function ($Y_{gen,r}$ = 0,012), i.e. the first place in our rating, which belongs to the leather «Vodogray». We have also found the biggest value of this parameter ($Y_{gen,r}$ = 0,063), i.e. the last place in our rating, and it goes to the Russian leather for the uppers of shoes of combined tanning methods; it is used for producing the shoes for military and work purposes (Table 2).

Table 2. The rating of natural leather of various kinds according to the multi-criteria compromise optimization technique.

Position	Yj						$\Sigma(1\text{-Djr})^2$	$Y_{ger,r}$
	TS	SF	L10	VP	WR	H		
1	19.3	18.2	33	4.3	180	18	0.180	*0.012*
2	15.9	14.4	19	2.5	37	12	5.143	0.063
3	17.7	16.3	20	4.1	35	13	2.932	0.048
4	15.0	14.0	27	5.0	30	11	4.197	0.057
5	19.0	17.9	30	5.3	30	20	1.056	0.029

Another independent technique, graphical-numeral, has been used in order to assess the quality of the natural leather materials. According to this technique, the area of the shape built identifies the quality level of a given material (P_i), and a correlation between the shape area for this material S_i and the shape area for the reference standard material S_e equals the complex quality assessment of the material under test. ($K_i = S_i : S_e$) (Table 3, Fig. 1).

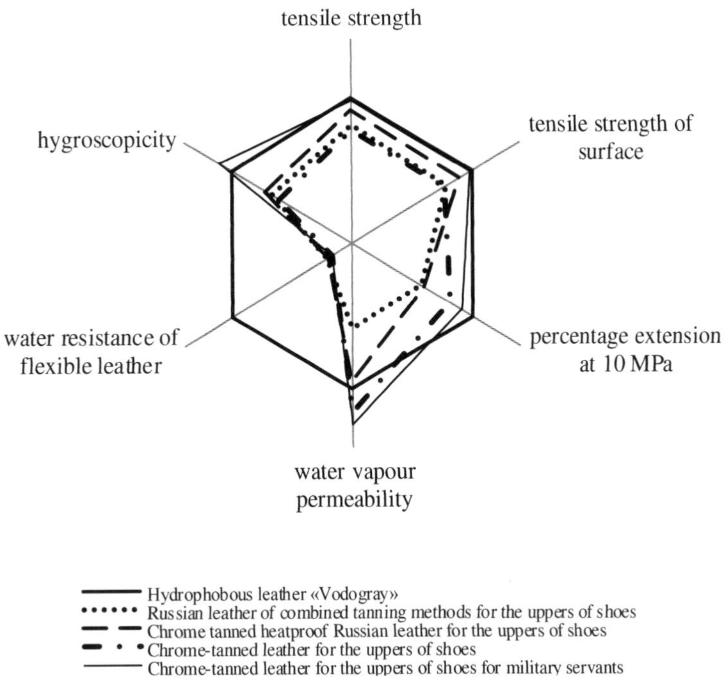

Figure 1. Making the complex assessment of the natural leather qualities

As a result of using the graphical-numerical technique (Table 3, Fig. 1) we have found the correlation between a shape area for each type of natural leather and a shape area for the reference standard leather (hydrophobic leather « Vodogray »): the Russian leather of combined tanning methods for the uppers of shoes – 43,7%, the chrome tanned heatproof Russian leather for the uppers of shoes – 50,9%, the chrome-tanned leather for the uppers of shoes – 59,7%, the chrome-tanned leather for the uppers of shoes for military servants – 93,3%.

Table 3. The complex quality assessment of the natural leather.

Index	Position									
	1		2		3		4		5	
	P_i=P_e	K_i	P_i	K_i	P_i	K_i	P_i	K_i	P_i	K_i
TS, 10 MPa	1.93	1.00	1.59	0.82	1.77	0.92	1.5	0.78	1.9	0.98
SF, 10 MPa	1.82	1.00	1.44	0.79	1.63	0.90	1.4	0.77	1.79	0.98
L10, %	33	1.00	19	0.58	20	0.61	27	0.82	30	0.91
VP, mg/sm^2·hour	4.3	1.00	2.5	0.58	4.1	0.95	5	1.16	5.32	1.24
WR, min	180	1.00	37	0.21	35	0.19	30	0.17	30	0.17
H, %	18	1.00	12	0.67	13	0.72	11	0.61	20	1.11
Si, sm^2	52.5		23.0		26.7		31.4		49.0	

According to the results of the multi-criteria compromise optimization, the natural hydrophobic leather « Vodogray » has been chosen the reference standard material. It has the best value of the generalized objective function ($Y_{gen,r}$ = 0,012), which is conditioned by its excellent parameters of reliability and longevity (tensile strength 1,93 MPa, tensile strength of surface 1,82 MPa, percentage extension at 10 MPa 33%), as well as by its ergonomic and hygienic qualities (water vapour permeability 4.3 mg/cm^2·hour; water resistance of flexible leather 180 minutes; hygroscopicity 20 %).

It should be noted that there the correlation has been found between the indices of the generalized objective function and a polygon area. Its correctness has been proved by the value of approximation validity: for the natural leather R^2 = 0.8880, which is > 0.7500 (Fig. 2).

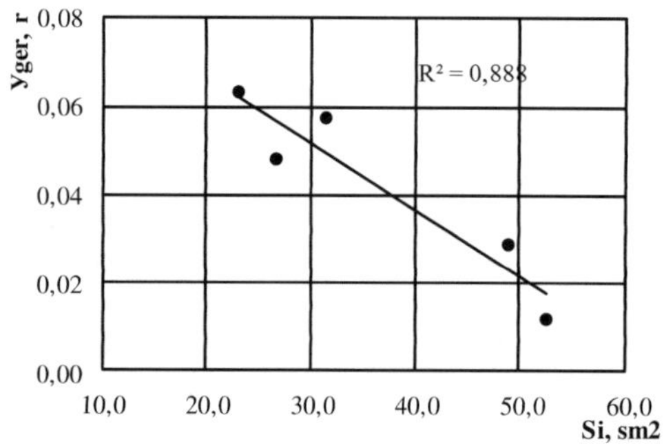

Figure 2. The correlation between the index of the generalized objective function and a polygon area

4. Conclusions

Using the statistic processing of the most important parameters for shoe materials with the help of the two independent techniques: the multi-criteria compromise optimization in the form of the generalized objective function and the complex assessment in the form of the graphical-numerical «polygon» technique we have provided the rationale behind using those polymeric materials of the natural origin which are the most capable of providing the high quality and ergonomics of purpose-made shoes: the hydrophobic leather «Vodogray» and the chrome-tanned leather for the uppers of shoes for military servants.

References

[1] Cao H, Wool R P, Bonanno P, Dan Q, Kramer J and Lipschitz S 2014 *International journal of Fashion Design, Technology and Education* **7** 21-30
[2] Hole L. G., Whittaker R. E. 1971 *Journal of materials science* **6** 1-15
[3] Downham, Rory P, Sears S, Vaughn G 2015 *Journal of Forensic Identification* **65** 138

[4] Tihonova N V, Jukovskaya T V, Mahotkina L Yu 2010 *Bulletin of the Technological University* **9** 369–372 (in Russian)
[5] Karkalića R M., Radulovićb J R., Jovanovićc D B 2017*Vojnotehnički glasnik / Military technical courier* 65 948–967.
[6] Sänchez-Navarro M.M, Cuesta-Garrote N, Arän- Ais F and Orgiles-Barcelö C 2011 *Journal of Dispersion Science and Technology* **32:12** 1722-1727
[7] Ibragimov A T, Maksudova U M, Raphikov A S 2016 *European journal of analytical and applied chemistry* **2** 9–12
[8] Nikitina L L 2012 *Bulletin of the Technological University* **15** 121–124 (in Russian)
[9] Tihonova N V, Abdulin I Sh,. Mahotkina L Yu, Jukovskaya T V 2010 *Bulletin of the Technological University* **9** 369–372 (in Russian)
[10] Deb A K, Shaikh Md A A, Sultan Md Z, Raf Md I H 2017 *Leather and Footwear Journal* **17(2)** 97-106
[11] Jucyte A, Valeika V, Sirvaityte J, Beleska K, Valeikiene V 2016 *Key Engineering Materials* **706** 89-93
[12] Valeika V, Beleska K, Sirvaityte J. 2012 *Brazilian Journal of Chemical Engineering* **29(2)**
[13] Nikonova A, Andreyeva O, Maistrenko L: 2016 *Proceedings of the 6th International Conference "ICAMS 2016 Advanced Materials and Systems"* 369-374
[14] Marukhlenko M. Mokrousova O, Okhmat O 2017 *Solid State Phenomena* **267** 52-57
[15] Maruhlenko M O, Palamar VA and Mokrousova O R 2016 *IOP Conference Series: Materials Science and Engineering* **111** conf. 1
[16] The hydrophobic leather «Vodogray»: http://chinbar.kiev.ua
[17] State Standard 485-82. Yuft Upper Leather. Specifications
[18] Industrial Standard 17-317-74. Heat-resistant Yuft Upper Leather
[19] State Standard 939-88. Upper leather
[20] State Standard 28422-89. Upper leather for military men. Specifications
[21] ISO 3376:2011. Leather. Physical and mechanical tests. Determination of tensile strength and percentage extension
[22] ISO 14268:2012. Leather. Physical and mechanical tests. Determination of water vapour permeability
[23] ISO 5403-1:2011. Leather. Determination of water resistance of flexible leather. Part 1: Repeated linear compression (penetrometer)
[24] ISO 2417:2002. Leather. Physical and mechanical tests. Determination of the static absorption of water
[25] Gorbachev A A et al 2007 Basis for the creation of modern production technologies (Kyiv: KNUTD) 189

IOP Conf. Series: Materials Science and Engineering **500** (2019) 012024 doi:10.1088/1757-899X/500/1/012024

A preliminary study on sodium hyaluronate loaded polyvinyl alcohol nanofiber webs obtained via roller electrospinning

Jegina S[1], Salaka L[1], Kukle S[1], Livkisa D[2], Gravitis J[3]

[1]Institute of Design Technologies, Riga Technical University, Riga, Latvia

[2]Laboratory of Bioanalytical methods and Biodosimetry, University of Latvia

[3]Latvian State Institute of Wood Chemistry, Riga, Latvia

E-mail: Sandra.Jegina@rtu.lv

Abstract. The aim of this study was to investigate whether it is possible to integrate cosmetic grade sodium hyaluronate or hyaluronic acid (HA) into polyvinyl alcohol (PVA) nanofiber matrix, as well as to analyse how different concentrations of HA effects the viscosity and conductivity of the spinning solution, fiber diameters, morphology and mechanical properties of the nanowebs.

1. Introduction

In biomedical field hyaluronic acid/sodium hyaluronate (HA) is widely used because of its properties- it is a nontoxic, biodegradable, biocompatible natural linear polysaccharide. HA can be chemically modified due to the presence of multiple acid and hydroxyl groups in the molecule [1]. In medical applications HA plays a significant role in wound healing, tissue engineering, drug delivery and immunomodulation [2].

HA belongs to the extracellular matrix (ECM) molecules. Recently, it is appreciated that ECM molecules that lie between cells, in addition to providing a constructive framework, they exert major effects on cellular function. Although ECM molecules appear amorphous by light microscopy, they form a highly organized structure, comprising mainly of glycosaminoglycan (GAG) with a unique capacity to bind and retain water molecules [3], proteoglycans, growth factors and structural proteins such as collagens. Yet, the predominant component of the skin ECM is HA [4].

In dermatology and cosmetology topical delivery of molecules into the human skin is one of the main issues and as studies have shown- HA with low molecular weight (20-300 kDa) passes through the outermost layer of the epidermis compared to the impermeability of high molecular weight HA (1000-1400 kDa) [5].

Besides effective consumer products with HA and therapeutic injections [6], HA has been studied also in terms of electrospinning to expand its applications using polyamide solution in formic acid as matrix with HA 0.04 %wt to prepare nano fibres by needleless spinning [7].

A number of studies have proven the efficient use of PVA to develop electro spun nano fibrous mats as a novel dressing matrix for controlled release of drugs and biopolymers for regenerative medicine [8-13]. Considering that polyvinyl alcohol (PVA) is a biocompatible and nontoxic polymer as confirmed by numerous studies including our previous experiments, in this study it is used as matrix to investigate spin ability and obtained web properties with different HA proportions in spinning solutions.

2. Materials and methods

Cosmetic grade low molecular weight (50 kDa) HA was bought in a local biocosmetic store "BB Factory", while PVA (Mowiol® 18-88, Mw ~130,000, Sigma-Aldrich) bought in Labochema Latvija" Ltd.

Both HA and PVA were dissolved in distilled water in different temperatures- HA in +20°C and PVA in 120-135°C, respectively. While the applied stirring speed was the same for both ingredients (1000 rpm), stirring time differed- 2 hours for PVA, 30 minutes for HA as HA dissolves relatively fast in water due to its well-known hydrophilic properties.

Five concentrations of HA were prepared and mixed into 10 wt% PVA solution (Table 1). To ensure stability of the bioactive ingredients of HA, PVA was cooled down to +20°C after dissolving process. When both solutions had the same temperature, they were mixed together using magnetic stirrer. Stirring time- 1 hour, speed- 1000 rpm, no heating was applied.

Table 1. Variants of polyvinyl alcohol (PVA) and PVA/hyaluronic acid (HA) spinning solutions

Composition	HA content wt%
PVA (10 wt%)	0
PVA (10 wt%)/HA	0.,04; 0.06; 0.07; 0.08; 0.1;

Electroconductivity and viscosity measured before electrospinning to prognose the success of nanofiber acquisition. It is known that viscosity values should be in the range 100-2000 mPa-s for solution to be spin able [14, 15].

Roller or needleless electrospinning was performed on Nanospider™ LAB 200 (Elmarco, Czech Republic) with following parameters- distance between electrodes 14 cm, roller speed- 2 rpm, applied voltage 65 kV, temperature- +20°C.

Fiber diameters were measured and morphology of the nanoweb was evaluated on atomic force microscope (AFM) Dimension Edge Veeco (Bruker, USA) using silicone cantilever OTESPA-R3, f0=300kHz, k=26N/m in tapping mode. 100 measurements of each sample were taken for nanofiber diameter analysis.

Fourier Transform Infra-Red (FTIR) spectroscopy was used to determinate HA in obtained nanofiber mats using equipment Thermo Scientific „Nicolet iS50 FT-IR".

Mechanical properties were tested on Instron Universal Tester Model 2519-107 (deformation speed- 1 mm/min., applied load- 1 N) and the results were collected through program Instron Bluehill. 10 samples (size 10 mm x 30 mm) of each composite were prepared, thickness was measured before placing them on a paper frame (Figure 1) for testing [13]. The paper frame ensure samples tightly fixed and do not damaged by the equipment's grippers in the testing process.

1.1 %wt HA and additional 5.2 %wt HA nano webs were tested on two cell lines- 3T3 mouse fibroblasts and HaCaT human keratinocytes [16]. Method- Neutral Red Uptake (NRU) assay.

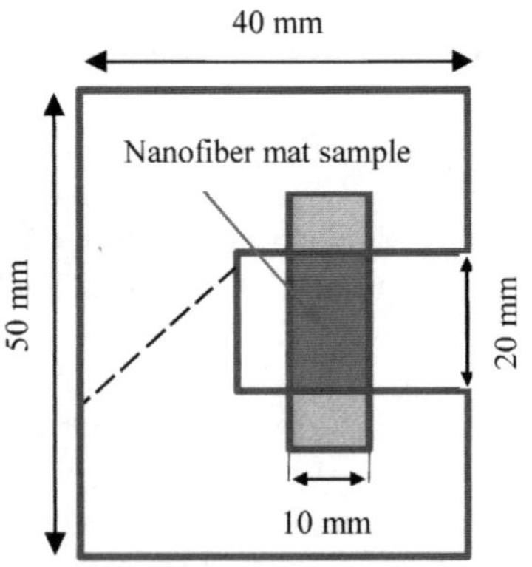

Figure 1. Preparation of nanofiber mat sample in Instron Universal Tester (Model 2519-107) [13]

3. Results and discussion

3.1. Viscosity and conductivity

Graphics of Figure 2 demonstrate that HA does not influence solution electro conductivity, which varies within the standard error, but viscosity proportionally increases with the HA concentration increase (equation (1)) from 796 to 1137 mPa-s, exceeding PVA solution viscosity (597 mPa-s) by 90%.

$$Yv = 542{,}9 + 6170\,X, \qquad R^2 = 0.94, \tag{1}$$

Where: Yv - viscosity, mPa-s; X - HA wt% of spinning solution.

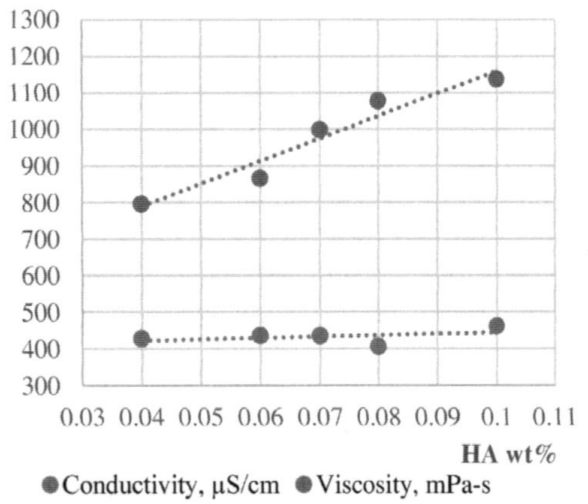

Figure 2. Properties of spinning solution

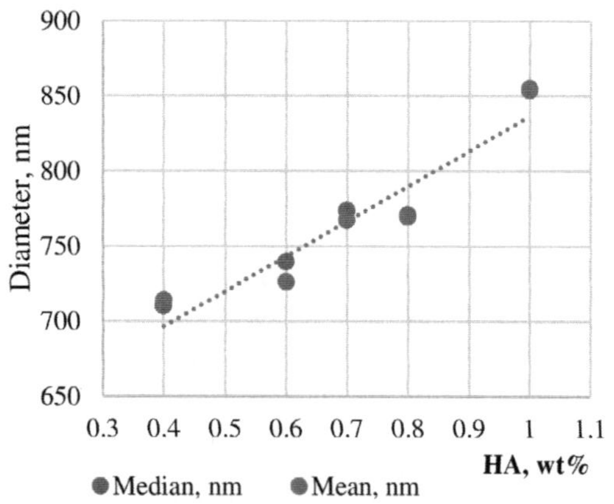

Figure 3. Fibers median and mean diameters

3.2. Diameters of nanofibers

Nanofiber diameter is the most important structural characteristic and its role in drug release process is parameters of electrospinning process [15]. Parameters of electrospinning process (applied voltage, distance between electrodes, relative humidity, temperature, etc.) can influence nanofibers diameters remarkably [16].

Median and average diameters of the nanofiber webs shown in Figure 3. Diameters tend to increase by added HA concentration and can be briefly described by equation (2).

$$Y_D = 602.87 + 233.64\,x, \qquad R^2 = 0.89 \tag{2}$$

IOP Conf. Series: Materials Science and Engineering **500** (2019) 012024 doi:10.1088/1757-899X/500/1/012024

3.3. *Morphology of nanofiber mats*

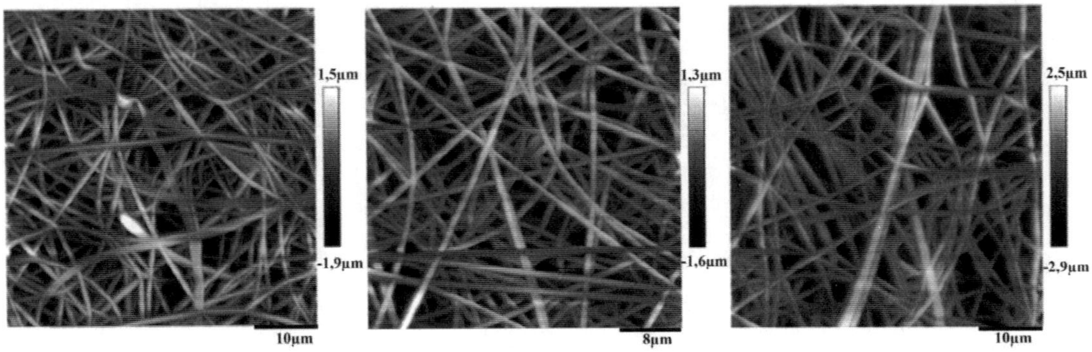

a **b** **c**

Figure 4. AFM images of nanofiber mats- **a** (PVA/0.4 wt% HA), **b** (PVA/0.8 wt% HA), **c** (PVA/1.1wt% HA)

As shown in AFM images (Figure 4), 0.8 wt% HA nanofiber mat has even and dense surface compared to 0.4 and 1 wt% HA samples. This could explain why 0.8 wt% sample demonstrated the best mechanical properties compared to other concentrations.

3.4. *Mechanical properties of nanofiber mats*

Table 2. Characteristics of nanofiber webs

	Web thickness, µm		**Web surface roughness, µm**	**Agglutinated fibres**
	Mean	**-/+**		
PVA 10 wt%	91.8	14.0	3.5	+
HA 0.4 wt%	147.7	6.3	3.4	+++
HA 0.6 wt%	113.5	8.9	4.1	++
HA 0.7 wt%	75.1	8.8	2.3	-
HA 0.8 wt%	91.1	13.2	2.9	+
HA 1.1 wt%	102.3	10.8	5.4	++++

Although all samples were electro spun within the same period (5 minutes), thickness and surface roughness of the nanofiber webs varied, as well as the amount of agglutinated nanofibers (Table 2). As seen from graphs of Figure 5, nonlinear variation nature of variants thickness and surface roughness with a HA in a range from 0.6 – 1.1 wt % are quite similar. Though surface roughness of the HA 0.4 wt percentage web is close to the relevant indicator of net PVA, the wed thickness differ substantially. The high share of agglutinated fibres in the web cross-section could be considered as a reason of difference (Table 2).

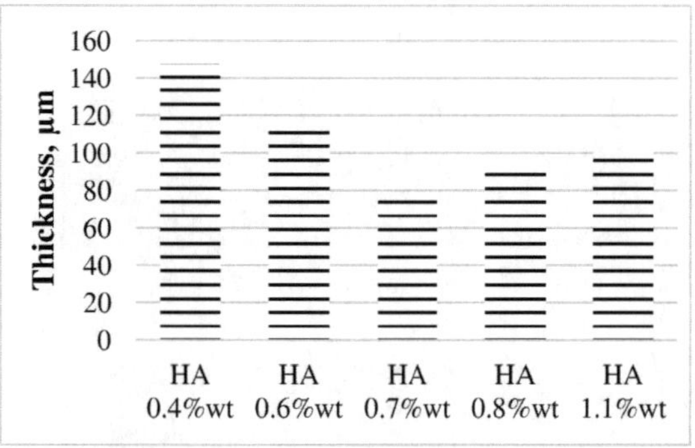

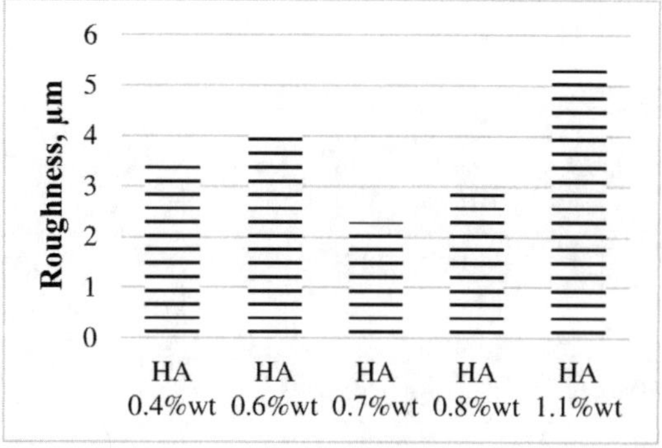

Figure 5. Variations of web variants thickness (left) and surface roughness (right)

Tensile stress and strain at break graphs (Figure 6) show non-linear dynamics depending on HA concentration in fibers, reaching the maximum value at 0,8 wt% HA (1,76 MPa, 17,84%), which is almost equal to the pure PVA sample (1,76 MPa, 18,65%) corresponding values. There is a linear relationship between tensile strain and stress where HA share in fibres fall in a range 0.6 – 0.8 %wt (Figure 6, right).

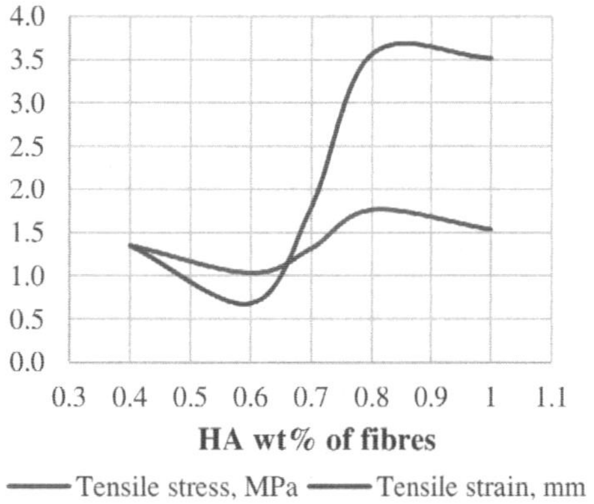

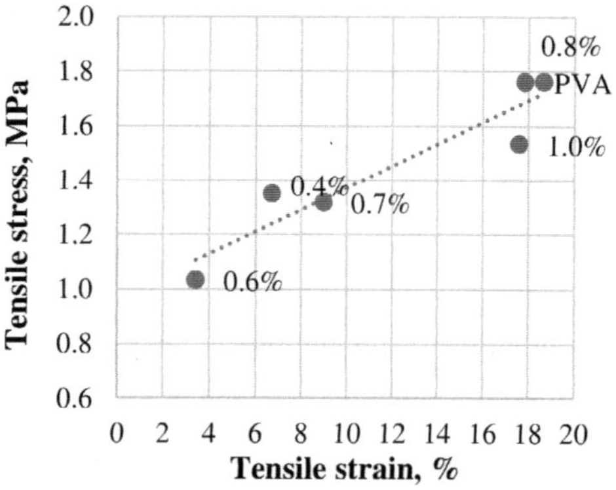

Figure 6. Mechanical properties of nanofiber webs

3.5. Fourier transform infrared (FTIR) spectroscopy

FTIR spectroscopy of composite PVA and HA nanofiber webs revealed different from net PVA absorption intensity in a frequency range 1557 – 2956 cm-1 and 3176-3462 cm-1(Figure 7).

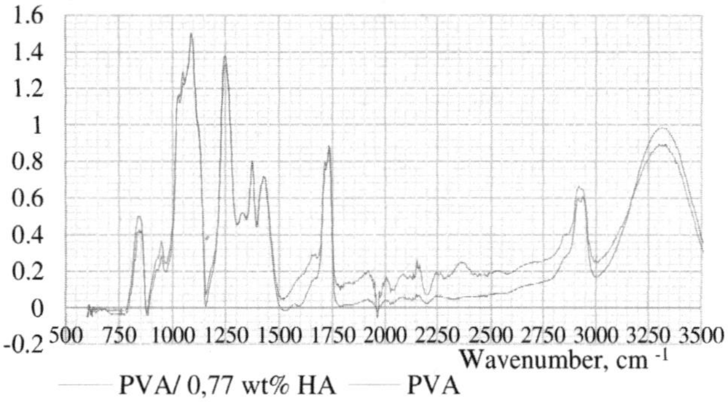

Figure 7. FTIR spectra of net PVA and PVA/HA 0.8 wt% webs

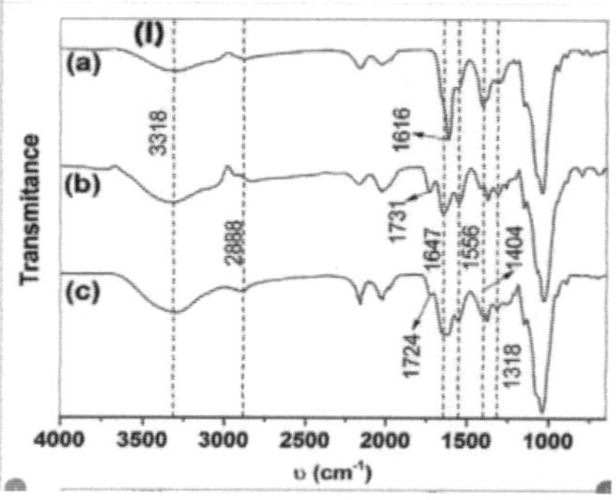

Figure 8. FTIR spectra of HA (a) and modified HA hydrogels[17]

Characteristic saccharide peaks identified in the sodium hyaluronate sample [17, 18]: at 3318 cm−1 (υ(O-H), stretching band), 2888 cm−1 (υ(aliphatic C-H), stretching band), 1616 cm−1 (Amide II), 1556 cm−1 (δ(-NH2), bending) fall in observed ranges of experimental web graphs. Peak at frequency 1054 cm−1 which refers to skeletal vibrations of C-O and C-C stretching, as well at 1404 cm−1 (δ(C-H) bending) partially overlapping with the bands of PVA (C-C; C-H2; C-H; C-OH) and therefore not easily identifiable

Further in-depth studies of cosmetic grade HA are necessary to compare IR values to medical grade HA which has been studied extensively. Pharmaceutical grade HA has to be used when the final product developed for tissue engineering, injections and treatment of open wounds. Cosmetic grade HA is suitable for beauty products such as facial mask, skin care cream, lotion, gel, etc.

3.6. Microbiological test results

Biological activity of the composites were tested on two types of cells- 3T3 mouse fibroblast cells and HaCaT human keratinocytes (Figure 9). As 1 wt% concentration showed no positive activity, additional concentration of HA (5 wt%) was prepared and tested. Results show that 5 wt% HA has stronger positive effects on 3T3 cells compared to HaCaT.

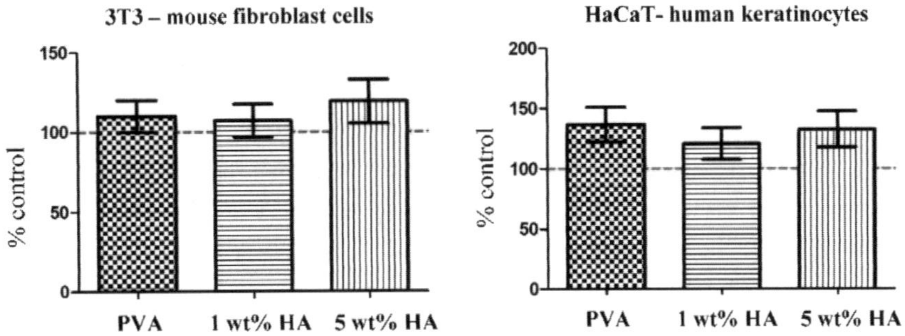

Figure 9. Microbiological test results of nanofiber mats

4. Conclusion

This study shows that spinning solutions based on PVA 10 wt% with the HA additive in a concentration range (0,04 to 0,5) wt% can be successfully electrospun. Viscosity of the spinning solution tends to increase linearly following HA concentration, but conductivity does not have the same dynamics.

Test results of mechanical properties revealed non-linear dynamics depending on HA concentration in fibers. FTIR analysis displayed new absorption peaks in all composites. Nanofiber diameters increased by added HA concentration.

5 wt% HA nanofiber mat sample showed better cell activity compared to 1 wt% HA. In further studies PVA concentration under 10 % wt should be evaluated and combined with a higher concentration of HA to keep appropriate viscosity and ensure better biological activity on cells.

References

[1] Larrañeta E, Henry M, Irwin N J et al, 2018, Synthesis and characterization of hyaluronic acid hydrogels crosslinked using a solvent-free process for potential biomedical applications, Carbohydr. Polym. 2018, 181, 1194–1205, doi:10.1016/j.carbpol.2017.12.015

[2] Zamboni F, Vieira S, Reis R L, Miguel Oliveira J, Collins M N, 2018, The potential of hyaluronic acid in immunoprotection and immunomodulation: Chemistry, processing and function. Prog. Mater. Sci. 2018, 97, 97–122, doi:10.1016/j.pmatsci.2018.04.003.

[3] Baumann L. Skin ageing and its treatment. J Pathol. 2007;211:241–51. doi: 10.1002/path.2098

[4] Eleni Papakonstantinou, Michael Roth, and George Karakiulakis. Hyaluronic acid: A key molecule in skin aging. Dermatoendocrinol. 2012 Jul 1; 4(3): 253–258.

doi: 10.4161/derm.21923

[5] Essendoubi M, Gobinet C , Reynaud R , Angiboust J F, Manfait M and Piot O, 2016, Human skin penetration of hyaluronic acid of different molecular weights as probed by Raman spectroscopy. Skin Res Technol, 22: 55-62. doi:10.1111/srt.12228

[6] Nobile V, Buonocore D, Michelotti A, Marzatico F, 2014, Anti-aging and filling efficacy of six types hyaluronic acid based dermo-cosmetic treatment: double blind, randomized clinical trial of efficacy and safety. Journal of Cosmetic Dermatology. 2014;13(4):277-287. doi:10.1111/jocd.12120.

[7] Milašius, R. , Ryklin, D. , Yasinskaya, N. , Yeutushenka, A. , Ragaišiene, A. , Rukuižiene, Ž. , Mikučioniene, D. Development of an Electrospun Nanofibrous Web with Hyaluronic Acid. Fibres & Textiles in Eastern Europe. 2017 | Nr 5 (125) | 8—12.

[8] Marcolin C, 2017, Electrospinning of biopolymers for regenerative medicine, PhD thesis, Biomaterial group of the Department of Chemistry, Materials and Chemical Engineering of Politecnico di Milano, http://hdl.handle.net/10589/132688.

[9] Jannesari M, Varshosaz J, Morshed M, Zamani M, 2011, Composite poly(vinyl alcohol)/poly(vinyl acetate) electrospun nanofibrous mats as a novel wound dressing matrix for controlled release of drugs. International Journal of Nanomedicine, 6:993-1003. doi:10.2147/IJN.S17595.

[10] Seham Abdelhady, Khaled M. Honsy, Mallesh Kurakula. Electro Spun- Nanofibrous Mats: A Modern Wound Dressing Matrix with a Potential of Drug Delivery and Therapeutics. Journal of Engineered Fibers and Fabrics.Volume 10, Issue 4 – 2015, 179-193.

[11] Enayati MS, Behzad T, Sajkiewicz P, Rafienia M, Bagheri R, Ghasemi-Mobarakeh L, Kolbuk D, Pahlevanneshan Z, Bonakdar SH. 2018, Development of electrospun poly (vinyl alcohol)-based bionanocomposite scaffolds for bone tissue engineering. J Biomed Mater Res Part A 2018:106A:1111–1120.

[12] Jegina S, Sutka A, Kukle S and Zelca Z, 2016, The effect of sodium chlorophyllin on polyvinyl alcohol electrospun nanofiber diameters (CYSENI 2016) (Kaunas: Lithuanian Energy Institute), pp 234 – 240.

[13] Sutka A, 2015, Technologies, Structure and Properties of Lignocellulosic Nanofibre Bicomponent Filaments. PhD Thesis. Rīga: [RTU], pp 80.

[14] Hrib J, Sirc J, Hobzova R et al, 2015, Nanofibers for drug delivery – incorporation and release of model molecules, influence of molecular weight and polymer structure, Beilstein Journal of Nanotechnology, vol. 6, pp 1939-1945.

[15] Yener F, 2014, New methods in the study of roller electrospinning mechanism. Summary of the thesis. Liberec: Technical University of Liberec, http://www.ft.tul.cz/document/240 .

[16] Olschläger V, Schrader A, Hockertz S. Comparison of primary human fibroblasts and keratinocytes with immortalized cell lines regarding their sensitivity to sodium dodecyl sulfate in a neutral red uptake cytotoxicity assay. Arzneimittelforschung. 2009;59(3):146-52. doi: 10.1055/s-0031-1296378.

[17] L. Grundelova, A. Gregorova, A. Mracek, R. Vicha, P. Smolka, A. Minarik, Viscoelastic and mechanical properties of hyaluronan films and hydrogels modified by carbodiimide, Carbohyd. Polym. 119 (2015) 142–148

[18] Javier Perez Quinones, Johanna Jokinen, Salli Keinänen, Carlos Peniche Covas, Oliver Brüggemann, Dmitri Ossipov. Self-assembled hyaluronic acid-testosterone nanocarriers for delivery of anticancer drugs. Macromolecular Nanotechnology. European Polymer Journal · December 2017 DOI: 10.1016/j.eurpolymj.2017.12.043

IOP Conf. Series: Materials Science and Engineering **500** (2019) 012025 doi:10.1088/1757-899X/500/1/012025

Surface Nanomechanics of Coatings and Hydrogels

Per M. Claesson, [a, b, *] Illia Dobryden,[a] Yunjuan He [a] and Gen Li [a]

[a] KTH Royal Institute of Technology, School of Engineering Sciences in Chemistry, Biotechnology and Health, Department of Chemistry, Division of Surface and Corrosion Science, Drottning Kristinas väg 51, SE-100 44 Stockholm, Sweden.
[b] RISE Research Institutes of Sweden, Division of Bioscience and Materials, SE-114 86 Stockholm, Sweden.
[*] Corresponding author. E-mail: percl@kth.se

Abstract
Due to the increasing use of nanostructured materials and thin coatings as barrier materials, it has become of high importance to measure and understand material properties on the nm to 100 nm length scales. In this article we demonstrate and discuss how atomic force microscopy techniques can be used to this end. It is demonstrated that the classical analysis based on the assumption of a purely elastic material response is a fair approximation for relatively stiff coatings (elastic modulus order of GPa), whereas viscous responses must be considered for soft materials (apparent modulus order of MPa) such as hydrogels.

1. Introduction

Very small probes are needed in order to determine local mechanical responses on the nm to 100 nm length scales. Such small probes are offered using the Atomic Force Microscopy (AFM) technique, where sharp tips with end radius of a few nms are commonly employed. This approach has successfully been used to determine variations in mechanical response over single core-shell nanoparticles [1], and to evaluate mechanical property variations across the interphase (interfacial region) between nanoparticles and polymer matrix in nanocomposites [2]. When performing such measurements one can choose between utilizing common single frequency methods and more sophisticated multi-frequency approaches.

Single frequency AFM modes include Peak Force Tapping (PeakForce QNM), Quantitative Imaging, QI, and Force Volume Mapping, where a complete force curve is acquired at each image pixel. Common to these modes is that the operator decides the maximum force applied, but they differ in terms of the tapping frequency used and in how nanomechanical information is extracted from the force curves. As an example we illustrate how the captured force curve is analysed, see Figure 1 left panel, but we note that all these images cannot be extracted in all modes. Importantly, several images are obtained simultaneously over the same surface region. The surface topography is shown in one image and the other images are reconstructed from the force curves. The adhesion force image is extracted from the retraction force curve and displays the maximum attractive force experienced upon separation. The adhesion energy image is the integral of the difference between the zero force line and the attractive part of the separation curve, the energy dissipation image is the integral of the difference between the approach and separation curves and the surface stiffness is extracted from the slope of the separation curve. The surface deformation (indentation depth) at each pixel can also be extracted as illustrated in Figure 1.

It is also possible to evaluate the local elastic modulus of the sample surface by fitting a contact mechanics model to a part of the retrace force curve. Of these quantities the surface stiffness and elastic modulus are most easily compared to quantities determined by e.g. indentation methods. We note that when evaluating the elastic modulus using contact mechanics models it is implicitly assumed that the response is purely elastic. As will be shown in this article, using a hydrogel sample, this is far from always correct since the viscous response of such soft materials is significant.

One way to analyse viscoelastic systems is to use multi-frequency dynamic AFM methods. Such approaches provide more information on the surface nanomechanical response, but they are also less

IOP Conf. Series: Materials Science and Engineering **500** (2019) 012025 doi:10.1088/1757-899X/500/1/012025

intuitive since they do not directly give a force vs. tip position curve. Recently, one such method, Intermodulation AFM (ImAFM) [3], has been utilized to gain detailed information on how both the surface of a soft sample and the cantilever tip moves in response to long-range and contact surface forces. In particular, it was shown that the common assumption of a static surface is incorrect for soft materials [4].

2. Results and Discussions

In this section we describe and discuss nanomechanical data obtained for two systems. The first is a relatively hard polymer coating designed for corrosion protection applications [5], whereas the other sample is a soft hydrogel of crosslinked polyHEMA. In both cases we illustrate typical surface-tip force curves and discuss how these are analyzed to provide nanomechanical information.

2.1. Nanomechanics of a corrosion protective coating

PTE Coatings (Gamleby, Sweden) provided the water-born hydroxyacrylic-melamine copolymer coating used in this example. The coating, with main components being a hydroxyacrylate resin mixed with the cross-linker hexakis(methoxymethyl)melamine (HMMM), was heat cured at 180 °C using different curing times in the range 2 – 30 minutes. More details on the corrosion protection performance after different curing times, and nanomechanical properties of this coating system can be found in our recent article [5].

A typical force curve between the hydroxyacrylic-melamine coating and the probing tip is illustrated in Figure 1 (middle). It is from such data sets that local nanomechanical properties are determined. We note that in this case the hysteresis between force curves measured on approach and on separation is small (except for a tip-surface adhesion seen in the retrace curve). This suggests that the deformation of the surface due to compression by the tip is predominantly elastic, and thus it is a good approximation to analyze the repulsive part of the force curve using an elastic contact mechanics mode. In this case, due to the small probe radius and the low adhesion, we utilized the DMT model [6].

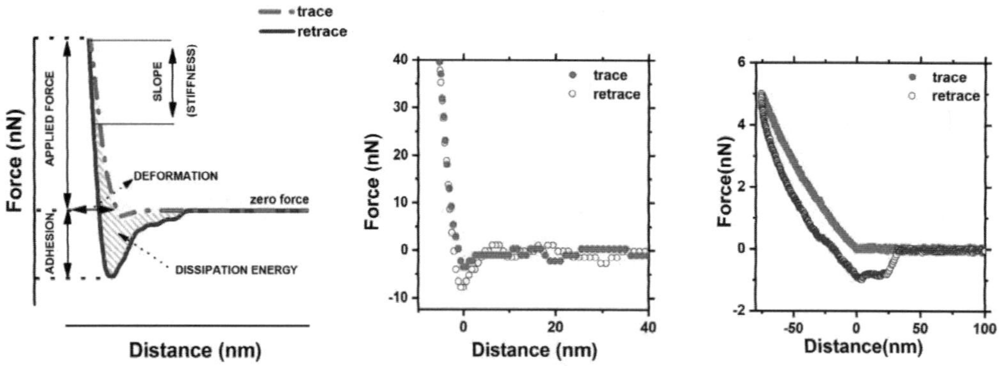

Figure 1. Typical force curves between an AFM tip and various samples. Left: Polystyrene sample in air. This panel illustrates how different nanomechanical properties are evaluated. Middle: The hydroxyacrylic-melamine coating measured in air on approach (trace) and retraction (retrace). Right: A cross-linked polyHEMA sample measured in water. Note the large hysteresis between forces measured on approach (trace) and separation (retrace) for this sample.

The topography, elastic modulus and deformation maps recorded simultaneously over the same surface area are illustrated for three curing conditions in Figure 2. We note that a relatively homogeneous surface layer is formed, with only small variations in height, elastic modulus and surface deformation. The curing time does, however, affect the mechanical properties significantly and the sample cured for 2 minutes is significantly softer than those cured for 10 or 30 minutes. This is due to a significantly lower conversion of the cross-linking reactions at the shorter curing time [5].

In our previous report we showed that even though the surface layer looks homogeneous and the corrosion protection performance is good, weak spots are present in the coating as can be observed after prolonged exposure times in corrosive environments [5].

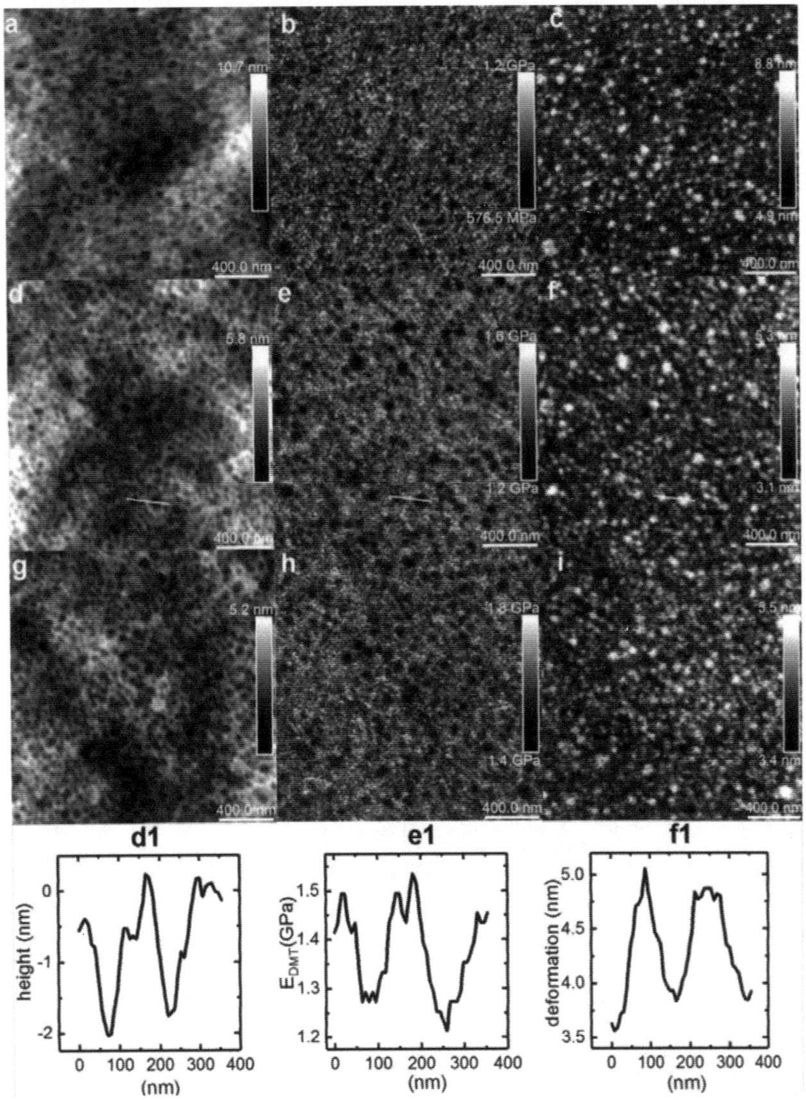

Figure 2. Topography (left), elastic modulus (middle) and deformation (right) images measured with AFM for coatings cured for 2 min (a-c), 10 min (d-f) and 30 min (g-i) min. The cross-section topography, modulus and deformation data for the sample cured for 10 min are taken along the white lines shown in panels d-f, respectively. The data were recorded in air. From reference 5, with permission. DOI: https://doi.org/10.1016/j.apsusc.2018.06.284. Creative Commons user license: https://creativecommons.org/licenses/by-nc-nd/4.0/

It is worth pointing out that the deformation used in these AFM-based nanomechanical measurements is small, in our case 3-5 nm. Thus, in these types of measurements it is the surface nanomechanical properties that are determined, and these may differ from bulk mechanical properties even for perfectly elastic solids. There are several reasons for this. The degree of cross-linking at the surface may be lower than that in bulk due to e.g. depletion of cross-linking agents at the interface or reaction inhibition by oxygen, which would make the surface layer softer than the bulk material. On the other hand, surface tension effects will counteract deformation and render the surface layer stiffer than the bulk material. This feature is often referred to as a skin effect and it is suggested to be due to non-linear stress-strain relations caused by the small size of the probing tip [7,8]. Thus, it is not uncommon to observe an indentation depth dependence on the elastic modulus [9, 10].

2.2. Nanowear of a corrosion protective coating

A protective coating can only fulfill its function as long as it remains intact, and thus the wear resistance of the coating is essential for its function. Below we report data obtained after 35 days exposure to corrosive environment, 0.1 M NaCl, at which point the corrosion protection experiment was terminated. In order to investigate the initiation of wear we have developed a two-step AFM based procedure. In the first step the tip scans the surface in contact mode, and here different loads are used over different surface areas. During this step the lateral force on the tip is also recorded, as it allows determination of friction and observation of stick-slip phenomena. The second step involves gently recording the surface topography over a slightly larger area that includes the worn area. Examples of such data are shown in Figure 3. As reported in detail in our previous article it was observed that wear at low loads was due to smooth plastic deformation, followed at higher loads by plastic deformation giving rise to topographical ripple structures. At even higher loads abrasive wear was predominant. It was also shown that the topographical ripple structures observed after wear had the same periodicity as the stick-slip periodicity as observed during wear [5]. Thus, a direct correlation between stick-slip and plastic surface topography changes was observed for this polymer coating.

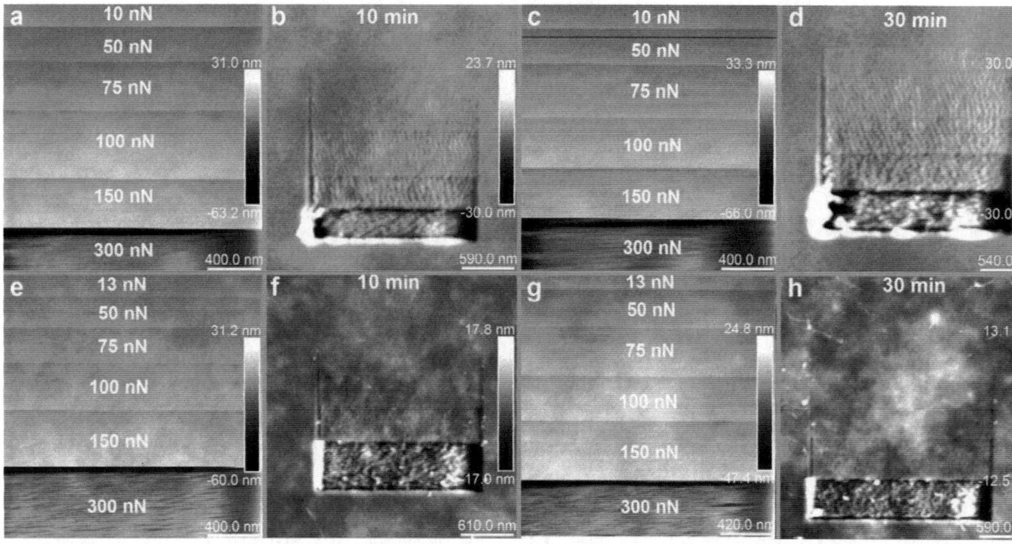

Figure 3. Wear experiments conducted with an AFM tip on a sample cured for 10 min before (a,b) and after (e,f) exposure to 0.1 M NaCl for 35 days, as well as on a sample cured for 30 min before (c,d) and after (g,h) exposure to 0.1 M NaCl for 35 days. The load was increased step wise as marked in the images. Contact mode topography images under the indicated loads, recorded during the wear experiment, are shown in panels a, c, e and g. Peak Force topographical images recorded after the wear experiment are shown in panels b, d, f and h. From reference 5, with permission. DOI: https://doi.org/10.1016/j.apsusc.2018.06.284. Creative Commons user license: https://creativecommons.org/licenses/by-nc-nd/4.0/

2.3. Nanomechanics of a hydrogel surface

The nanomechanical response for a thermoplastic nanocomposite, where the coating stiffness varies with temperature, was recently reported [11] and illustrates that coatings with varying stiffness can be studied by means of AFM. In the next example we consider a very soft material, a cross-linked polyHEMA hydrogel studied in aqueous solution where it is fully hydrated. In air, the hydrogel is significantly harder and shows a surface nanomechanical response that is humidity dependent as will be reported in a forth-coming publication. A typical tip-sample force curve for this case is illustrated in Figure 1 (right). Here we observe a large hysteresis between forces measured on approach and separation, signifying that viscous effects are of importance during the time scale of the measurements. We also note an adhesion on separation that arises from stretching of polymer chains that bind to the tip.

In order to investigate the nanomechanical response of this sample we adopted an experimental approach similar to that used for studying nanowear in the previous example. We probe different surface areas with different loads and evaluate the load-dependent nanomechanical response. In the example below we illustrate the surface topography (Figure 4, left) and how the apparent elastic modulus varies with indentation depth (applied load) (Figure 4, right). We use the term apparent modulus to acknowledge that the mechanical response is not elastic but rather viscoelastic.

We note that the apparent modulus increases with increasing indentation depth (Figure 4, right). Note that the indentation depth for this soft material is significantly larger than for the polymer coating discussed above. The load dependence of the apparent modulus suggests that the hydrogel sample is less cross-linked at the surface, and there are also some dangling polymer tails as observed in the force curve (Figure 1, right). At higher compression the indentation is deeper and the response of the cross-linked hydrogel material further below the surface becomes more important, but viscous effects still preclude determination of the elastic modulus. Thus, for such soft samples as hydrogels viscous responses cannot be ignored. For this reason no single value of apparent modulus (or, alternatively, surface stiffness) can characterize the mechanical response. Instead it is more appropriate to report the load dependence of these properties at a defined measuring speed. Viscous effects also have to be taken into account when performing indentation measurements with other techniques and evaluating the data according to the Oliver-Parr methodology [12]. It is our opinion that multi-frequency techniques, such as ImAFM, that separate viscous and elastic responses have an advantage over single frequency AFM techniques for studying such samples. One example of such a study can be found in a recent article [4].

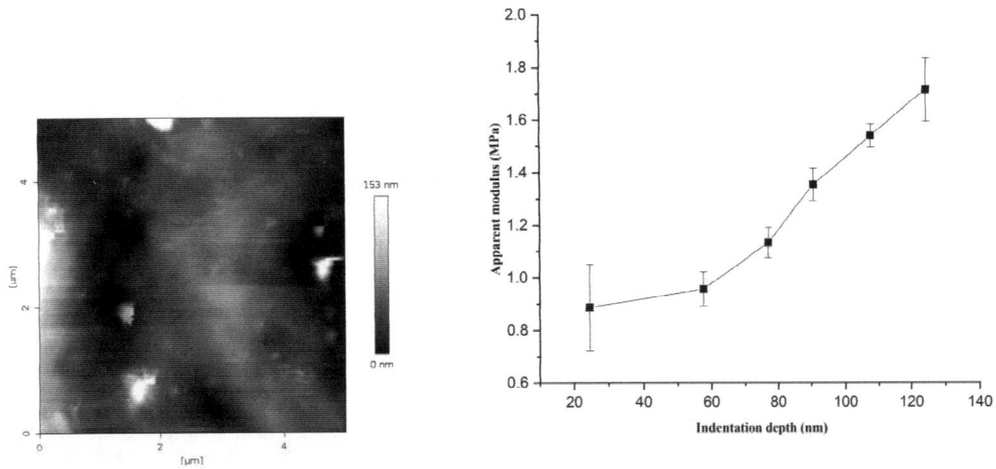

Figure 4. Topography (5x5 μm^2) of a cross-linked polyHEMA sample (left) and the apparent modulus as a function of indentation depth (right) determined using a constant driving speed of 5 $\mu m/s$.

3. Conclusions

AFM-based techniques can be used for measuring force vs. distance data, and from these it is possible to extract local nanomechanical surface properties. The data evaluation is relatively straightforward for samples showing a predominantly elastic response, and here elastic modulus and surface stiffness can be evaluated using classical contact mechanics theories. For soft samples, special protocols need to be adopted to allow determination of the indentation (or load) dependence of the extracted nanomechanical data. This is due to the importance of viscous effects for such samples. AFM-based techniques are also of great importance for investigating the initiation of wear and wear mechanisms under different load and shear conditions, and we have suggested an experimental protocol for such measurements.

4. Acknowledgement

Yunjuan He and Gen Li acknowledge the China Scholarship Council (CSC) for financial support for their PhD study at KTH. PTE coatings AB Gamleby (Västervik, Sweden) is acknowledged for providing the polymer coating. This work was supported by the Swedish Research Council, VR [2015-05080].

5. References

1. Dunér, G., Thormann, E., Dedinaite, A., Claesson, P.M., Matyjaszewski, K., and Tilton, R.D., Soft Matter, 8, 312 (2012)
2. Huang, H., Dobryden, I., Thorén, P.-A., Ejenstam, L., Pan, J., Fielden, M.L., Haviland, D.B., and Claesson, P.M., Composite Science and Technology, 150, 111 (2017)
3. Platz, D., Forchheimer, D., Tholén, E.A., and Haviland, D.B. Nat. Commun. 4, 1320 (2013)
4. Thorén, P.-A., Borgani, R., Forchheimer, D., Dobryden, I., Claesson, P.M., Kassa, H.G., Leclére, P., Wang, Y., Jaeger, H.M., Haviland, D.B., Phys. Rev. Applied, 10, 024017 (2018)
5. He, Y., Dobryden, I., Pan, J., Ahniyaz, A., Deltin, T., Corkery, R.W., and Claesson, P.M., Appl. Surface Sci., 457, 548 (2018)
6. Derjaguin, B.V., Muller, V.M., and Toporov, Yu. P., J. Colloid Interface Sci. 53, 314 (1975)
7. Dokukin, M.E. and I. Sokolov, Macromolecules 45, 4277 (2012)
8. Dokukin, M.E. and I. Sokolov, Langmuir 28, 16060 (2012)
9. Alderighi M, Ierardi V, Fuso F, Allegrini M, Solaro R., Nanotechnology, 20, 235703 (2009)
10. Dimitriadis, E.K., Horkay, F., Maresca, J., Kachar, B., and Chadwick, R.S., Biophys J, 82, 2798 (2002)
11. Huang, H., Dobryden, I., Ihrner, N., Johansson, M., Ma, H., Pan, J., and Claesson, P.M., J. Colloid Interface Sci. 494, 204 (2017)
12. Oliver, W.C., and Parr, G.M., J. Mater. Res. 19, 3 (2004)

IOP Conf. Series: Materials Science and Engineering **500** (2019) 012026 doi:10.1088/1757-899X/500/1/012026

Properties of pre-modified linen fabric dyed with reactive dyes

Bernava A [1], Reihmane S [2]

[1] Institute of Polymer Materials, Riga Technical University, Paula Valdena 3/7, Riga, LV-1048, Latvia, aina.bernava@rtu.lv

[2] Institute of Polymer Materials, Riga Technical University, Paula Valdena 3/7, Riga, LV-1048, Latvia, skaidrite.reihmane@rtu.lv

Abstract. Reactive dyes are one of the most popular dyes for cotton goods and other cellulosic fibres. The main investigations for application of reactive dyes for cellulosic fibres are devoted to cotton textiles. In this paper the research results of investigation of colour properties, colour and rubbing fastness, mechanical properties of pre-modified with alkali and enzyme linen fabric dyed with colour triad of Cibacron reactive dyes are presented.
The investigation results show that the main factor influencing the colour properties of linen fabric dyed with examined reactive dyes is used dye. The modifiers use increases the rubbing fastness of dyed linen textile. The modifiers influence on colour and colour properties before and after light test is selective for used dye, but not significant.

1. Introduction

Cellulosic fibres can be dyed with more types of dyes than other ones – mostly they are water- soluble dyes applied by one batch process [1]. Reactive dyes having been developed in the 1950s, in market from 1956 are very popular for dyeing and printing of cellulosic textiles [2]. Reactive dyes are today the most popular and most developed dyes for natural fibers due to its high level of washing fastness, and versatility of application [3, 4, 5, 6]. They are the only textile colorants designed to form covalent bond, with the substrate during the application process [1, 7, 8]. The functional groups presented in the structure of reactive dye molecule have distinct influence on dyeing behaviour [9].
Conventional immersion application processes employed for reactive dyes customarily involve the use of significant amounts of added inorganic electrolyte to promote dye uptake; indeed, traditional dyeing theory dictates that either NaCl or Na_2SO_4 must be added to a reactive dye dye bath in order to achieve adequate dyeing [8]. The major attempts on reduction or elimination of salt content in dye bath can be summarized under three categories; reduction of liquor ratio during the dyeing process, modification of dyes, such as usage of polyfunctional dyes [10] like Cibacron LS, Novacron LS, Teegafix HE reactive dyes and modification of cotton, using chemicals, compounds, enzymes, nanoparticles, ultrasonic waves, plasma, gamma, ozone treatments [11]. With the increasingly view on health, "Green" and environmentally friendly clothing obtain the popularity [12]. The textile industry is now facing many constraints. The creation of huge water and air pollutions and cost enhancement due to high energy consumption are two major constraints. The waste minimisation by process control can reduce the problem to a large extent, and the toxic chemicals can be replaced by eco-friendly enzymes and biopolymers [13].
The main investigation for application of reactive dyes for cellulosic fibres is devoted to cotton textiles [1, 3, 6, 13].

With the trend of appearance natural, comfortable yet elegant fabrics, linen and linen-blended fabrics accept acquired authority and added in acceptability attributable to the fabric qualities of comfort, hygiene, and elegance recently. Linen reduces γ-radiation nearly by bisected and protects the human organism towards sun radiation. It is an array of nice actual for advantageous applications due to its transfer and permeability to air. Heat conductivity of linen is five times as high as that of wool and 19 times as that of silk, additionally, the tensile strength of linen thread is twice as high as that of cotton and three times that of wool. Dye ability is the main barrier that unable its wider applications in the real life [12].

2. Materials and Methods

The local bleached, plain weave 100% linen fabric with density of 167,7 g/m^2 (Ltd, Larelini, Latvia) was used.

Reactive dyes (Huntsman International LLC): Cibacron Yellow F-4G - single monofluorotriazine, azo class dye (final exhaustion 65%, fixation 52%, washing-off properties–very good), Cibacron Red FN-R heterobifunciol monofluorotriazine–sulfatoetilsulphone, azo class dye, (final exhaustion 82%, fixation 72%, washing-off properties– very good) and Cibacron Blue FN-R heterobifunciol monofluorotriazine–sulfatoetilsulphone dye final exhaustion 92%, fixation 77%, washing-off properties –very good) [13] were applied.

2.1. Methods of Production

Before the modification linen fabric was washed in water at 75 ± 2 0C temperature for 15 min, after it treated in distilled water solution of washing agent Felosan NOF (designation VAV) 2 g/L at 98±2 °C temperature for 1 h. The rinsing in warm /cold water followed (designation of sample U).

2.1.1. Mercerization. The mercerization of fabric without tension with NaOH solution (200 g/L; liquid ratio VM 50) at 16±2 °C temperature for 1 min with followed rinse and neutralization with acetic acid solution 10 g/L at 16±2 °C temperature for 1 min and final rinse was done (designation of sample A).

2.1.2. Enzyme pre – treating. The pre-treating of fabric with 4% solution of enzyme Beisol PRO and 1% Felosan NOF (VM 25) at 55 ± 2 °C temperature for 15 ± 1 min with following rinsing in warm /cold water was managed (designation of sample E).

2.1.3. Dyeing Technology. The dyeing (Figure 1) was realized according to the recommendation of the Cibacron Reactive dyes producer.

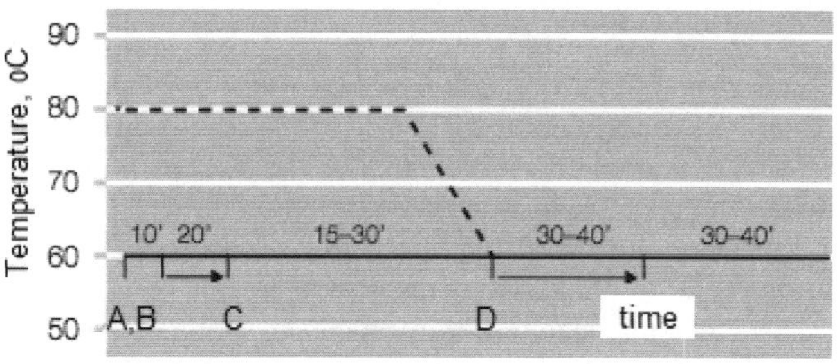

Figure 1. CIBACRON F/FN method of dyeing with alkali agent for 100 % cellulose. A –auxiliaries, B – dye, C – salt, D –alkali agent [13].

The pad dyeing was subjected; using solutions of 2% Cibacron dyes (VM 50, NaCl – 60 g/L; $Na_2 CO_3$ –20 g/L and Felosan NOF –2 g/L). After dyeing the treating with Felosan NOF 2g/L (VM 50) at 60 – 80 ± 2 °C temperature and rinse with hot/ cold water was realized.

The general principle of bachwise or exhaust dyeing with reactive dyes is to encourage as mush dye as possible to more into fibre over a period of 30-45 min by adding a high concentration of salt to the neutral dye bath. During this period, the dye is not reacting with the fibre, but most of dye is exhausted onto the fibre before the alkali is added, thus providing the best possible conditions for fixation before the dye- fibre reaction initiated. When the levelling process is completed, alkali is added and fixation continues for further 30-60 min [1]. The dye molecule has a reactive group which forms a chemical bond with the OH-group of the cellulose fibre. The dyeing process is divided in the migration phase and the fixing phase. The migration phase can be changed in time and temperature, depending on the material and machinery available.
The fixing phase is very susceptible to changes of temperature. It begins with the addition of alkali (soda). Deviations of the prescribed temperature (already at deviations of 5 °C) result that the final dye is much lighter: at too high temperatures due to hydrolysis and at too low temperatures due to an insufficient fixing. Mercerized cotton fabrics require 40% less colorant for the same hue than non-mercerized [14].

Table 1. Designation of samples, modification method and used dye

Sample	**UR**	**AR**	**ER**	**UB**	**AB**	**EB**	**UY**	**AY**	**EY**
Modification	-	Alkali	Enzyme	-	Alkali	Enzyme	-	Alkali	Enzyme
Used dye	Reactive red Cibacron			Reactive blue Cibacron			Reactive yellow Cibacron		

2.2. Methods of Testing

2.2.1. Colour properties. The tests were performed with Easy Colour QA device (Pocket Spec. Technologies Inc. USA), using RGB (red, green and blue) light system in CIELAB-76 colour space on both sides of samples as in [15]. Lightness difference (Δ L), common colour difference (Δ E), hue difference (Δ H) and chroma (S) is a measure of intensity or saturation of colour. Calculations were made following ISO 105-JO3:2009 [16].

2.2.2. Colour Fastness. The colour fastness of dyed samples were tested in light camber (Q- SUN, Xenon Test Chamber, mod. Xe-1-B, Q-LAB) for 72 hours at temperature 55 0C, intensity of radiance 1, 1 W/m2, according ISO 105-B02:2014 [17]. Xenon lamps emit UV, visible light, and infrared in the same way the sun does, and can therefore provide very accurate data [18]. After testing the main colour properties were examined.

2.2.3. Rubbing Fastness. The color fastness to dry and wet rubbing according of LVS EN ISO 105-X12:2016 [19] was assessed with Croc meter 238 A of SDLA (Shirley Development Laboratories Atlas Inc. USA). The highest rubbing fastness represents 5.

2.2.4. Mechanical Testing. The samples (width – 2,5 cm, length – 16 cm;) in warp and weft direction with INSTRON dynamometer (IntronLtd, UK, between clamps 100 mm, test speed 100,0 mm / min), according LVS EN 13934-1:2013 [20] were tested.

3. Results and Discussion

3.1. Colour properties.

The examination of obtained experimental results (Table 2) shows that colour properties of dyed textiles depend on the used dye. The lightness (L) and saturation (S) of unmodified samples forms the line UB < UR < UJ. The use of modifier NaOH causes small decrease of L and S for UB and UR. The enzymes treatment is not significant.
The results of ΔL, ΔE and ΔH once more affirm the influence of used dyes on colour properties. The colour properties characteristics differ for UY, in comparison with other dyes. This observation can be explained with the Cibacron Yellow F-4G dye chemical structure.

The influence of modifiers is different. For example, the colour difference (ΔE) with alkali modification is higher for UJ and UB samples than for UR, but influence of enzyme modification is observed only for UR sample.

The saturation (S) of UJ samples is higher and influence of alkali modification is more notable, than for other dyes. The lowest hue difference (ΔH) is observed for untreated and alkali treated UR samples, while for enzyme treated samples increase (5-7 times) of ΔH is observed. For UB and UJ samples the influence of modification methods is insignificant.

3.2. Colour fastness.

After colour fastness test to light the same colour properties were examined (Table 2). As expected, in all cases they are different in comparison with initial samples. It is understandable, that after light influence the lightness (L) of samples are growing and saturation (S) diminishing. The comparison of obtained results (ΔL, ΔE and ΔH) show common changes – some colour properties are reducing depending of used dye. The modifiers have not significant influence on colour properties after light test, only small nuances are observed, for example, for UR and UB samples a small increase of lightness after enzyme modification is observed, the reduce of saturation is noticed for all samples, but for UB samples decrease of saturation is less than for others.

The highest colour difference (ΔE) is for UY samples, while influence of modification was insignificant.

Table 2. Colour properties, colour and rubbing fastness properties of dyed samples

	UR	**AR**	**ER**	**UB**	**AB**	**EB**	**UY**	**AY**	**EY**
	Colour properties								
Lightness, L	39.98	35.60	41.01	33.42	26.35	32.42	79.58	79.37	79.75
Saturation, S	57.94	56.14	58.41	36.46	35.76	36.79	72.88	79.6	75.47
Lightness difference, ΔL	-43.74	-48.12	-42.71	-50.29	-57.37	-51.3	-4.13	-4.35	-3.97
Colour difference, ΔE	67.94	69.43	72.48	62.05	67.31	63.00	73.33	79.78	75.83
Hue difference, ΔH	5.24	3.53	26.54	20.14	19.16	20.07	30.24	30.98	30.64
	Colour properties after light fastness test								
Lightness, L,	45.24	36.61	45.90	33.56	26.90	35.48	81.31	80.31	81.60
Saturation, S	52.82	53.57	52.76	32.82	32.69	33.03	61.43	70.5	65.4
Lightness difference, ΔL	-38.49	-45.80	-36.51	-47.17	-55.51	-46.94	-2.41	-2.11	-0.11
Colour difference, ΔE	60.58	67.09	60.49	57.32	63.65	56.80	61.60	71.37	66.40
Hue difference, ΔH	5.34	0.22	1.32	19.36	13.34	14.56	27.20	27.21	26.56
	Rubbing fastness								
Dry	3/4	4/5	4/5	4	4	4	4/5	4/5	4
Wet	3/4	4	4	4/5	4/5	4/5	4/5	5	4/5

3.3 Rubbing fastness.

The fastness properties on dry and wet rubbing alike to colour properties depends on the used dye. The small increase (Table 2) of rubbing fastness is observed for modified samples.

3.4. Mechanical properties.

The obtained experimental results of mechanical properties of researched linen fabric are presented in Table 3. The error of measurement is within 2 - 3%.

In all cases the tensile strength is higher in warp direction in comparison of weft. As expecting, the alkali treatment without tension (samples U, A) causes the some decrease of tensile strength and significant increase of elongation. Such treatment of linen threads or fabrics can be used for creation of new textiles with higher elasticity.

The influence of enzyme modified textiles on mechanical properties and used dye is not significant.

Table 3. Mechanical properties of dyed fabric.

Sample	Maximum load, N		Extension at maximum load, %	
	Warp direction	Weft direction	Warp direction	Weft direction
U	370.8	275.7	9.6	14.0
A	296.3	294.7	38.9	22.9
E	360.9	274.4	10.3	13.2
UR	356.7	305.6	11.2	15.8
AR	353.5	258.5	37.5	21.3
ER	353.1	273.2	10.1	14.4
UB	371.6	277.6	10.0	14.5
AB	386.1	252.5	41.5	20.2
EB	357.6	272.1	10.3	10.8
UY	391.7	277.7	9.7	15.3
AY	341.4	266.9	38.7	19.5
EY	367.9	273.1	9.4	11.5

4. Conclusion

- The main factor influencing the colour properties of linen fabric dyed with examined reactive dyes is used dye
- The modifiers influence on colour and colour properties after light test is selective for used dye, but not significant
- The modifiers use increase the rubbing fastness of dyed linen textile
- The use of alkali treatment of linen textile without tension causes the significant increase of elongation characteristics of fabric.

5. References

[1] Koh J, 2011, Dyeing of cellulosic fibres, *Handbook of textile and industrial dyeing* **2**: *Application of dyes* Clark M, Woodhead Publishing Limited, 129-146.

[2] Chattopadhyay D P, 2011, *Chemistry of dyeing, Handbook of textile and industrial dyeing* **1**: *Application of dyes*, Clark M, Woodhead Publishing Limited, 150- 183.

[3] Farouk R, Gaffer H E, 2013, Simultaneous dyeing and antibacterial finishing for cotton cellulose using a new reactive dye, *Carboh. Pol.*, 138-142.

[4] Burkinshaw S M, Salihu G, 2017, The role of auxiliaries in the immersion dyeing of textile fibres: Part **6** Analysis of conventional models that describe the manner by which inorganic electrolytes promote reactive dye uptake on cellulosic fibres. *Dyes and Pigm.*, In Press.

[5] Burkinshaw S M, Salihu G, 2017, The role of auxiliaries in the immersion dyeing of textile fibres: Part **9** Practical aspects of the role of inorganic electrolytes in dyeing cellulosic fibres with pure reactive dyes. *Dyes and Pigm.*, In Press.

[6] Arivithamani N, Giri Dev V R, 2017, Industrial scale salt-free reactive dyeing of cationized cotton fabric with different reactive dye chemistry, *Carboh. Pol.* **174**, 137-145.

[7] Ojstrsek A, Doliska A, Fakin D, 2008, Analysis of reactive dyestuffs and their hydrolysis by capillary electrophoresis, *Analyt. Sc.*, 1581-1587.

[8] Burkinshaw S M, Salihu G, 2018, The role of auxiliaries in the immersion dyeing of textile fibres: Part **10** The influence of inorganic electrolyte on the wash-off of reactive dyes, *Dyes and Pigm,* **149**, 652-661

[9] Hunger K, 2003, *Industrial Dyes: Chemistry, Properties, Applications*, Wiley-CVH, New-York, USA, 113-118.

[10] Tang X, Zhong W, Zhuang D, Li C and Liu Y, 2013, Application of New Environmentally Friendly Materials in Clothing, Appl. *Mech. and Mat.* **295-298, 2,** 389-392.

[11] Choudhury A K R, 2017, Various eco-friendly finishes, *Principles of Textile Finishing*, Woodhead Publishing, Elsevier, 467–525.

[12] Hadj Taieb A, Msahli S, Sakli F, 2010, A New Index for Evaluating the Mechanical Comfort of Linen Fabric, *J. of Nat. Fib.* **7, 4,** 251-266.

[13] Ciba Specialty Chemicals, Textile Dyes, CIBACRON Warm, Exhaust Dying, CIBACRON F/FN, 1999/2000, http://www.huntsman.com/textile_effects (28.08.2018.)

[14] Von Lerber K, Dyeing with CIBACRON® F - An excerpt from "Färbedreiecke", http://shop.kremerpigments.com/media/pdf/345110-345170e.pdf (28.08.2018.)

[15] Bernava A, Reihmane S, 2018 Influence of Modification Method on Colour Properties of Linen Fabric Dyed with Direct Dyes. *Proc. of the Est. Ac. of Sc.*, **67, 2,** 131-137.

[16] LVS EN ISO 105-J03:2010, Textiles Tests for colour fastness, Part J03: Calculation of colour differences, https://www.lvs.lv/lv/library/read/27430 (16.02.2018.)

[17] LVS EN ISO 105-B02:2014, Textiles - Tests for colour fastness - Part B02: Colour fastness to artificial light: Xenon arc fading lamp test, https://www.lvs.lv/lv/library/read/55603 (16.02.2018.)

[18] Lucas J, 2001, Keep your true colours: lightfastness and weatherability testing, **4-5**, 13-16.

[19] LVS EN ISO 105- X12:2016, Textiles -Tests for colour fastness - Part X12: Colour fastness to rubbing. https://www.lvs.lv/lv/library/read (16.02.2018.)

[20] LVS EN ISO 13934-1:2013 Textiles - Tensile properties of fabrics - Part 1: Determination of maximum force and elongation at maximum force using the strip method https://www.lvs.lv/lv/library/read/33971 (16.02.2018.)

IOP Conf. Series: Materials Science and Engineering **500** (2019) 012027 doi:10.1088/1757-899X/500/1/012027

The manufacture and properties of nonwoven composites from fibres mix

Bernava A[1], Reihmane S[2] and Merijs – Meri R[3]

[1] Institute of Polymer Materials, Riga Technical University, Paula Valdena 3/7, Riga, LV-1048, Latvia, aina.bernava@rtu.lv

[2] Institute of Polymer Materials, Riga Technical University, Paula Valdena 3/7, Riga, LV-1048, Latvia, skaidrite.reihmane@rtu.lv

[3]Institute of Polymer Materials, Riga Technical University, Paula Valdena 3/7, Riga, LV-1048, Latvia, remo.merijs-meri@rtu.lv

Abstract. The application of nonwoven composites expands each year. The heat-bonded nonwovens (NW) are used for geotextile production due to the simple production cycle and low processing cost. To create the geotextile with desirable properties the mix of fibres are used. The aim of this study is to use recycled hemp, polyethylene terephthalate and polypropylene fibres for the development of new materials and technologies for nonwoven material composites (NWC) production. The series of NWC from mix of natural and synthetic fibres and series with one or two sides thermoplastic net reinforcing are manufactured. The NWC characterization parameters, water penetration, air permeability and results of physico - mechanical properties are presented.

1. Introduction

The nonwoven (NW) textile industry has grown as a broad array of engineered fibers and polymer-based products by high-speed, low-cost, innovative, value-added, fully automated processes and the ability to recycle textile products into useful products [1]. The common thermoplastic composites market is projected to grow up for 8.34% between 2017 and 2022. Increased use of thermoplastic composites in the transportation and aerospace & defence applications due to various properties offered by them, which include ease of recyclability, low curing time, high strength, and increased rigidity, among others are the factors driving the growth of the thermoplastic composites market across the globe [2]. The nonwoven composite (NWC) geotextiles are fibrous three-dimensional structures, used for different applications such as reinforcement, separation, filtration and drainage instead of coarse-grained soil due to their easy installation and gain [3]. Geotextile are generally made from a limited number of polymers (polypropylene, polyethylene and polyester). The three main properties which are required and specified for geotextile are its mechanical responses, filtration ability and chemical resistance [4]. The Rural Support Service of Latvia has planned financial support for the rebuilding and restoration of melioration systems, as well as for the construction or conversion of areas at production facilities [5, 6]. The 147 km long single water supply system in Riga City also requires continuous maintenance and construction of new facilities, according to the rules for the use and maintenance of the Riga City Hydrographic Network [7].

In view of the above facts, the development of new materials for geotextile is up to date. In the previous studies the optimal composition of fibres mix and the production condition of nonwovens' composites was identified [8, 9].

The aim of this research is to investigate the properties and the factors affecting NWC from fibers mix without and with reinforcing. The NWC series of the local hemp fibres variety "Bialobrzeskie" of Kraslava district, recycled polyethylene terephthalate and polypropylene fibres mix as well as series of composites with one or two side thermoplastic net reinforcing is manufactured and properties tested.

2. Materials and Methods

For NW web production the local hemp fibres variety "Bialobrzeskie" (HF; length 50-60 ± 2 mm, Ø 15 - 50 ηm, linear density 30 tex) [10], recycled polyethylene terephthalate fibres (PET; length 64 ± 1 mm, linear density 6.7 dtex, producer Starlinger & Co. Gesellschaft m.b.H., Uzbekistan [11]) and polypropylene fibres (PP; length 12 ± 1 mm; Ø 18.7 ηm, linear density 7.7 dtex; producer Bautech [12]) are used. The thermoplastic PP net (mesh size 20 × 20 mm, thickness 0.37 mm, producer Tenax SPA, Italy) is applied for NW web reinforcement.

2.1. Methods of Production

The untreated hemp fibres are stiff and on the used laboratory carding machine it is not possible to produce a satisfactory NW web [6]. Therefore, after tests, alkali treatment method of hemp (100 g/L at 20±1 ^{0}C temperature for 2 min) is selected [8, 9]. A neutralization of pre-treated fibres with acetic acid (2 g/L) at 19 ± 1 ^{0}C temperature for 15 min is done.

The NW web (Table 1) is made from fibrous blend (HF- 59 %; PET- 23 %; PP- 18 %) using laboratory carding machine MESDAN 337A (delivery velocity 10-15 m/min). After NW web formation the samples (160 x 160 mm; mass 2.6 g (designation of sample A) and 3.8 g (designation of sample B) are cut out and prepared for NWCs production in a frame without and with PP net reinforcement from one side (designation of samples A1, B1,) and both sides (designation of samples A2, B2). For thermal bonding of NWC the laboratory press Labtech Engineering ASTM LP-S-50/S at the required pressure 26 ± 1 MPa and temperature 160 ± 2 ^{0}C is used.

2.2. Methods of Testing

Mass per unit area and thickness of NWCs are calculated according LVS EN 12127:2001 [13] and LVS EN ISO 9073-2:2001 [14] respectively.

The tensile strength and elongation at break at maximum force is determined on universal testing machine Zwick/Roell BDO-FB020TN, according LVS EN ISO 13934-1-2001 [15]. Ten samples of NWC (50 x 150 mm) are tested, based on the experience gained in previous research [8, 9]. The resistance to water penetration of NWC is verified by hydrostatic pressure test according LVS EN ISO 9073-16:2009 [16]. For determination of the air permeability LVS EN ISO 9073-15:2008 [17] is applied. Ten NWC samples are tested from both sides and the average values of water penetration and air permeability are calculated.

3. Results and Discussion

Series of composites without netting and with one or both sides net reinforcing were manufactured and tested. The results of NWC parameters are summarised in the Table 1. Before explaining the obtained results, it must be mention, that NW samples have free arrangement of fibers and they are made manually, therefore irregularity in fibers location of further NWC is observed which can an influence on the obtained results of examined properties.

3.1. NWC characteristic and properties

The NWC thickness and mass without netting (Table1; samples A, B). depends on mass of the NW. The use of reinforcement decreases the thickness of one side reinforced NWC – about 16 % (A1) and 10 % (B1) due to additional influence of thermoplastic PP net on NW web bonding. Both sides reinforcement in comparison with one side is not significant. As expected, the use of reinforcing net increases the mass per area of NWC and depends on reinforcing net amount (one side, both sides).

Table 1. Mass of NW web, number of reinforcement layers, characterization parameters, water penetration and air permeability of NWC

Sample	Mass of NW web		Layers of net		NWC characterization parameters		Water penetration, L/min	Air permeability, mm/s
	2.6± 0.01 g	3.8± 0.01g	1	2	Thickness, ± 0.01 mm	Mass per area, ± 0.5 g/m^2		
A	x				0.38	108.9	61.0	9.5
A1	x		x		0.32	115.6	55.9	16.0
A2	x			x	0.34	122.7	71.7	14.9
B		x			0.40	153.5	79.0	3.9
B1		x	x		0.36	164.4	66.3	5.0
B2		x		x	0.35	176.0	75.0	5.4

3.2. Water penetration and air permeability

The water penetration - significant property for geotextiles (Table 1) for examined NWC without reinforcing is higher for samples with higher initial mass of web (A, B samples). With one layer netting, the water penetration decreases for both series samples. It can be explained with NWC thickness decrease and fibers crowding in the NW web after reinforcing. The double side netting (samples A2; B2) causes increase of water penetration in comparison with one-sided reinforcement (A1, B1). The wetting processes influence on water penetration is possible.

The air permeability (Table 1) is higher (59 %) for samples with lower web initial mass and mass per area (A series). The reinforcement of NWC increases the air permeability in all cases.
The expected correlation between the water penetration and air permeability is not confirmed.

3.3. Mechanical properties

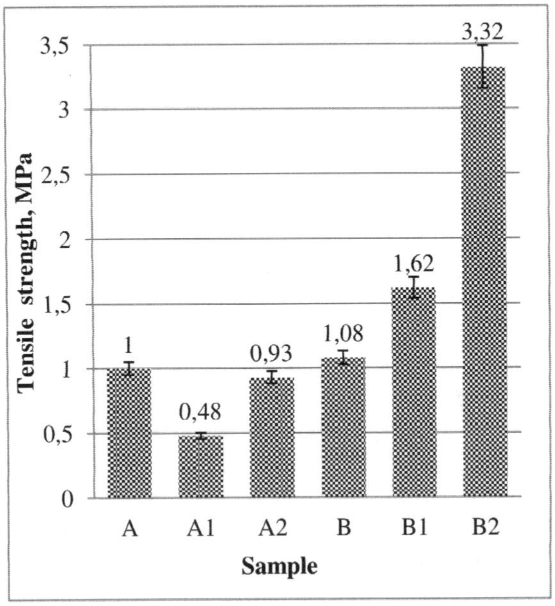

Figure 1. Tensile strength of NWC

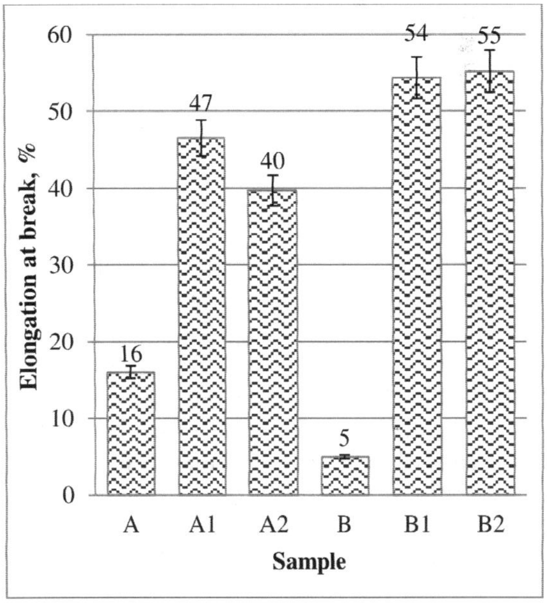

Figure 2. Elongation at break of NWC

The physico-mechanical properties of NWC are important for starting an application of NWC. The influence of thickness and mass per area of NWC on tensile strength (Figure 1; samples A, B) without reinforcing is not significant. The significant increases of tensile strength, after reinforcement is observe for B series - 33 % for one-sided (sample B1), 70 % for double-sided (sample B2).

The elongation characteristics grow remarkably (60 - 66 % for A series and 81 % for B series) in all cases due to use of reinforcement (Figure 2). The influence of reinforcing method (one or both sides netting) on the elongation at break is smaller -15 % (A series) and 2 % (B series).

4. Conclusions

- The thickness and mass per area of NWC without reinforcing depends on initial mass of NW web - 5 % and 30 % respectively.
- The use of reinforcement decreases NWC thickness (one side netting about 16 % (A1) - 10 % (B1); both side netting 11 % (A2) - 13 % (B2)
- The use of netting causes the increase of NWC mass per area; one side netting about 6 % (A1) - 7 % (B1), double-sided netting 12 % (A2) - 13 % (B2).
- The increase of water penetration for NWC with higher web initial mass, 23 % (without netting), 16 % (with 1 play netting) and 5 % (with both sides netting), is observed.
- The air permeability is higher (59 %) for samples with lower web initial mass and mass per area. The reinforcement of NWC increases the air permeability in all cases.
- The physico-mechanical properties depends on NWC characterizing properties and reinforcing method; the increase of tensile strength 70 % for B series double sided NWC is achieved.
- The use of reinforcement causes significant increase of elongation at break; 66 % (A series) and 81 % (B series).

5. Acknowledgements
The Financial Support of NRP of Latvia, project IMIS2, is greatly acknowledged.

6. References

[1] Pourdeyhimi B 2016 *Nonw. Text. World Com.* **3 - 4** 32 – 35.
[2] Thermoplastic Composites Market Fiber - Global Forecast to 2022 2017 *Markets and Markets Rep. Code ASDR-432582* **Nov** (8.12.2017).
[3] Lamy E, Lassabatere L, Bechet B, Andrieu H 2013 Effect of a nonwoven geotextile on solute and colloid transport in porous media under both saturated and unsaturated conditions, *Geot. and Geom.* **36** 55 – 65.
[4] *Handbook of technical textiles* 2004 Ed. Horrock A R and Anand S C Woodhead publishing limited 358 - 372.
[5] Lauku attīstības programmas (LAP) investīciju pasākumi 2014 - 2020 4 3 Atbalsts ieguldījumiem infrastruktūrā, http://www.lad.gov.lv/lv/atbalsta-veidi/projekti-un-investicijas/lap-investiciju-pasakumi/4-3-atbalsts-ieguldijumiem-infrastruktura (30.08.2018).
[6] Latvijas lauku attīstības programma (LAP) 2014. -2020. Zemkopības ministrija, Rēzekne 19.03.2015.,http://www.lad.gov.lv/files/lap_prezentacija_19032015.pdf
[7] Meliorācija,http://mvd.riga.lv/nozares/vides-parvalde/melioracija (30.08.2018).
[8] Bernava A, Reihmane S, Bitenieks J, Maniņš M 2017 The nonwovens properties made from hybrid fibres, *Key Eng. Mat.* **721**, 53 – 57.
[9] Bernava A, Reihmane S, Bitenieks J, Merijs - Meri R 2017 The properties of mesh reinforced nonwoven composites *Solid State Phenom.* **267** 63 - 67.
[10] Baltiņa I, Zamuška Z, Stramkale V, Strazds G, 2011, Physical properties of Latvian hemp fibres *Environment Technology. Resources, Proc. of the 8th Intern. Sc. and Pract. Conf* **2** 237 -243.
[11] *Baucon polypropylene fibres* © 2018 http://www.bautech.eu/en/products/concrete-floor/baucon.html (13 11.2018.)

[12] © *Starlinger &Co GmbH Recycling line – reco STAR PET (non food)* https www.starlinger.com/en/recycling/recostar-product-line/recostar-pet-non-food (13.11.2018.)
[13] LVS EN 12127:2001 *Textiles - Fabrics - Determination of mass per unit area using small samples*, https://www.lvs.lv/lv/products/4020, (13.06.2018).
[14] LVS EN ISO 9073-2:2001 *Test methods for nonwovens - Part 2: Determination of thickness* https://www.lvs.lv/lv/products/4616 (13.06.2018).
[15] LVS EN ISO 13934-1:2001 *Textiles - Tensile properties of fabrics - Part 1: Determination of maximum force and elongation at maximum force using the strip method* https://www.lvs.lv/lv/products/5094 (13.06.2018).
[16] LVS EN ISO 9073-16:2009 *Textiles - Test methods for nonwovens - Part 16: Determination of resistance to penetration by water (hydrostatic pressure) (ISO 9073-16:2007)* https://www.lvs.lv/lv/products/25312, (13.06.2018).
[17] LVS EN ISO 9073-15:2008 *Textiles - Test methods for nonwovens - Part 15: Determination of air permeability* https://ww.lvs.lv/lv/products/24642 (13.06.2018).

IOP Conf. Series: Materials Science and Engineering **500** (2019) 012028 doi:10.1088/1757-899X/500/1/012028

Study of print paste composition for natural and synthetic textiles. Part 1: Printing of cellulosic fabrics

A Borisova

Institute of Polymer Materials, Chair of Polymer Materials Technology, Riga Technical University, Riga, Latvia

E-mail: anna.borisova@rtu.lv

Abstract. Reactive dyes have proven to be one of the most commonly applied classes of modern synthetic dyes. Unlike other dye classes for cellulosic textiles dyeing and printing, the reactive dyes provide excellent wash-fastness for a very wide shade gamut. One of the most important components of print paste, especially for reactive dye printing of cellulosic textiles, is urea. Despite urea's great importance in adjustment of dye distribution in the paste, it can cause environmental problems. In this study, the application of polyethylene glycol (PEG) (10, 30 and 50 g/100 g) was examined for urea's replacement in reactive dye print paste. The optimal concentration of sodium bicarbonate, as pH buffer, was found. Testing results of colour fastness and colour measurements approved PEG application possibility.

1. Introduction
Textile printing is the most diverse and important of the methods used for introducing colour and design to textiles. It is a process, which combines a design idea, one or more colorants, and a textile substrate, using a technique for applying the print paste with some precision.

The composition of dye-based print paste is a complex combination of multiply constituents [1]. Among water, dye, thickener and pH buffers, the paste contains fixation agents and other auxiliaries. One of the most important components of print paste, especially for reactive dye printing of cellulosic textiles, is urea, which is used as a swelling agent during the steaming process, dye solubilizing and disaggregation agent. However, because of high nitrogen content, it can cause environmental problems. Thus, investigation in the field of optimization of textiles printing technology towards cleaner process is of current importance.

In continuation of the investigation of twill weave 50% cotton/ 50% polyester blended fabric one-step one-bath dyeing technology elaboration using a single dye class [2], the present study covers 100% cotton (Part 1) and 100% polyester (Part 2) fabric print pastes' recipe elaboration for direct printing method. The main goal of the research was to avoid the application of conventionally used constituent – urea, replacing it with polyethylene glycol.

2. Materials and methods
2.1. *Printing technology*
Prior to printing, the cotton fabric was scoured in non-ionic liquor solution Felosan NOF (1 g/L) (CHT Bezema, Switzerland) at 55-60°C for 60 min, at liquor ratio 1:30, rinsed and dried at ambient conditions.

In this study, reactive dye-based print paste was applied for 100% cotton fabric (100 g/m^2 surface density, and 0.11 mm average thickness). Print paste contained of distilled water, reactive dye Cibacron Navy W-B (CHT Bezema, Switzerland), sodium alginate (George Weil & Sons Ltd, UK) as thickener, sodium bicarbonate (Sigma-Aldrich Co., USA) as pH buffer, sodium 3-nitrobenzenesulphonate (Sigma-Aldrich Co., USA) as resisting salt, urea (Merck KgaA, Germany) as dye solubilizing agent, and polyethylene glycol (M_w = 400) (Kremer Pigmente, Germany) as substituent of urea. Examined print paste recipes are shown in Table 1.

Table 1. Recipes of screen print paste for cotton fabric [1, 3]

Print paste constituents	Print paste composition, g/100 g paste	
Reactive dye	5	
Urea	10, 30, 50	-
PEG	-	10, 30, 50
$NaHCO_3$	1, 1.5, 2	
3-nitrobenzenesulphonate	1	
Sodium alginate thickener	up to 100 g	

Each printed sample was labelled, depending on urea or PEG, and $NaHCO_3$ concentration. Examples of sample labelling are given in Table 2. Each series consisted of 3 parallel samples; total number of samples was 54.

Table 2. Examples of sample labelling

Sample label	Urea concentration, x g/100 g paste	PEG concentration, x g/100 g paste	$NaHCO_3$, x g/100 g paste
U-10-1	10	-	1
U-30-1.5	30	-	1.5
P-10-1	-	10	1
P-30-1.5	-	30	1.5

The direct 'all-in' method was applied for cotton fabric printing at the laboratory level (Fig. 1).

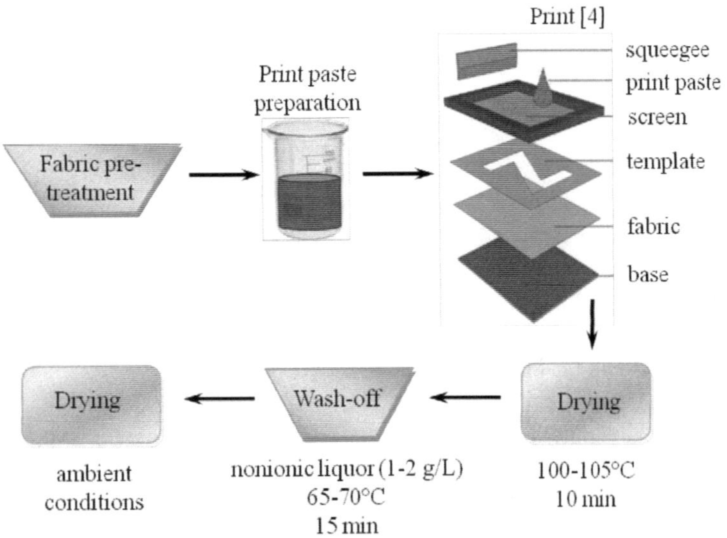

Figure 1. Flow-chart of applied screen printing on cotton fabric

IOP Conf. Series: Materials Science and Engineering **500** (2019) 012028 doi:10.1088/1757-899X/500/1/012028

2.2. Testing methods

Colour measurements of printed area, colour fastness to rubbing, washing and artificial light were determined according to national and international standards, as described in [2]. Organoleptic estimation of printing quality, using point scale from 1 (poor) to 5 (excellent) for printed fabric handle, paste spread homogeneity, penetration and sharpness of contours was evaluated.

3. Results and discussion

Table 3 shows colorimetric data of printed cotton samples in dependence of urea or PEG, and sodium bicarbonate concentration. It can be seen that increasing sodium bicarbonate concentration in the case of urea-based print paste, samples proved to have darker shades (L* values) (see Table 3). This correlation conforms to findings of Xie K et al. [5].

Table 3. Colorimetric data of printed cotton fabric samples

Sample	L*	a*	b*	C^*_{ab}
U-10-1	60.58	-0.72	-17.41	17.42
U-10-1.5	55.32	-1.67	-18.96	19.04
U-10-2	26.81	0.12	-16.22	16.22
U-30-1	48.72	-2.25	-20.09	20.21
U-30-1.5	29.09	-0.50	-17.26	17.27
U-30-2	29.31	-0.22	-18.04	18.04
U-50-1	57.62	-1.38	-17.63	17.69
U-50-1.5	55.07	-1.35	-18.48	18.53
U-50-2	21.09	1.82	-11.41	11.55
P-10-1	29.88	1.03	-19.11	19.14
P-10-1.5	30.93	0.87	-19.47	19.49
P-10-2	39.93	-1.61	-21.63	18.46
P-30-1	32.84	-1.11	-20.40	20.43
P-30-1.5	48.97	-2.61	-20.72	20.88
P-30-2	28.39	0.60	-18.28	18.29
P-50-1	73.30	2.01	-11.53	11.70
P-50-1.5	37.59	-1.77	-21.05	21.12
P-50-2	66.82	-0.44	-15.85	15.85

Increasing urea and sodium bicarbonate concentration in the print paste, colour fastness to artificial light, dry/wet rubbing and washing increased (except U-10-2 and U-50-1.5 samples), though, print handle properties deteriorated (see Table 4). For traditional, urea-containing print paste, the combination of urea (50 g/100 g) and sodium bicarbonate (2 g/100 g) was found to be an optimal one, though U-50-2 samples were the stiffest among all printed cotton samples.

Table 5 shows that in the case of PEG-containing paste it was found, that with increase of PEG concentration for samples with equal sodium bicarbonate amount, colour fastness to artificial light and washing decreased (higher ΔE^*_{ab} values). The correlation between sodium bicarbonate concentration and colour fastness parameters couldn't be found. Though, results of colour rub fastness were comparable for urea- and PEG-based pastes (see Table 4 and Table 5).

Table 4. Printed cotton fabric colour fastness, visual and handle parameters in dependence of urea-containing paste recipe

Sample	Total colour differences ΔE^*_{ab}		Colour fastness to dry | wet rubbing	Organoleptic estimation			
	72 h of artificial light exposure	5-cycle washing		Paste spread	Paste penetration	Sharpness of contours	Print handle (stiffness)
U-10-1	18.06	3.65	5 | 5	3	1	2	5
U-10-1.5	14.08	3.62	4 | 5	3	1	3	4
U-10-2	36.31	12.33	5 | 4	3	3	2	3
U-30-1	8.64	12.43	5 | 5	2	2	2	4
U-30-1.5	7.55	2.02	5 | 4	2	2	2	3
U-30-2	7.01	7.96	5 | 4	3	3	2	3
U-50-1	15.3	6.59	5 | 5	3	1	1	4
U-50-1.5	17.33	6.05	5 | 4	3	3	2	4
U-50-2	3.33	3.30	4 | 4	4	5	4	2

For PEG-containing print paste, the combination of PEG (30 g/100 g) and sodium bicarbonate (2 g/100 g) proved to be possible substituent of urea-containing paste. It can be seen from Table 5, that in comparison with optimal urea-based recipe, i.e. U-50-2 sample, PEG-containing print paste provided lighter colour (L* value was higher for 26%), lower light fastness (ΔE^*_{ab} value was higher for 57%), but better wash-fastness (ΔE^*_{ab} value was lower for 12%), print contour sharpness and handle properties. Colour rub fastness results of P-30-2 sample were comparable with Zhang et al reported [6].

Table 5. Printed cotton fabric colour fastness, visual and handle parameters in dependence of PEG-containing paste recipe

Sample	Total colour differences ΔE^*_{ab}		Colour fastness to dry | wet rubbing	Organoleptic estimation			
	72 h of artificial light exposure	5-cycle washing		Paste spread	Paste penetration	Sharpness of contours	Print handle (stiffness)
P-10-1	7.90	1.88	5 | 4	4	4	4	3
P-10-1.5	12.20	3.42	4 | 4	2	3	3	3
P-10-2	7.65	6.55	5 | 4	2	2	3	4
P-30-1	13.94	2.74	5 | 4	3	3	3	3
P-30-1.5	12.68	1.61	5 | 4	4	3	3	5
P-30-2	7.70	2.90	5 | 4	4	5	5	3
P-50-1	13.15	3.52	5 | 5	4	3	3	4
P-50-1.5	15.08	4.97	5 | 4	4	4	4	4
P-50-2	14.15	2.14	5 | 4	5	3	5	4

4. Conclusions

The present study aimed to decrease environmental impact of direct cotton printing by replacement of urea, as reactive dye solubilizing and disaggregation agent, with PEG.

Print paste with urea (50 g/100 g) and sodium bicarbonate (2 g/100 g) proved to provide darker colour, better colour fastness to artificial light and 5-cycle washing, but the stiffest handle among all printed cotton samples in this study.

As alternative, print paste with PEG (30 g/100 g) and sodium bicarbonate (2 g/100 g) can be recommended, providing better wash-fastness, sharpness of print contours and handle properties for cotton fabric.

References

[1] Kinnersly-Taylor J 2011 *Dyeing and screen-printing on textiles* (London: A & C Black) pp 70-92
[2] Borisova A 2018 *Key Eng Mater* **762** 390
[3] Miles L W C 2003 *Textile printing. Revised second edition.* (Bradford: SDC) pp 158-161
[4] Ryan V *Screen printing* www.technologystudent.com (accessed: 30-08-2018)
[5] Xie K, Liu H and Wang X 2009 *Carbohyd Polym* **78** 538
[6] Zhang H, Goa A, Song X and Hou A 2016 *J Clean Prod* **124** 126

IOP Conf. Series: Materials Science and Engineering **500** (2019) 012029 doi:10.1088/1757-899X/500/1/012029

Study of print paste composition for natural and synthetic textiles. Part 2: Printing of polyester fabrics

A Borisova

Institute of Polymer Materials, Chair of Polymer Materials Technology, Riga Technical University, Riga, Latvia

E-mail: anna.borisova@rtu.lv

Abstract. Urea, as dye disaggregation and solubilizing agent, plays an important role in direct printing process, though it has negative environmental impact due to ammonia nitrogen emission. Not only cotton reactive dye printing involves application of urea. Also disperse dye print pastes consists of it, mainly applied by professional artists and designers. In this study, the possibility of urea replacement with polyethylene glycol (PEG) for polyester fabric printing was examined. The action of glycerine, as additional dye dissolving agent, was also explored. With variation of urea and PEG concentration (25, 50 and 100 g/1000 g paste), and addition of glycerine (20 g/1000 g paste), an impact on results of colour measurements and fastness was analysed. With application of PEG to polyester printed fabrics higher colour fastness to washing and artificial light can be obtained.

1. Introduction

The dye-sublimation printing is the most common method of synthetic textiles decoration worldwide. According to recent global investigation from Smithers Pira [1], growth in the amount of dye-sublimation printed materials is forecasted to be 18.4% across the next five years. Despite overall positive aspects of dye sublimation, it has some technical restrictions [2-4]. Besides, there will always be a large group of professional artists and designers, who implement their ideas, using different printing techniques, e.g. direct style, which covers textile printing of positive image, applying one or more print pastes [5].

In continuation of the research of cotton/polyester one-step one-bath dyeing with disperse dyes [6], this study covers investigation of urea, as dye disaggregation and solubilizing agent, possible replacement with less hazardous chemical - polyethylene glycol, for polyester screen-printing method. For textile design projects, glycerine is recommended as dye dissolving agent with or without combination of urea [5]. Thus, the action of addition of glycerine was also examined. This study aimed print paste recipe elaboration with less negative environmental impact, avoiding ammonium nitrogen emission, caused by application of urea in print pastes.

2. Materials and methods

2.1. Pre-treatment and surface modification

Prior to printing, the polyester fabric was scoured in solution at liquor ratio 1:30, at 90-95°C for 60 min, containing sodium hydroxide (5 g/L) (PPH Stanlab Sp.j, Poland), sodium hexametaphosphate (0.5 g/L) (Chempur, Poland) and non-ionic liquor Felosan NOF (1 g/L) (CHT Bezema, Switzerland)

according to [7]. Following rinse, samples were neutralized in 30% acetic acid (1 mL/L) at ambient conditions for 10 min, rinsed again and dried.

2.2. Printing technology

In this study, disperse dye-based print paste was applied for 100% polyester fabric (125 g/m^2 surface density, and 0.2 mm average thickness). Print paste contained of distilled water, disperse dye Itosperse Blue Rap and dispersing agent Itosolt LJ 557 (LJ Specialities Ltd., UK), sodium alginate (George Weil & Sons Ltd, UK) as thickener, glycerine (Sigma-Aldrich Co., USA) as dye dissolving agent, sodium 3-nitrobenzenesulphonate (Sigma-Aldrich Co., USA) as resisting salt, urea (Merck KgaA, Germany) as dye dissolving agent, and polyethylene glycol (M_w = 400) (Kremer Pigmente, Germany) as possible substituent of urea. Examined print paste recipes are shown in Table 1.

Table 1. Recipes of screen print paste for polyester fabric [5, 8]

Print paste constituents	Print paste composition, g/1000 g paste			
Disperse dye	5			
Dispersing agent	10			
Urea	25, 50, 100	25, 50, 100	-	-
PEG	-	-	25, 50, 100	25, 50, 100
Glycerine	-	20	-	20
3-nitrobenzenesulphonate	10			
40°C distilled water	150			
Sodium alginate thickener	up to 1000 g			

Each printed sample was labelled, depending on urea or PEG concentration, and glycerine presence. Examples of sample labelling are given in Table 2. Each series consisted of 3 parallel samples; total number of samples was 36.

Table 2. Examples of sample labelling

Sample label	Urea concentration, x g/1000 g paste	PEG concentration, x g/1000 g paste	Glycerine, x g/1000 g paste
U-25-G	25	-	20
U-50	50	-	-
P-25-G	-	25	20
P-50	-	50	-

The direct 'all-in' method was applied for polyester fabric screen printing at the laboratory level, following by drying at 100-105°C for 10 min and thermal fixation at 190°C for 1 min. Samples were rinsed and scoured in non-ionic liquor Felosan NOF (1-2 g/L) at 65-70°C for 15 min, and dried at ambient conditions.

2.3. Testing methods

All testing procedures were implemented according to national and international standards, as described in Part 1 of the present work. Additionally, non-printed area's cross-staining after 5-cycle washing was examined using grey scale, according to ISO105-A03:1993.

IOP Conf. Series: Materials Science and Engineering **500** (2019) 012029 doi:10.1088/1757-899X/500/1/012029

3. Results and discussion

Table 3 shows colour measurement results for printed samples in dependence of urea and PEG concentration, and glycerine addition. Comparing urea- and PEG-containing print paste samples, traditional application of urea resulted darker fabric colouration (lower L* values).

Obtained results indicate that concentration of urea almost didn't influence lightness (L*) and chroma (C^*_{ab}) values. Addition of glycerine to urea-containing print paste led to slightly darker, but less saturated shades.

In the case of PEG-based print paste, with an increase of PEG concentration, colour lightness values increased for 16% in average. Samples printed with glycerine addition had lighter (higher L* values) shades (see Table 3) and less homogenous spread of print paste (see Table 4).

Table 3. Colorimetric data of printed polyester fabric samples

Sample	L*	a*	b*	C^*_{ab}
U-25	47.26	5.42	-36.26	36.67
U-50	46.57	6.44	-35.66	36.24
U-100	48.11	5.52	-34.87	35.30
U-25-G	43.63	6.30	-36.35	36.89
U-50-G	49.31	4.09	-34.59	34.83
U-100-G	48.81	4.78	-34.46	34.79
P-25	45.23	5.37	-28.73	29.23
P-50	54.68	3.00	-30.55	30.70
P-100	53.36	2.83	-30.94	31.07
P-25-G	50.96	5.24	-34.80	35.19
P-50-G	52.15	3.50	-31.91	32.10
P-100-G	51.78	3.86	-32.28	32.51

Correlation of total colour differences (ΔE^*_{ab}) with print paste recipe is shown in Figure 1. With the increase of urea concentration in print paste, print colour fastness to 5-cycle washing improved (lower ΔE^*_{ab} values) for 68% in average.

As stated above, glycerine is recommended as additional or separate dye disaggregation agent. Results in Figure 1 indicate that application of glycerine in urea- and PEG-containing print paste tended to reduce its colour light fastness. That can be explained with too amplified dye solubilizing action, which led to weak dye bonding with fibre.

Analysing PEG concentration impact, samples printed with PEG (100 g/1000 g - sample P-100) had lower for 65% in average light fastness values and moderate paste spread homogeneity in comparison with P-25 and P-50 samples. Application of glycerine almost didn't change colour wash-fastness results for PEG-containing print pastes (see Table 4).

Samples printed with PEG (25 and 50 g/1000 g) proved to have better colour fastness to washing and artificial light. That can be attributed to similarities in chemical composition of PEG and polyester, i.e. one of the main components of polyester is ethylene glycol. Polyglycol is water soluble, but majority of it tends to be trapped inside the fibre structure [9], which can be the possible reason of fibre blocking from dye desorption.

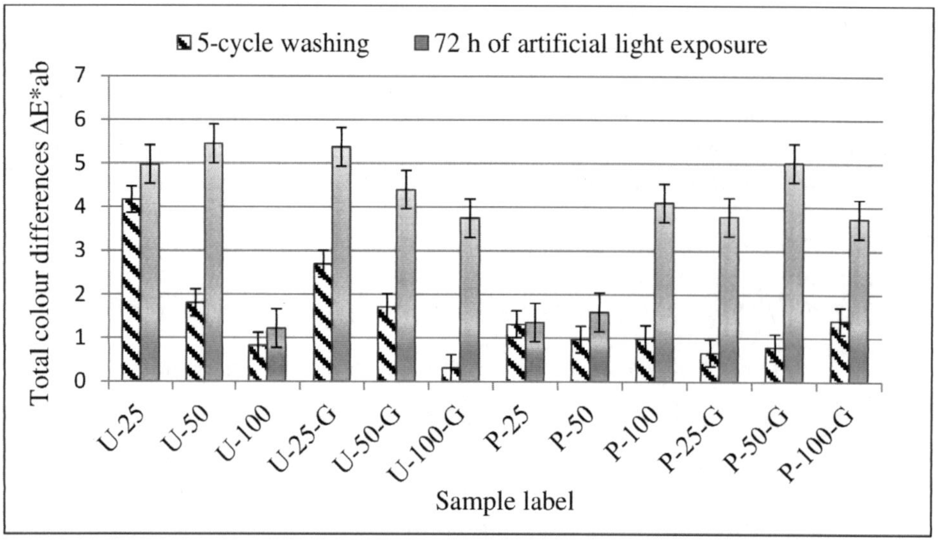

Figure 1. Total colour differences of printed samples after 5-cycle washing and 72 h of artificial light exposure

Table 4. Printed polyester fabric visual and handle parameters in dependence of paste recipe

Sample	Non-printed area's cross-staining	Organoleptic estimation			
		Paste spread	Paste penetration	Sharpness of contours	Print handle (stiffness)
U-25	5	3	5	5	4
U-50	5	4	5	5	4
U-100	5	4	5	5	5
U-25-G	5	3	5	5	4
U-50-G	5	2	5	5	5
U-100-G	5	2	5	5	5
P-25	3/4	5	5	5	4
P-50	4/5	5	5	5	5
P-100	4/5	3	5	5	5
P-25-G	4/5	3	5	5	5
P-50-G	4/5	4	5	5	4
P-100-G	4/5	3	5	5	5

Despite excellent results of colour fastness to dry and wet rubbing for all printed samples in this study, cross-staining tests showed that PEG application led to non-printed area's tinting (see Table 4). However, addition of glycerine to PEG-containing print paste slightly improved cross-staining results. In the case of urea-containing paste, concentration of urea and addition of glycerine didn't affect these parameters.

4. Conclusions

This study aimed to examine possibility of urea replacement with PEG in print paste for polyester fabric.

Achieved results show that application of PEG is possible in concentration up to 50 g/1000 g paste. These samples showed better results of colour fastness to 5-cycle washing and artificial light, print handle and paste spread properties. PEG concentration in the amount of 50 g/1000 g paste can be recommended as substitution of urea.

Addition of glycerine to urea- and PEG-containing print pastes decreased colour light fastness and didn't improve colorimetric parameters. Thus, application of glycerine in print pastes for polyester can't be recommended.

Achieved results are promising as an alternative cost-effective and less environmentally hazardous print paste option and can be suggested for application in textile printing design projects.

References

[1] Information from www.smitherspira.com *Five factors pushing dye sublimation print market towards €3 billion* (accessed on 12-09-2018)
[2] Wang L, Cui S, Ni H, Wu M and Wang W 2018 *Prog Org Coat* **123** 75
[3] Shakoor A, Baig G A, Tausif M and Gilani S Q Z 2017 *Dyes Pigments* **136** 865
[4] Cie C 2015 *Ink jet textile printing* (Cambridge: Woodhead Publishing) pp 125-137
[5] Kinnersly-Taylor J 2011 *Dyeing and screen-printing on textiles* (London: A & C Black) p 70-92
[6] Borisova A 2018 *Key Eng Mater* **762** 385
[7] Krichevsky G E 2000 *Khimicheskaya tehnologya tekstilnikh materialov. Tom 1* (in Russian) (Moscow: Moscow State University) pp 315-323
[8] Miles L W C 2003 *Textile printing. Revised second edition.* (Bradford: SDC) pp 172-176
[9] East A J 2005 *Synthetic fibres: nylon, polyester, acrylic, polyolefin* (Cambridge: Woodhead Publishing) pp 95-166

IOP Conf. Series: Materials Science and Engineering **500** (2019) 012030 doi:10.1088/1757-899X/500/1/012030

How synergistic aqueous lubrication is mediated by natural and synthetic molecular aggregates

A Dėdinaitė[1,2] and PM Claesson[1,2]

[1]KTH Royal Institute of Technology, School of Engineering Sciences in Chemistry, Biotechnology and Health, Department of Chemistry, Surface and Corrosion Science, Drottning Kristinas väg 51, SE-10044 Stockholm, Sweden
[2]RISE Research Institutes of Sweden, Materials and Surfaces, SE-114 86 Stockholm, Sweden
E-mail: andra@kth.se

Abstract. Nature lubricates in aqueous environment, and thus the example of a human synovial joint with its seamless function has been a fascination for scientists since the times of the birth of modern science. Here, inspired by nature, we investigate the mechanistic function of three different types of synergistic molecular aggregates. Firstly, we show how simple phospholipids lubricate hydrophilic model surfaces of silica and how this lubrication is facilitated further by the presence of an anionic polysaccharide, hyaluronan, due to the enhanced surface build-up of lubricant material. Next, we mimic natural polylectrolyte-surfactant aggregation by employing a highly positively charged polyelectrolyte and anionic surfactant that strongly associate both in the bulk and at the surfaces by building structured aggregates that lubricate due to hydration lubrication. This occurs despite of the presence of strong attraction between the lubricated surfaces. This is an example of synergistic lubrication due to particular internal structural arrangement of the aggregates. Finally, we investigate the case of synergistic lubrication due to preferential surface ordering of two biological polyelectrolytes, cartilage oligomeric matrix protein and lubricin, that leads to favourable lubrication.

1. Introduction

Since the beginning of modern science inquisitive minds admired and wondered about the structure and function of the human synovial joint. The surgeon William Hunter wrote in Philosophical transactions of the Royal Society in 1743: "*The fabric of the joints in the human body is a subject of so much the more entertaining, as it must strike every one that considers it attentively with an idea of fine mechanical composition. Wherever the motion of one bone upon another is requisite, there we find an excellent apparatus for rendering that motion safe and free*"[1]. In other words, the cartilage surfaces covering the bones are perfectly lubricated by the molecules and molecular aggregates of the synovial fluid. Today we start getting a grasp on what molecular constructs and their aggregates that are responsible for the lubrication of cartilage. However, despite the appreciable age of the considered topic, until recently we find a heated debate as to what is "the vital active ingredient in synovial fluid (SF) that imparts effective boundary lubrication"[2] of synovial surfaces. To give a perspective, while Hills was convinced that it is simple phospholipids that lubricate the surfaces of the cartilage [3, 4], Jay et al. argued for the case of a glycoprotein, lubricin [5, 6]. The ingenious machinery of Nature has blended multiple components, macromolecular and small molecular weight, in both cartilage and synovial fluid [7]. We argue that it is for a reason that aggregates of molecules work far more efficiently as lubricants than the individual components. There are multiple modalities by which the lubrication synergies of the components may be achieved. This applies so much to biological as to synthetic molecular aggregates. In the following we will illustrate this by showing three such cases of synergistic action of the self-assembled molecular aggregates in lubrication. In the first case we employ phospholipids and hyaluronan for friction control on silica surfaces. In the second case we investigate cationic polyelectrolyte and anionic surfactant aggregates on silica surfaces. And in the third case we study associating biological glycoproteins at mildly hydrophobic polymeric surfaces.

2. Experimental

2.1 Materials

Hyaluronan, HA, is a linear polysaccharide consisting of a repeating disaccharide unit of β1-4-D-glucuronic acid and β1-3-N-acetyl-D-glucosamine [8], Figure 1a, left. The hyaluronan used was of relatively low molecular weight (average $M_w = 6.2 \times 10^5$ g mol^{-1} as determined by asymmetric flow flow-field fractionation). Dipalmitoylphosphatidylcholine, DPPC, is a molecule with a zwitterionic headgroup of phosphatidylcholine and a double C16 hydrocarbon tail group, Figure 1a, right. The solubility of DPPC in water is very low and the DPPC molecules aggregate in layered structures when in contact with water and assemble at interfaces [9]. The cationic polyelectrolyes used were poly[3-(2-methyl propionamido)-propyl]trimethylammonium chloride (polyMAPTAC) with a molecular weight of 480 000 g mol^{-1} or poly[2-(propionyloxy)ethyl]trimethylammonium chloride (polyCMA) with a mean molecular weight of 1.5×10^6 g mol^{-1}. The monomeric segments of the polyelectroytes are shown in Figure 1b, top. The surfactant used was anionic sodium dodecyl sulphate, SDS, Figure 1b, bottom. Lubricin was isolated from synovial fluid collected from a pool of patients diagnosed with rheumatoid arthritis (Figure 1c, left), and a monomeric form of recombinant human cartilage oligomeric matrix protein (COMP)/trrombospondin-5 with predicted molecular mass of 81.8 kDa (Figure 1c, right) was used.

a.
b.
c.

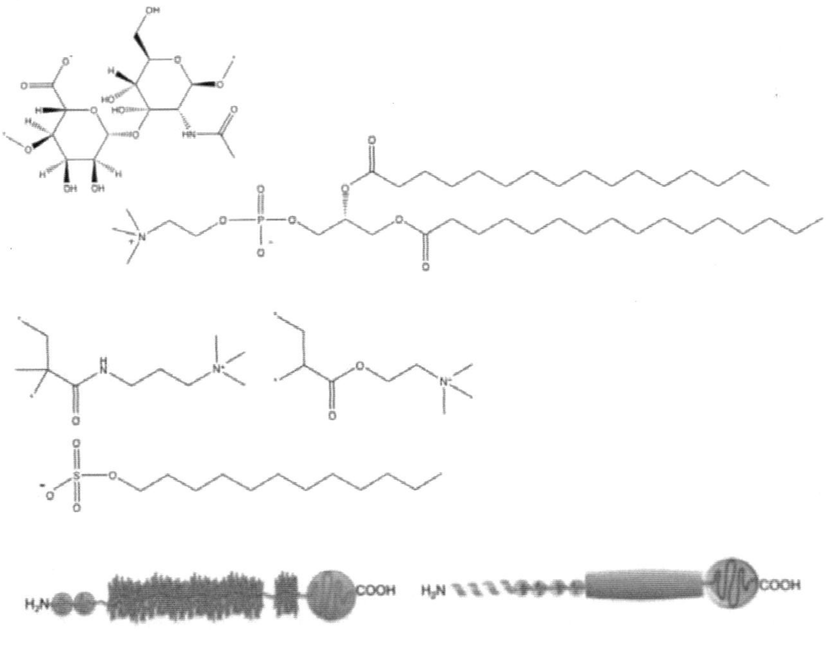

Figure 1. a. Molecular structures of a unit of HA (left) and DPPC (right), b. monomeric units of polyMAPTAC (top left) and polyCMA (top right), and SDS (bottom), c. schematic representation of lubricin (left) and COMP (right).

2.2 Surfaces and force measurements.

Silica or mica was employed as a smooth hydrophilic surface. The silica surface is suitable for phospholipid bilayer formation [10] and subsequent HA adsorption. Silica and mica are negatively charged. Poly(methylmethacrylate), polyMMA, was used as a surface for lubricin and COMP adsorption. This was motivated by the slightly hydrophobic character of polyMMA (contact angle of water on polyMMA is 68°) that promotes adsorption of molecules with hydrophobic chain regions.

Normal and friction forces were measured between the surfaces by the atomic forces microscope – colloidal probe technique that is described in detail before [11]. Measurements were performed in a fused silica liquid cell. Forces were measured between coated a silica surface and a silica sphere with diameter of ~7 μm, or between a coated polymethylmethacrylate (polyMMA) surface and a polyMMA sphere with a diameter of ~10 μm.

3. Results and discussion

3.1 DPPC and HA

Forces acting between pure DPPC bilayers in 155 mM NaCl solution (physiological ionic strength) at 47 °C (above chain melting temperature) have been reported [12] and are shown in the inset of Figure 2 left. They are characterized by a steep repulsion appearing at short separations upon approach and separation, which is understood as being due to hydration and protrusion forces characteristic for phospholipid bilayers [13]. Adsorbing HA on the DPPC bilayers results in a longer-range repulsive force and a week pull-off force upon separation.

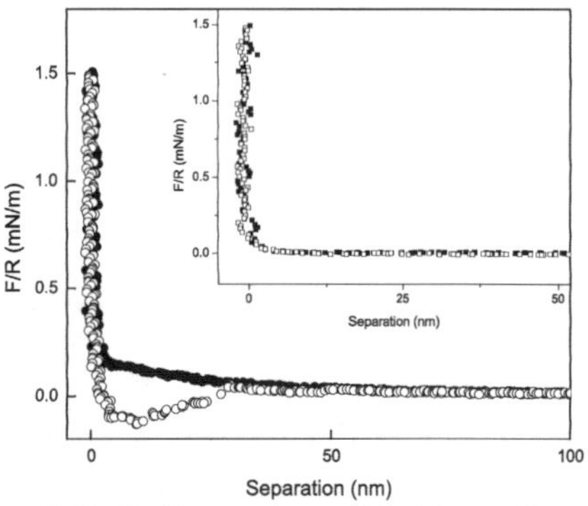

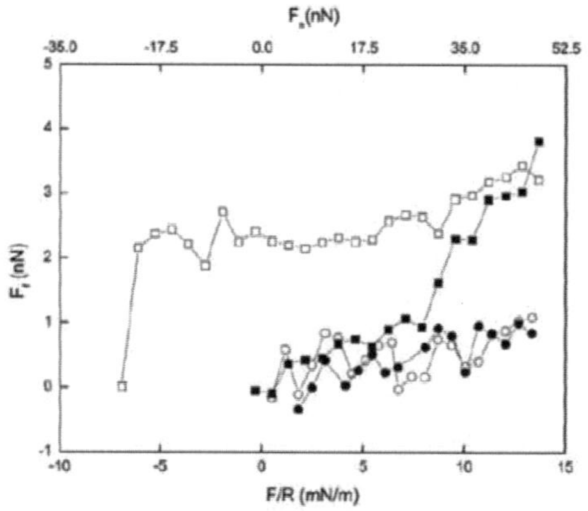

Figure 2. Left: Force normalized by radius as function of surface separation. Forces measured between silica surfaces covered by the DPPC bilayer with HA adsorbed on top of it. Inset: Forces measured between two silica surfaces covered by the pure DPPC bilayer. Right: Friction force as a function of load between silica surfaces coated by a DPPC bilayer (● loading and ○ unloading) and between DPPC bilayers with hyaluronan adsorbed on top (■ loading and □ unloading). Reproduced from ref. 12 with permission from The Royal Society of Chemistry.

Friction forces between DPPC bilayers immersed in 155 mM NaCl solution are very low, with friction coefficient of 0.01, and of very high load bearing capacity, sustaining pressures of at least around 68 MPa (Figure 2, right). Friction forces between DPPC layers with HA adsorbed are comparable to these between pure DPPC bilayers until a load of about 27nN is reached (in our experimental set-up corresponding to a pressure of 56 MPa) with a friction coefficient of 0.03. Upon further increase of the load, the friction force is rapidly increasing, thus we witness a disintegration of the DPPC – HA layer, and no recovery of the layer occurs on separation (Figure 2, right).

When the HA and DPPC are mixed in bulk solution at the ratio of 0.5 mg/mL DPPC and 0.5 mg/mL HA and are allowed to adsorb on silica surfaces for 40 minutes followed by rinsing, the force curves reflect the fact that the adsorbed layer is heterogeneous and there can be several adsorption layers or aggregates expelled from between the surfaces upon compression [14] (Figure 3, left.) Interestingly, the average friction force measured with such mixed layers adsorbed on the surfaces is low and the value of the friction coefficient is as low as 0.006, well comparable to that reported in synovial joints [15]. Thus, we suggest that the initially inhomogeneous layer re-arrange upon load and shear into phospholipid bilayers or multilayers separated by tenaciously attached though easily sheared water layers. The role of HA in this case is synergistic in that it facilitates the collection of lubricating phospholipids at the surfaces.

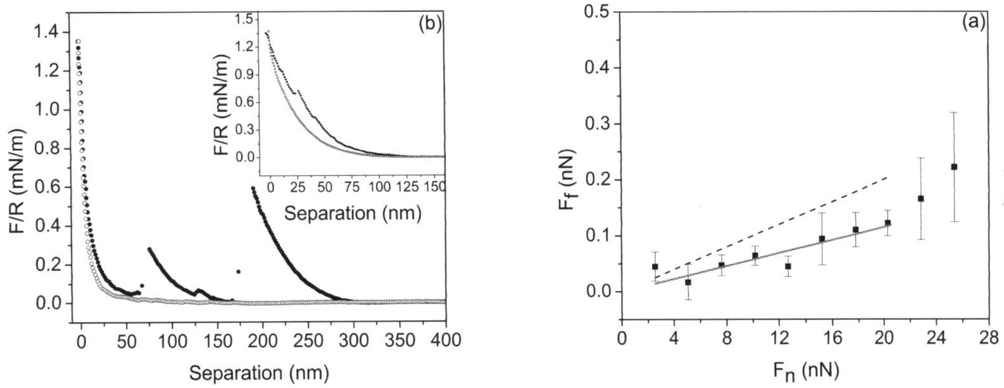

Figure 3. Left: Long-range repulsive forces measured upon first compression of the layers (•). Inset: the 10th approach (•) and separation (○). Right: Friction force as a function of load between two silica surfaces carrying a mixed DPPC-HA adsorption layer formed from solutions containing 0.5 mg/mL DPPC and 0.5 mg/mL HA in 155 mM NaCl and rinsed with 155 mM NaCl. The solid line has a slope of 0.006 and the dashed line of 0.01. The measurements were performed at 47 °C. Reproduced with permission from ref. [14] (doi.org/10.1016/j.jcis.2016.10.091) under terms of https://creativecommons.org/licenses/by-nc-nd/4.0/

3.2 polyCMA/polyMAPTAC and SDS

Cationic methacrylate-based polymers, such as polyCMA and polyMAPTAC, readily adsorb on negatively charged surfaces to compensate for their charge, see [16] and references therein. They also readily associate with anionic surfactants, as SDS. The association results in spontaneously organized structures both in bulk [16] and at interfaces [17] with the cylindrical SDS micelles in hexagonal arrangement that are stabilized by the polyelectrolyte wrapped around them [16]. The surface forces mediated by the polyCMA/polyMAPTAC and SDS aggregates are peculiar. They reflect the layering of cylindrical SDS aggregates at the surfaces and the forces oscillate from repulsive to attractive with the periodicity of that of an SDS micellar diameter of 40 Å [18] (Figure 4, left). Note, that very strong normal alternating repulsive and attractive forces operate between the surfaces. Similar normal forces can be detected also at silica surfaces pre-coated with polyMAPTAC and immersed in 1 cmc SDS solution [19]. The friction forces measured between polyMAPTAC coated surfaces in the presence of 1 cmc SDS solution in 1 x 10^{-4} M KBr are shown in Figure 4, right. These data are exciting as the friction force is very low up to a load of 9 mN m^{-1} (corresponding to a pressure of about 20 MPa), except for two transient peaks occurring at normalized forces of 3 - 4 mN m^{-1} and 8 mN m^{-1}. This friction peaks are due to restructuring of the aggregates under compression and shear. Thus, it is clear that aggregation structures mediate good surface lubrication – low friction combined with high load bearing capacity along with self-healing despite the presence of strong normal attractive forces. Thus, we encounter a case of synergistic lubrication brought about due to the structural arrangement of the aggregates that enable free flow of interspacing water layers that surround the surfactant headgroups. Note, that the individual components by themselves in this case are not good lubricants for negatively charged surfaces: SDS, being negatively charged, does not adsorb on negatively charged surfaces at all [16, 17 and references therein], and polyCMA/polyMAPTAC that do adsorb on negatively charged surfaces to compensate for their charge, generate strong bridging between the surfaces and the friction between such surfaces coated with individual polyelectrolytes is high [19].

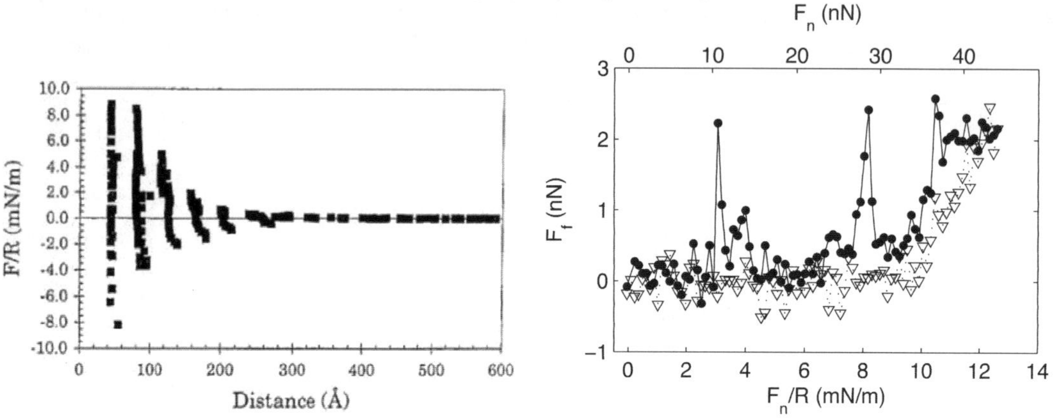

Figure 4. Left: Force normalized by radius as a function of surface separation determined using Surface Forces Apparatus. They were measured between mica surfaces pre-coated with polyCMA across a solution containing 1 cmc SDS and 10^{-4} M KBr. Reprinted with permission from ref. [18]. Copyright © 2000 American Chemical Society. Right: Friction force as a function of load between one mica and one silica surfaces pre-coated with polyMAPTAC and immersed in solution of 1 cmc SDS in 10^{-4} M KBr. Filled symbols designate compression and open symbols represent retraction of the surfaces. Reproduced from ref. [19] with permission from The Royal Society of Chemistry.

3.3 Lubricin and COMP

Lubricin is one of the biomacromolecules that has been suggested as being imperative to synovial joint lubrication. Indeed, lubricins molecular architecture with a dense highly hydrated polysaccharide brush in the core of the molecule and two anchoring naked protein endings (Figure 1e) appears to be conducive to good lubrication. Thus, lubricin's ability to lubricate model surfaces has been investigated but the results have indicated poor lubrication performance of lubricin alone [20]. One reason for this is that lubricin does not adsorb strongly enough onto the model surfaces used, and other molecules that anchor and structure lubricin into an ordered layer are needed [21], [22-24].

COMP is one of the biomolecules of cartilage that promotes anchoring of lubricin in a fashion that facilitates lubrication. Flowers et al. identified COMP-lubricin complexes in arthritic synovial fluid and found evidence for non-covalent binding between the C-terminal of COMP and the N-terminal of lubricin [25]. This is important, as it seems that COMP is the protein in the cartilage that is able to induce attachment and orientation of lubricin such that its lubrication ability is expressed optimally. Some data obtained for the COMP-lubricin synergistic pair are presented in Figure 5 [26]. It is seen in the inset of Figure 5 that the friction force when lubricin is directly adsorbed on polyMMA is high, and lubricin is eroded from the surface during shearing, as can be understood from the hysteresis between the measured curves on loading and unloading. The attachment of COMP on polyMMA is robust and no hysteresis is observed in the friction curves, but the friction is rather high. However, when lubricin is adsorbed on COMP-coated polyMMA surfaces, a durable layer is formed and no erosion is observed up to a pressure of 7 MPa along with significant reduction in friction (friction coefficient is 0.06 at the highest load applied). Thus there is orientational attachment synergy between

COMP and lubricin that enhances lubricins frictional performance that is inherent in its molecular architecture.

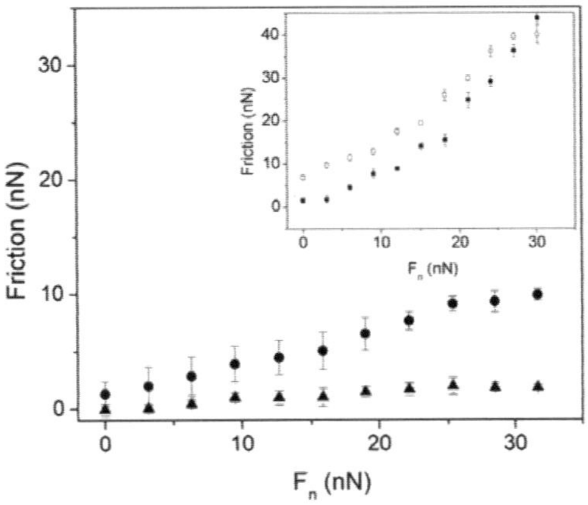

Figure 5. Main figure: Friction as function of load measured between two COMP-coated polyMMA surfaces (●), and COMP-lubricin coated surfaces (▲). Inset: Friction forces measured between two lubricin-coated polyMMA surfaces (loading ■, and unloading □). Reproduced with permission from ref. [27] (doi.org/10.1016/j.jcis.2017.02.007) under terms of https://creativecommons.org/licenses/by-nc-nd/4.0/

4. Conclusion

There are different ways to achieve surface lubrication in aqueous environment as good as that in mammalian joints. When using potent lubricants as phospholipids, it is imperative to supply an ample amount of the lubricant to the surface. Biopolymers, such as HA, can do this by operating as carriers for the lubricant and thus assuring the surface reservoir of the lubricant. To achieve good surface lubrication we do not even need to have a substance that is a good lubricant by itself. As long as the mixture of the components is self-assembled into aggregates that have a capacity to accommodate water layers that are tenaciously bound to the aggregates yet mobile along the surface, lubrication synergy is achieved. This is the mechanism of hydration lubrication. And finally, macromolecules with a blocky structure that provide hydration of the surface can also be used and here we find synergies in how one macromolecule (e.g. COMP) can aid adsorption of another (e.g. lubricin) that results in low friction.

References

[1] W. Hunter, VI. Of the structure and diseases of articu-lating cartilages, Philosophical Transactions 42(470) (1743) 514-521.
[2] B.A. Hills, Identity of the joint lubricant, The Journal of Rheumatology 29(1) (2002) 200-201.
[3] I.M. Schwarz, B.A. Hills, Surface-active phospholipid as the lubricating component of lubricin, British Journal of Rheumatology 37(1) (1998) 21-26.
[4] B.A. Hills, Surface-active phospholipid: a Pandora's box of clinical applications. Part I. The lung and air spaces, Internal Medicine Journal 32(4) (2002) 170-178.
[5] G.D. Jay, K.A. Waller, The biology of Lubricin: Near frictionless joint motion, Matrix Biology 39 (2014) 17-24.
[6] G.D. Jay, D.A. Harris, C.-J. Cha, Boundary lubrication by lubricin is mediated by O-linked beta(1-3)Gal-GalNAc oligosaccharides, Glycoconjugate Journal 18 (2001) 807-815.
[7] S.M. McNary, K.A. Athanasiou, A.H. Reddi, Engineering Lubrication in Articular Cartilage, Tissue Engineering, Part B: Reviews 18(2) (2012) 88-100.

[8] I.H. Park, E.J. Choi, J. Polymer 37 (1996) 313.
[9] S.Z. Can, C.F. Chang, R.A. Walker, Spontaneous formation of DPPC monolayers at aqueous/vapor interfaces and the impact of charged surfactants, Biochimica et Biophysica Acta (BBA) - Biomembranes 1778(10) (2008) 2368-2377.
[10] P.S. Cremer, S.G. Boxer, Formation and Spreading of Lipid Bilayers on Planar Glass Supports, The Journal of Physical Chemistry B 103(13) (1999) 2554-2559.
[11] T. Pettersson, A. Dedinaite, Normal and Friction Forces between Mucin and Mucin-Chitosan Layers in Absence and Presence of SDS, J. Colloid Interface Sci. 324(1-2) (2008) 246-256.
[12] C. Liu, M. Wang, J. An, E. Thormann, A. Dedinaite, Hyaluronan and phospholipids in boundary lubrication, Soft Matter 8(40) (2012) 10241-10244.
[13] J. Israelachvili, The interactions between lipid bilayers and biological membranes, Intermolecular and surface forces1992, pp. 395-421.
[14] A. Raj, M. Wang, T. Zander, D.C.F. Wieland, X. Liu, J. An, V.M. Garamus, R. Willumeit-Romer, M.L. Fielden, P.M. Claesson, A. Dedinaite, Lubrication Synergy: Mixture of Hyaluronan and Dipalmitoylphosphatidylcholine (DPPC) Vesicles, Journal of Colloid And Interface Science 488 (2017) 225-233.
[15] J. Klein, Molecular mechanisms of synovial joint lubrication, Proceedings of the Institution of Mechanical Engineers Part J-Journal of Engineering Tribology 220(J8) (2006) 691-710.
[16] M. Bergström, U.R.M. Kjellin, P.M. Claesson, J.S. Pedersen, M.M. Nielsen, A Small-Angle X-ray Scattering Study of Complexes Formed in Mixtures of a Cationic Polyelectrolyte and an Anionic Surfactant, J. Phys. Chem. B 106 (2002) 11412-11419.
[17] P.M. Claesson, A. Dedinaite, E. Blomberg, V.G. Sergeyev, Polyelectrolyte-Surfactant Association at Solid Surfaces, Ber. Bunsenges. Phys. Chem. 100 (1996) 1008-1013.
[18] A. Dedinaite, P.M. Claesson, M. Bergström, Polyelectrolyte-Surfactant Layers: Adsorption of Preformed Aggregates versus Adsorption of Surfactant to Preadsorbed Polyelectrolyte, Langmuir 16(12) (2000) 5257-5266.
[19] A. Dedinaite, T. Pettersson, B. Mohanty, P.M. Claesson, Lubrication by organized soft matter, Soft Matter 6(7) (2010) 1520-1526.
[20] B. Zappone, M. Ruths, G.W. Green, G.D. Jay, J. Israelachvili, Adsorption, Lubrication, and Wear of Lubricin on Model Surfaces: Polymer Brush-Like Behaviour of a Glycoprotein, Biophys. J. 92 (2007) 1693-1708.
[21] R.C.A. Eguiluz, S.G. Cook, C.N. Brown, F. Wu, N.J. Pacifici, L.J. Bonassar, D. Gourdon, Fibronectin mediates enhanced wear protection of lubricin during shear, Biomacromolecules 16(9) (2015) 2884-2894.
[22] S.E. Majd, R. Kuijer, A. Kowitsch, T. Groth, T.A. Schmidt, P.K. Sharma, Both hyaluronan and collagen type II keep proteoglycan 4 (lubricin) at the cartilage surface in a condition that provides low friction during boundary lubrication, Langmuir 30(48) (2014) 14566-72.
[23] D.P. Chang, F. Guilak, G.D. Jay, S. Zauscher, Interaction of lubricin with type II collagen surfaces: Adsorption, friction, and normal forces, Journal of Biomechanics 47(3) (2014) 659-666.
[24] S. Das, X. Banquy, B. Zappone, G.W. Greene, G.D. Jay, J.N. Israelachvili, Synergistic Interactions between Grafted Hyaluronic Acid and Lubricin Provide Enhanced Wear Protection and Lubrication, Biomacromolecules 14(5) (2013) 1669-1677.
[25] S.A. Flowers, S. Kalamajski, L. Ali, L.I. Björkman, J.R. Raj, A. Aspberg, N.G. Karlsson, C. Jin, Cartilage oligomeric matrix protein (COMP) forms protein complexes with synovial lubricin via non-covalent and covalnt interactions, Osteoarthritis Cartilage 25(9) (2017) 1496 - 1504.
[26] A. Raj, M. Wang, C. Liu, L. Ali, N.G. Karlsson, P.M. Claesson, A. Dedinaite, Molecular synergy in biolubrication: The role of cartilage oligomeric matrix protein (COMP) in surface-structuring of lubricin, Journal of Colloid and Interface Science 495 (2017) 200-206.

IOP Conf. Series: Materials Science and Engineering **500** (2019) 012031 doi:10.1088/1757-899X/500/1/012031

Thermal properties of polylactide / recycled lignin and cellulose filler biocomposites

O Platnieks, S Gaidukovs and A Barkane
Institute of Polymer Materials, Faculty of Materials Science and Applied Chemistry, Riga Technical University, P.Valdena Street 3/7, Riga, LV-1048, Latvia.

E-mail: oplatnieks@gmail.com, sergejs.gaidukovs@rtu.lv, barkaneanda@gmail.com

Bio-based polymer composites were prepared using poly-L-lactide (PLLA) as a matrix and three different types of natural fillers. Waste paper from regular office paper, cellulose from Tetra Pak® recycling industry and lignin waste from paper manufacturing industry were chosen as fillers. Density, crystallinity and melting/crystallization behavior of PLLA biocomposites loaded with different filler contents (20, 30, 40 wt.%) were compared and discussed.

1. Introduction

Polymer products based on renewable biomass feedstock including agricultural crops, forestry residues, biomass processing residues can form the basis for eco-efficient, sustainable and competitive solutions to substitute partially, and to some extent totally petroleum-based products on a technical and cost-performance values [1]. To further increase value and environmental impact of bio-economy, the use of sustainable polymer materials based on various bio-based oils, bio-based polymers derived from corn and starch and biopolymers lignin, cellulose are encouraged for further intensive investigations [2,3].
Cellulose and lignin are sustainable, biodegradable and relatively cheap materials for the bio-based materials preparation due to their availability from various waste streams, especially involving paper production industry. While cellulose is desirable filler due to its potential reinforcing properties to increase final mechanical characteristic of the composite's material [4].
We have chosen three filler types obtained as common waste products: waste paper filler (WPF) particles from recycled office paper, recycled cellulose filler (RCF) particles from Tetra Pak recycling industry and lignin waste filler (LWF) from paper production plant. Density measurements and differential scanning calorimetry analysis were performed for the obtained PLLA /recycled natural filler biocomposites.

2. Methodology

2.1. Materials and sample preparations

Poly-L-lactide with trademark PLLA-Biomer L9000 produced by Biomer® was chosen. The main parameters of PLLA are melt flow index 5.3 g/10 min, tensile strength 70 MPa, tensile deformation 2.4%, elastic modulus 3600 MPa.

LWF filler with the particles size bellow 0.25 mm, while WPF and RCF fillers with the particles size bellow 2 mm were used to produce biocomposites. RCF and WPF were used as received from the companies. Cellulose was separated from aluminum and polymer components during industrial processing by the recycling company. The PLLA and fillers were dried at 100 °C for 72 hours in a vacuum oven. Polymer biocomposites were prepared by blending with the thermoplastic mixer Brabender® Mixer 50EHT. PLLA biocomposites were processed with fillers contents of 20, 30, 40 wt%. PLLA biocomposites were obtained by melt blending at 170 °C for 5 min and screws velocity of 70 rpm. Pure PLLA sample was processed in the same conditions for comparison analysis.

2.2. Differential Scanning Calorimetry

Mettler DSC-1 instrument was used to investigate thermal properties. Specimens about 10 mg in weight were scanned in the temperature range from 25 to 250°C by using nitrogen as a purge gas. The DSC heating/cooling rate of 10°C/min was used. Enthalpies (ΔH), melting (T_m) and glass transition (T_g) temperatures were calculated from the experimental heating curves. Degree of crystallinity was calculated from the equation 1

$$\chi = \frac{\Delta H_{exp}}{\Delta H_{100\%}} \quad (1)$$

χ- degree of crystallinity of sample, where ΔH_{exp} is experimental melting enthalpy of crystalline phase, J/g; $\Delta H_{100\%}$ is theoretical melting enthalpy of PLLA crystalline phase, 93.7 J/g [5,6].

2.3. Density measurements

For density measurements a simple hydrostatic weighing equipment Sartorius YDK 01 was used. Ten measurements were performed for every sample composition and density was calculated from the equation 2

$$\rho = \frac{W_{(a)} \cdot (\rho_{(C_2H_5OH)} - 0.0012)}{0.99983 \cdot (W_{(a)} - W_{(C_2H_5OH)})} + 0.0012 \quad (2)$$

where, ρ – density of the sample (g/cm^3), $W_{(a)}$ – weight of the sample in air (g), $W_{(C_2H_5OH)}$– weight of the sample in ethanol (g), $\rho_{(C_2H_5OH)}$– density of ethanol (g/cm^3).

3. Results and discussions

3.1. Density measurements

Density data for PLLA /recycled lignin and cellulose filler biocomposites are compared with pure PLLA density 1.219 g/cm^3 and the results are summarized in Figure 1.

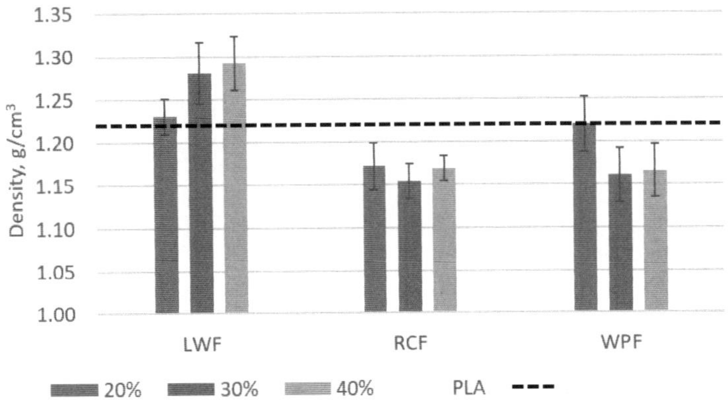

Figure 1. Density of PLLA biocomposites with lignin and cellulose fillers in comparison to pure PLLA

The lignin filler loading increases density of composites materials; while the cellulose fillers loading decreases the density of composites materials. For example, at the concentration of 40 wt % of the filler, the lignin composite density is 1.292 g/cm^3, but the density of the waste paper composite is only 1.165 g/cm^3. Observed density decrease can be explained with the uneven melt blending and formation of the voids in the composite material at high filler loading concentrations.

3.2. Differential Scanning Calorimetry

The second heating and cooling DSC thermograms of the processed PLLA compositions are shown in Figure 2. The characteristic parameters are calculated and summarized in the Table 1. The cooling and heating curves indicate that the addition of lignin and cellulose fillers slightly effect the glass transition temperature T_g of the polymer matrix. The T_g of the PLLA is 62.6 °C, while the T_g of the composite filled with 20 wt% of WPF is 64.7 °C. This could be attributed to the possible some interactions between hydroxyl groups of cellulose with polymer chains through hydrogen bonding development, which can reduce the overall flexibility of the polymer chain mobility [7,8].

The obtained results show that the melting temperature T_m of the composites with the fillers higher loading decreases. T_m of the pure PLLA is 168.8 °C, while the T_m of the PLLA biocomposites loaded with 40 wt % of RCF and WPF are 166.4 °C and 168.0 °C, respectively.

Table 1. Thermal properties obtained from second heating and cooling DSC thermograms

	Heating					**Cooling**		
	T_g (°C)	T_{cc} (°C)	χ_{cc} (%)	T_m (°C)	χ_m (%)	T_g (°C)	T_c (°C)	χ_c (%)
PLLA	58.6	98.3	23.4	168.8	53.1	62.6	95.4	4.8
LWF 20%	58.8	105.9	30.3	169.0	43.8	63.0	95.9	3.0
LWF 30%	58.8	103.3	20.8	168.7	48.9	62.3	95.9	7.0
LWF 40%	58.4	102.0	19.2	168.1	49.4	61.3	96.8	10.5
RCF 20%	57.9	98.6	7.9	168.7	48.3	62.9	100.3	25.0
RCF 30%	58.8	-	-	168.8	52.7	63.0	117.5	41.2
RCF 40%	58.0	-	-	166.4	47.8	-	128.1	41.3
WPF 20%	57.5	-	-	169.1	46.4	64.7	100.6	34.6
WPF 30%	57.4	-	-	169.0	42.6	-	103.9	35.4
WPF 40%	-	-	-	168.0	35.4	-	103.5	28.6

IOP Conf. Series: Materials Science and Engineering **500** (2019) 012031 doi:10.1088/1757-899X/500/1/012031

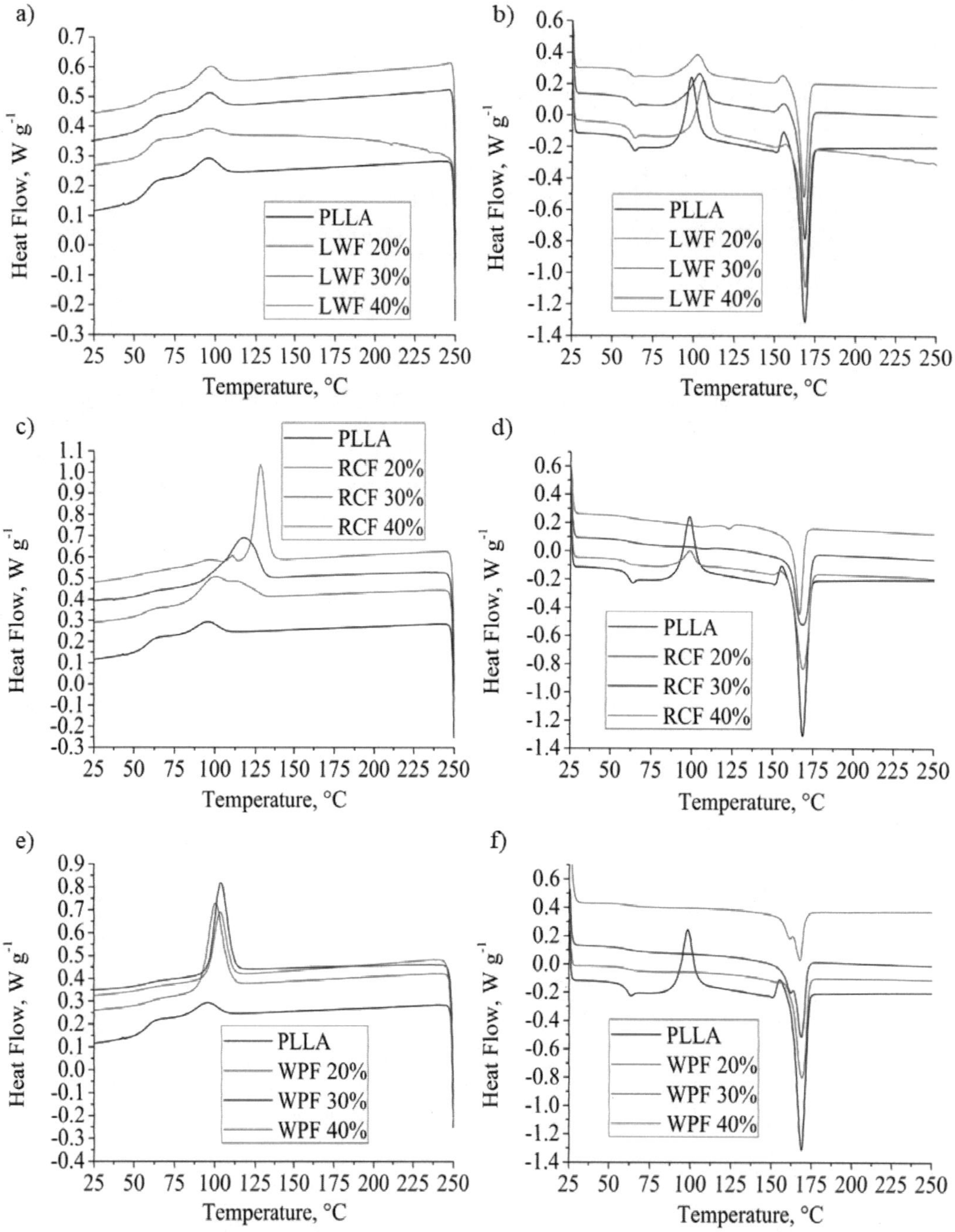

Figure 2. Heating and cooling thermograms of the biocomposites: a) cooling curves of LWF biocomposites, b) heating curves of LWF biocomposites, c) cooling curves of RCF biocomposites, d)

heating curves of RCF biocomposites, e) cooling curves of WPF biocomposites, f) heating curves of WPF biocomposites.

Most notable changes are observed for the melting and crystallization of the PLLA. Pure PLLA is characterized from the 2nd heating thermogram with the degree of the crystallinity (χ_m) – 53.1%, cold crystallization process (T_{cc}) at about 98.3 °C and degree of the cold crystallization (χ_{cc}) – 23.4%, but during cooling the polymer is characterized by almost amorphous structure and degree of the crystallinity (χ_c) – 4.8%. Lignin filler containing biocomposites also show the cold crystallization process during the cooling; while the other biocomposites show the absence of it. Crystallinity degree for the cold crystallization process decrease to 20.8% and 19.2% for 30 and 40 wt% loading of LWF. PLLA biocomposites with the cellulose based fillers RCF and WPF show almost no cold crystallization process. During cooling stage PLLA filled with the 40wt% of RCF and WPF fillers show degree of crystallinity 41.3% and 28.6%. Crystallization temperature of the PLLA biocomposites increase from 95.4 °C for the pure PLLA till 128.1 °C for the PLLA biocomposites filled with the 40 wt% of RCF. The increase of degree of crystallinity during cooling can be explained with nucleating effect of fillers towards PLLA matrix, which is especially strong for cellulose fillers RCF and WPF. This could be explained with very strong intermolecular interactions developed between the composite's components [7,8,9].

3. Conclusions

The PLLA biocomposites density increases after loading with lignin filler; while, PLLA biocomposites density decreases after loading with cellulose fillers. This behavior can be related to the formations of the voids at very high filler loading contents. Differential scanning calorimetry analysis show that cellulose fillers slightly effect the glass transition temperature, which decreases by almost 2 degrees. The fillers incorporation into the polymer matrix also increases the crystallinity degree of the PLLA biocomposites. Finally, it can be concluded that the cellulose fillers show more pronounced interaction with polymer matrix in comparison with the lignin filler, which relates to the nature of the fillers.

4. References

[1] Madbouly S, Zhang C, Kessler M R 2016 *Bio-Based Plant Oil Polymers and Composites* 10-217

[2] Gaidukova G, Ivdre A, Fridrihsone A, Verovkins A, Cabulis U and Gaidukovs S 2017 *Ind Crops Prod* **102** 133–143

[3] Garrison T F, Murawski A and Quirino R L 2016 *Polymers* **8(7)** 262

[4] Sutka A, Sutka A, Gaidukov S, Timusk M, Gravitis J and Kukle S 2015 *Holzforschung* **69(6)** 737-743

[5] Leng J, Kang N, Wang D Y, Wurm A and Schick C 2017 *A. Polymer (Guildf)* **108** 257-264

[6] Saeidlou S, Huneault M A, Li H and Park C B 2012 *Prog. Polym. Sci.* **37 (12)** 1657-1677

[7] Ghanbaria A, Tabarsaa T, Ashorib A, Shakeric A and Mashkour M 2018 *Carbohydr Polym.* **197** 305-311

[8] Lei W, Fanga C, Zhoua X, Li Y and Pub M 2018 *Carbohydr Polym.* **197** 385-394

[9] Gaidukov S, Danilenko I and Gaidukova G 2015 *Int J Polym Sci* art. no. 123469

IOP Conf. Series: Materials Science and Engineering **500** (2019) 012032 doi:10.1088/1757-899X/500/1/012032

Different epoxide compound influence on two component silyl-terminated polymer/epoxide systems mechanical, rheological and adhesion properties

Ritvars Berzins[1,2], Remo Merijs Meri[1], Janis Zicans[1]

[1]Institute of Polymer Materials, Faculty of Materials Sciences and Applied Chemistry, Riga Technical University, Latvia
[2]TENACHEM Ltd, Dobele, Latvia

remo.merijs-meri@rtu.lv

Abstract. Polyether silyl-terminated pre-polymer (SIL) and epoxy resin (EP) based two component systems are most perspective alternatives to the dominant polyurethane adhesive and sealant materials. This study focuses on the influence of the amount of epoxide component on the rheological, indentation, tensile and adhesion properties of SIL/EP composites within a whole range of the component weight to weight ratios. Results show that epoxide content and structure both considerably affect all the investigated properties of the material. Depending on its composition, the obtained materials demonstrate tensile strength in the range of 0.22-4.06 MPa, tensile deformation in the range of 25-1527 % and Shore A hardness in the range of 4-96.

1. Introduction

Sealant and adhesive market are constantly growing, because of new material development, increasing quality requirements and environmental issues. Despite wide range of adhesives in the market, only few materials (mostly polyurethane type) simultaneously can provide good adhesion to various substrates, high tensile strength (>3 MPa) and deformation (>100%). Previous publications [1-2] showed that two component silyl-terminated polymer /epoxy resin systems (SIL/EP) are able to achieve the above mentioned properties. Consequently, in the current research attention is paid to revealing the effects of various structurally different EP addition on the rheological, indentation, tensile and adhesion properties of SIL based two-component systems.

2. Experimental methods

Formulations of component A, based on SIL (SAX 520), and component B, based on structurally different EPs (bisphenol A epoxy resin (D.E.R. 331), hexion cycloaliphatic epoxy resin (Eponex 1510), tetra-functional aliphatic epoxy compound (D.E.R. 749), mono-functional aliphatic epoxy compound (D.E.R. 721)), are summarized in Table 1, where the particular epoxide groups containing raw material is labelled as epoxy compound.

A and B components were made by using 3-liter laboratory mixer TEJA Engineering. First, all liquid raw materials (resin/pre-polymer, plasticizer, adhesion promoters, drying agent) were stirred for 5 minutes at 1000 (dissolver)/ 5 (planetary) rpm, then fillers were added and the composition was stirred

20 minutes at 3500 (dissolver)/ 35 (planetary) rpm, then mixing was continued under vacuum for 30 minutes at 3500 (dissolver)/ 35 (planetary) rpm.

Table 1. Two-component SIL/EP formulations (in weight percentage)

Polymers percentage	100/0	90/10	80/20	70/30	60/40	50/50	40/60
A component							
SAX 520 (Kaneka Belgium)	40	36	32	28	24	20	16
Hexamol DINCH (BASF)	15.0	13.5	12.0	10.5	9.0	7.5	6.0
Dynsylan 1189 (Evonik)	1	1	1	1	1	1	1
Omycarb 1T (Omya)	17.13	15.00	13.80	12.00	10.30	8.60	6.80
Hakuenka CCR-S10 (Omya)	25.0	22.5	20.0	17.5	15.0	12.5	10.0
Dynasilan VTMO (Evonik)	1.0	0,9	0,8	0,7	0,6	0,5	0,4
Lupragen N600 (Air products)		0,62	1,33	2	2,66	3,33	4
B component							
Epoxy compound (see in text)	0	4	8	12	16	20	24
Hexamol DINCH (BASF)	0	1.5	3.0	4.5	6.0	7.5	9.0
Omycarb 1T (Omya)	0	1.70	2.38	3.69	4.92	6.14	7.45
Hakuenka CCR-S10 (Omya)	0	2.5	5.0	7.5	10.0	12.5	15.0
Tibcat 216 (Tibchemicals)	0.20	0.18	0.16	0.14	0.12	0.10	0.08
Water	0.67	0.60	0.53	0.47	0.40	0.33	0.27

SIL/EP compositions were mixed using SpeedMixer DAC 150 centrifugal laboratory mixer at room temperature 23 °C and 3000 rpm, casted in Teflon moulds and cured at standard conditions – 23±2 °C, 50 ±5% RH – for 1, 7 and 28 days.

Shore A hardness was measured after 1, 7 and 28 days of curing by using durometer (SCHMIDT) according to ISO 7619.

Rheological measurements were performed by using Bohlin CVO 100 rheometer. Instrument was equipped with 20 mm diameter spindle with plate-plate geometry (gap size 1000 μm). Tests at 25 °C were performed in oscillation mode – frequency 1 Hz and strain 0.006.

Tensile stress-strain measurements were done by using Zwick /Roell Z010 universal testing machine. Tests were made according to ISO 527 at test speed 100 mm/min. Tensile strength measurements were made after 1, 7 and 28 days of curing.

Lap shear test was performed by using different substrates: stainless steel, PVC, wood (ash tree). Test was made according to EN 1465. Specimen overlap area was 12.5*25.0 mm and material thickness was 0.2 mm.

3. Results and discussion

3.1. Mechanical properties.

After 1, 7 and 28 days of curing systems show wide range of tensile strength values. In previous publication [2] it was shown that the most effective SIL/EP ratio from adhesion and aging perspective is 80/20. In this publication we will pay special attention to the corresponding polymer composition.

Tensile strength data (Fig. 1) show that addition of even 10-20 wt.% of epoxy group containing compound can cardinally change material properties. Thus after 28 days of curing of the investigated systems at SIL/EP ratio 80/20 following tensile strength values are obtained: 3.05 MPa (for D.E.R. 331 containing systems), 2.75 MPa (for D.E.R. 749 containing systems), 1.75 MPa (for D.E.R. 721 containing systems) and 3.78 MPa (for Eponex 1510 containing systems). SIL/EP compositions with three of four EPs, namely D.E.R. 331, D.E.R. 749 and Eponex 1510, show similar tendencies throughout all hardening time. From the three previously mentioned, the systems with Eponex 1510 show some interesting properties. Because of alkyl type structure of Eponex 1510 it has lower reactivity comparing to bisphenol A type epoxy compounds (this effect is observed also from rheological data). Because of this effect Eponex 1510 containing system has different σ_{break}(SIL/EP ratio) relationship than that for the systems with D.E.R. 331 and D.E.R. 749. Continuous progress of the curing reaction between Eponex 1510 and SIL, especially in the systems with higher EP content, is confirmed by rising tensile strength value as curing time is increased: if after 1 day of curing minimum in σ_{break}(SIL/EP ratio) relationship is observed along SIL/EP ratio 70/30-40/60, than by increasing curing time this minimum range is getting narrower, namely tensile strength minimum is observed at SIL/EP ratio of 70/30 – 50/50 (7 days of curing) and 70/30 (28 days of curing). It can be hypothesized that by increasing curing time even more, tensile strength minimum will be lost and σ_{break}(SIL/EP ratio) relationship will show maximum values in the range 70/30- 40/60. In contradiction to other epoxy compounds, mono-functional D.E.R. 721 containing system show completely different trend of σ_{break}(SIL/EP ratio) relationship. Tensile strength decrement at EP contents above 10 wt.% of the systems with D.E.R. 721 can be explained with mono-functional chain of D.E.R. 721, preventing effective cross-linking within the system.

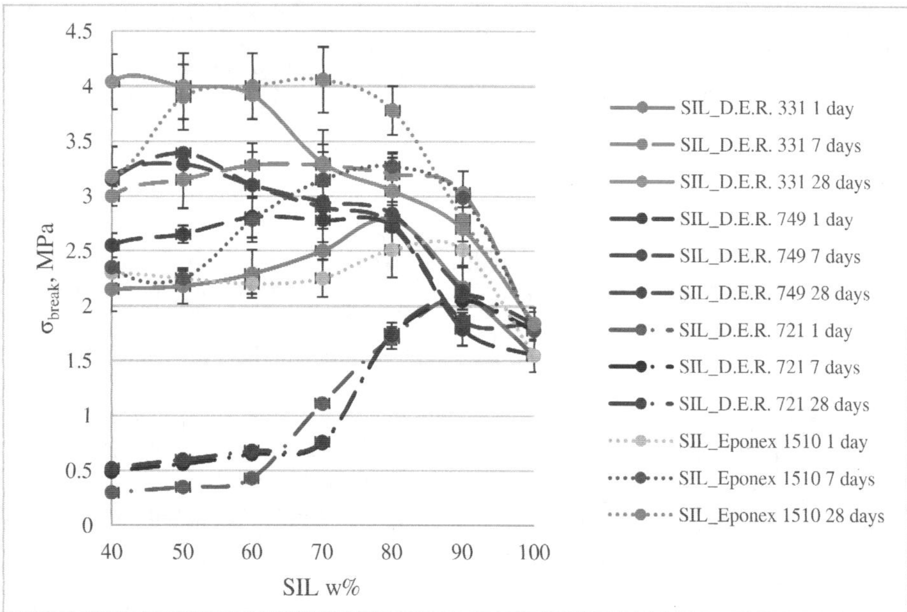

Figure 1. Tensile strength at break (σ_{break}) as a function of SIL weight content

Tensile deformation data make good correlation with tensile strength data. The systems, containing D.E.R. 331, 749 and Eponex 1510, show similar results after curing for 7 and 28 days. Considerably higher tensile deformation values of the system with Eponex 1510 after 1 day of curing are most probably explained with lower curing rate. Addition of D.E.R. 721 to SIL results in sharply increased tensile deformation values, especially after 1 day of curing of the 80/20 SIL/EP system. By increasing the curing time up to 7 days, tensile deformation values of D.E.R. 721 containing systems may even increase due to the increment of overall toughness of the system, especially at higher EP contents (60/40-40/60 SIL/EP ratio). However, at 28 days of curing this this effect is somewhat reduced, especially at lower EP contents (60/40-70/30 SIL/EP ratio), due to the development of tighter cross-linked network.

In comparison to tensile strength at break and ultimate elongation, showing complex non-monotonous relationships in respects to SIL/EP content, Shore A hardness as function of SIL/EP ratio of the investigated compositions is characterized by almost linear relationships. This could be probably attributed to the fact that tensile stress-strain behaviour characterizes materials in bulk, while indentation related Shore A hardness characterizes materials surface, where hardening occurs at the first. Interestingly, that in the case of D.E.R. 331, D.E.R. 749 and Eponex 1510 containing systems Shore A hardness increases along with EP content, concomitant in the case of D.E.R. 721 decrease in Shore A hardness values is observed. This could be explained with distinct curing dynamics in the case of mono-functional (D.E.R. 721) and bi- (D.E.R. 331 and Eponex 1510) or tetra-functional (D.E.R. 749) epoxy compounds. By increasing the content of mono-functional epoxy compound, which cures mainly by homo-polymerization, SIL/EP two-component system is not efficiently cross-linked, as indirectly confirmed also by low tensile strength at break and ultimate deformation values. In contradiction, in the case of bi- and tetra-functional EPs, effective cross-link network is developed resulting in increased values of Shore A hardness, concomitant to tensile strength at break and ultimate hardness. Increase in curing time causes increment in Shore A hardness of all the investigated systems, irrespectively from the epoxy compound used.

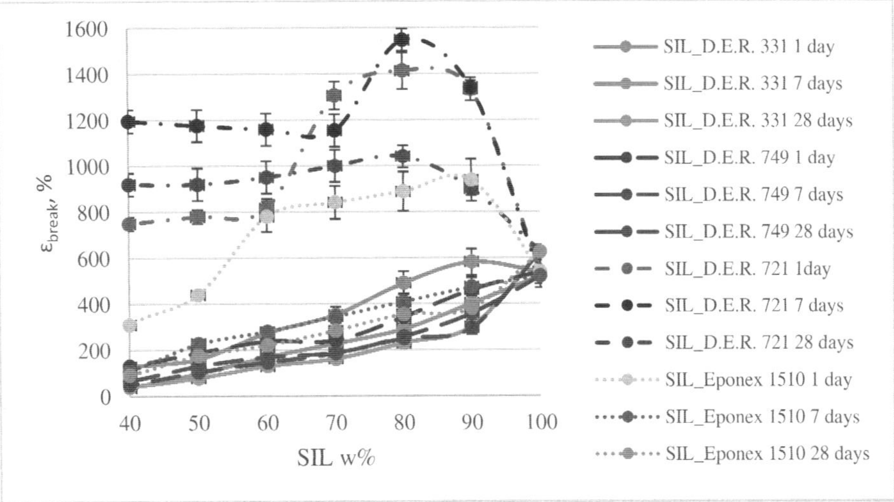

Figure 2. Tensile deformation at break (ε_{break}) as a function of SIL weight content

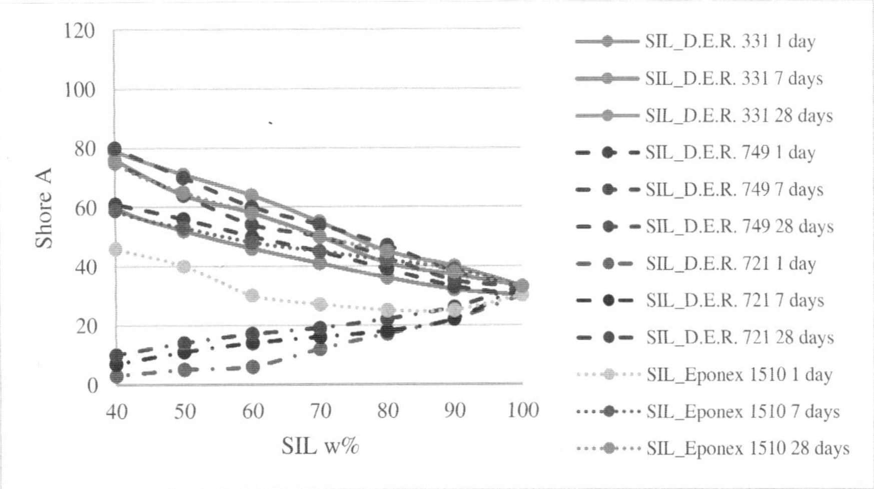

Figure 3. Shore A hardness value as a function of SIL weight content

3.2. Rheological characteristics.
Viscosity-time relationships of the investigated epoxy compounds and respective SIL/EP compositions at 80/20 ratio are demonstrated in Fig. 4 and Fig. 5 respectively. The composition with SIL/EP ratio 80/20 is chosen, because in previous publication [2] it was shown that this particular concentration was the most effective from adhesion perspective.

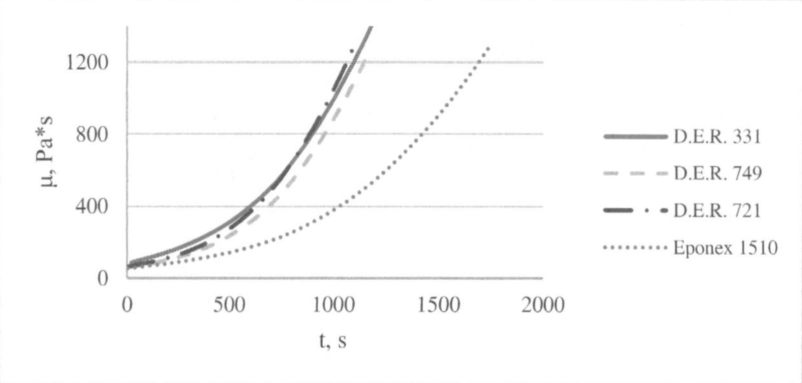

Figure 4. Viscosity change in time of two component SIL/EP system

Results show that three of four epoxide type raw materials (D.E.R. 331, D.E.R. 749 and D.E.R. 721), when mixed with SIL, influence viscosity growth of the respective two-component systems almost in the same way, testifying that these epoxy compounds influence SIL curing similiary. Viscosity growth rate of Eponex 1510 containing SIL is considerably smaller in comparison to the rest of the recipes, i.e., time at cross-over point of Eponex 1510 containing system is ca 1500 s in respect to ca 800-1000 s, characteristic for other investigated two-component systems. Besides it one can observe that cross-over moluli of D.E.R. 331, D.E.R. 749 and Eponex 1510 containing systems are higher than that of D.E.R. 721 containing system, denoting to the development of more resistant 3D network in the former case, which is also consistent with tensile strength at break values of the respective SIL/EP two-component systems.

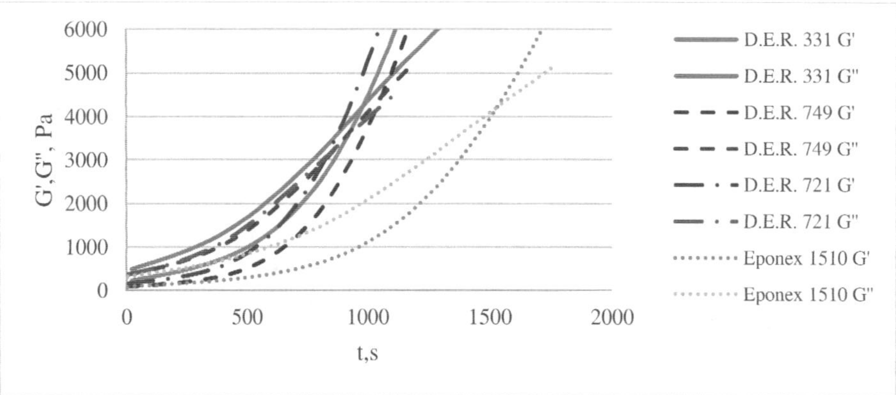

Figure 5. Elastic and viscous moduli growth in time of two component SIL/EP system

3.3. Lap shear test.
Lap shear test results show that addition of three (D.E.R. 331, D.E.R. 749 and Eponex 1510) of four EPs to SIL at ratio of 20/80 ensures good adhesion to stainless steel and wood. Eponex 1510 containing system possesses the highest shear strength values to the both previously mentioned substrates. Let's

mention that this composition demonstrates also the highest values of tensile strength at break and cross-over modulus. It is especially interesting to note that D.E.R. 749 containing system ensures also good adhesion to PVC after 28 days of curing, although its overall shear strength values to stainless steel and wood are somewhat lower in comparison to those with D.E.R. 331 and Eponex 1510. Better adhesion of D.E.R. 749 containing systems to PVC could be explained by considering its greater polarity in comparison to other epoxy compounds, considered in this research. It is also hypothesized that shear strength to stainless steel and wood as well as tensile strength at break of this high functionality (4) epoxy compound containing systems potentially could be increased by using structurally similar 3-functional epoxy compound due to the reduced density of the 3D network and hence increased evolution of tensile strength because of greater flexibility of the structure. In contradiction to other investigated systems the system with D.E.R. 721 shows low adhesion to all the substrates, which is one more indication that mono-functional epoxy compounds can't ensure the necessary adhesion to the substrates, highly attractive in the adhesives and sealants market, because they tend to block chain ends of the SIL.

Table 2. Lap shear test results of the investigated SIL/EP 80/20 compositions to various substrates

Substrate	Days	Epoxide type							
		D.E.R. 331		D.E.R. 749		D.E.R. 721		Eponex 1510	
		σ_{break} [MPa]	ε_{break} [%]	σ_{break} [MPa]	ε_{break} [%]	σ_{break} [MPa]	ε_{break} [%]	σ_{break} [MPa]	ε_{break} [%]
PVC	7	0.16	1.52	0.33	5.26	0.44	36.0	0.29	8.13
	28	0.21	9.00	3.51	42.0	0.50	32,0	0.38	4.69
Stainless steel	7	5.78	17.1	4.26	20.4	0.48	20,7	5.95	23.6
	28	5.50	14.5	4.34	27.1	0.47	30.0	5.81	36.0
Wood	7	4.20	21.1	3.32	32.4	0.28	16.0	4.43	36.4
	28	4.30	30.0	3.52	30.8	0.49	30.2	4.62	26.3

4. Conclusions

The influence of the addition of various structurally different epoxy compounds (mono-functional D.E.R. 721, bi-functional D.E.R. 331 and Eponex 1510 and tetra-functional D.E.R. 749) on the rheological, indentation, tensile and adhesion properties of polyether silyl-terminated pre-polymer two-component systems is investigated.

It has been revealed that:

- epoxide content and structure both significantly affect mechanical properties of the silyl-terminated polymer composite - values of the tensile strength at break change in the range of 0.22-4.06 MPa, tensile deformation in the range of 25-1527% and Shore A hardness in the range of 4-96;
- curing speed of the three of four epoxide compounds is practically the same, except of Eponex 1510, which cures slower;
- D.E.R. 749 containing composition ensures adhesion to all the tested substrates, including stainless steel, wood and polyvinylchloride, showing shear strength above 3 MPa after 28 days of curing;
- D.E.R. 721 containing system shows decreased adhesion to all the substrates, but because of high elongation values these compositions are perspective to create materials with exceptionally high flexibility.

References

[1] Bērziņš R, Merijs-Meri R, Zicāns J 2016 *Proc. Int. Conf. Riga Technical University 57th International Scientific Conference "Materials Science and Applied Chemistry"* (Riga: RTU Press) pp 31-35

[2] Berzins R, Merijs Meri R, Zicans J 2017 *Solid State Phenom* vol 267 (Zurich: Trans Tech Publications Inc.) pp 28-34

IOP Conf. Series: Materials Science and Engineering **500** (2019) 012033 doi:10.1088/1757-899X/500/1/012033

Effect of rheological properties of composition based on modified starch on film formation

I Resnytskiy [1], O Ishchenko [1], V Plavan [1], M Koliada [1], V Valeika[2]

[1] Department of Applied Ecology, Technologies of Polymers and Chemical Fibers, Kyiv National University of Technologies and Design, 2, Nemirovich-Danchenko str., Kyiv 01011, Ukraine
[2] Department of Physical and Inorganic Chemistry, Kaunas University of Technology, 73, K. Donelaicio str., Kaunas LT-44029, Lithuania

e-mail: ilya9res@gmail.com

This study investigated the use of carboxymethylated corn starch with sodium alginate for films preparation. The rheological properties as well as water absorption and mechanical properties of aqueous solutions and films of corn starch, carboxymethylated starch, sodium alginate and mixtures of starch and sodium alginate were studied. It was found that application of modified starch allows obtaining of water resistant films characterized by better physical-mechanical properties comparatively to the corn starch based films. It was found that mechanical strength of the formed films is due to the fact that alginate and carboxymethylated corn starch are hydrocolloids with an unbranched structure. This composition is characterized by stronger structure gels and higher viscosity of the solutions.

1. Introduction

Recently, in the field of medicinal preparations, obtaining of additional therapeutic properties by introducing other polymeric material in dressing systems became very popular [1, 2].

The ideal polymer coating for wounds and burns should be characterized by the following properties: protective (prevents the penetration of infection from the outside, prevents from external mechanical injury), absorptive and healing (coatings should block the development of infection in the wound, have anesthetic and hemostatic effects, and stimulate healing of wounds).

Classic polymers that are widely used in medicine as the main component of dressing agents are cellulose and other derivatives of polysaccharides. At the same time, the greatest interest are protective coatings based on natural hydrocolloids (agar-agar, pectin, carrageenan, starch, cellulose and alginates) [3].

Currently, among the new wound bandages, composite materials predominate and are used in the form of gels, films and applied by the method of "printing" on the textile basis of plaster [4].

Polymers that have antimicrobial properties represent a valuable alternative to conventional antibiotics and is currently gaining interest in coatings, personal care, active food packaging, and biomedical applications. Typical antimicrobial agents used with polymers are organic or inorganic acids, metals, alcohols, ammonium compounds or amines. However, due to the low molecular weight of these antimicrobial agents, their retention capacity has been reported to be poor causing them to leach out when applied directly to a substrate or polymer system, thereby inhibiting their antimicrobial performance. The interest in starch as an antimicrobial agent carrier is accrued from its film-forming properties, and high molecular weight.

Native starch is hydrophilic in nature, insoluble in water at room temperature. Furthermore, it cannot be melt-processed as it degrades under relatively low temperature and lacks mechanical integrity. These shortcomings have limited its use for polymer applications requiring mechanical strength and thermal stability, especially in the plastic industry where starch should be melt processed in most cases. Also, the poor solubility of starch in cold water limits its potential applications.

Thus, starch modification is desirable not only to mitigate these challenges but also to bring about other functional properties. Some of the properties that can be achieved via starch modification include thermal stability, hydrophobicity, amphiphilicity, paste clarity, mechanical strength, freeze-thaw stability and others [5-10]. Several starch modification processes are reported in literature including, physical processes [11, 12], chemical modifications [13–16], and enzymatic approaches [17–19] or their combinations.

This work aimed to study the possibility of using polysaccharides to develop a film having healing properties by casting the mixture of biopolymers: sodium alginate and carboxymethylated corn starch.

2. Materials and methods

2.1. Materials

All materials were supplied at Sigma Aldrich as following grades: Corn starch (CAS 9005-25-8), sodium alginate (CAS 9005-38-3), chloroacetic acid (CAS 79-11-8), acetic acid (CAS 64-19-7), sodium hydroxide (CAS 1310-73-2), and glycerol (CAS 56-81-5).

2.2. Preparation of carboxymethylated (modified) starch (Na-CS)

Composition recipe, %
Corn starch 13.7
Water 76.6
Chloroacetic acid 3.1
Sodium hydroxide 6.6

Corn starch had been dissolved in deionized water to a milk-like state, afterwards, pre-dissolved in water chloroacetic acid was added. After stirring, sodium hydroxide was added as 40 % solution. Synthesis was carried out at 50-60 °C for 30 minutes. After cooling down to 20 °C, the viscous solution was neutralized by 30 % acetic acid to pH 8.

2.3. Film formation

Films were produced by casting method of the colloidal solution. The resulting mixture has the appearance of a viscous solution, which was applied onto the glass plate. Then the samples were further dried at 40-50 °C, after, were thermostabilized at 90-110 °C.

2.4. Rheological properties

The rheological properties of the samples were performed on the rheometer "Brookfield" DV-III (USA) using a thermoplatform block with a temperature range of 23-25 °C.

2.5. Water absorption

For the study of water absorption samples 10×10 mm having 1 mm thickness we prepared, they were weighed to an accuracy of 0.0001 g. Measurements were made after a certain period of time (10, 20, 30, 40 minutes). Water absorption of the sample, X (%), was calculated by the formula:

$$X = (m - m_o) / m_o \cdot 100\% \quad (1)$$

where: m - mass of the sample at a certain time of stay in water, g;
m_o - mass of the dry specimen, g.

2.6. Mechanical properties

Tensile strength and elongation at break were determined by the ASTM D882-18, Standard Test Method for Tensile Properties of Thin Plastic Sheeting [20] at a humidity of 45%, 65%, 80%.

3. Results and discussion

In this work, the effect of the addition of starch and modified starch on the viscosity of sodium alginate was studied, the influence on the mechanical properties of films at 45%, 65%, and 80% humidity was determined and the water absorption of films based on the composition of biopolymers was determined. Concentrations were adjusted as well as ratios for sodium alginate and modified starch in compositions (table 1). The glycerin as plasticizer was used in order to improve the mechanical properties (elasticity) of the films.

Typically, for preparing of the compositions, 10-15 % aqueous solutions of polysaccharides are used. Studies of the low concentration solutions (3-7 %) are required for better understanding of the mechanism of changing of the starch viscosity by different shear stresses. Significance of the flow index of the two-component system (starch-sodium alginate, modified starch-sodium alginate) are within the range of 0.303-0.445 by the Ostwald-de-Waele equation.

Table 1. Composition of films based on starch and Na-CS with sodium alginate, %*

Sodium alginate	Corn starch	Na-CS
7.5	2.5	–
7.5	–	2.5
5.0	5.0	–
5.0	–	5.0
2.5	7.5	–
2.5	–	7.5
10	–	–
–	10	–
–	–	10

*Note: Deionized water, % - 88; Glycerol, % - 2.

The rheological properties of the material were determined by parameters of temperature, pressure, stress or shear rate using a rheometer. The obtained results (fig 1.) are used to assess the storage stability of the substance, melting temperature, solidification temperature, shear stability, molecular weight, quality of production, chemical, mechanical and thermal processing of this material.

It was found that the composition based on the studied polysaccharide refers to structured systems with anomalously viscous properties, i.e. when increasing shear stress, the effective viscosity decreases: the viscosity of such composition in the coordinates $\eta = f(\gamma)$ obeys a power law and is described by the Ostwald-de-Waele equation. The values of the flow index of the two-component compositions are in the range of 0.29-0.45 and have a complex of properties predicting the possibility of their effective use for the production of films by casting.

IOP Conf. Series: Materials Science and Engineering **500** (2019) 012033 doi:10.1088/1757-899X/500/1/012033

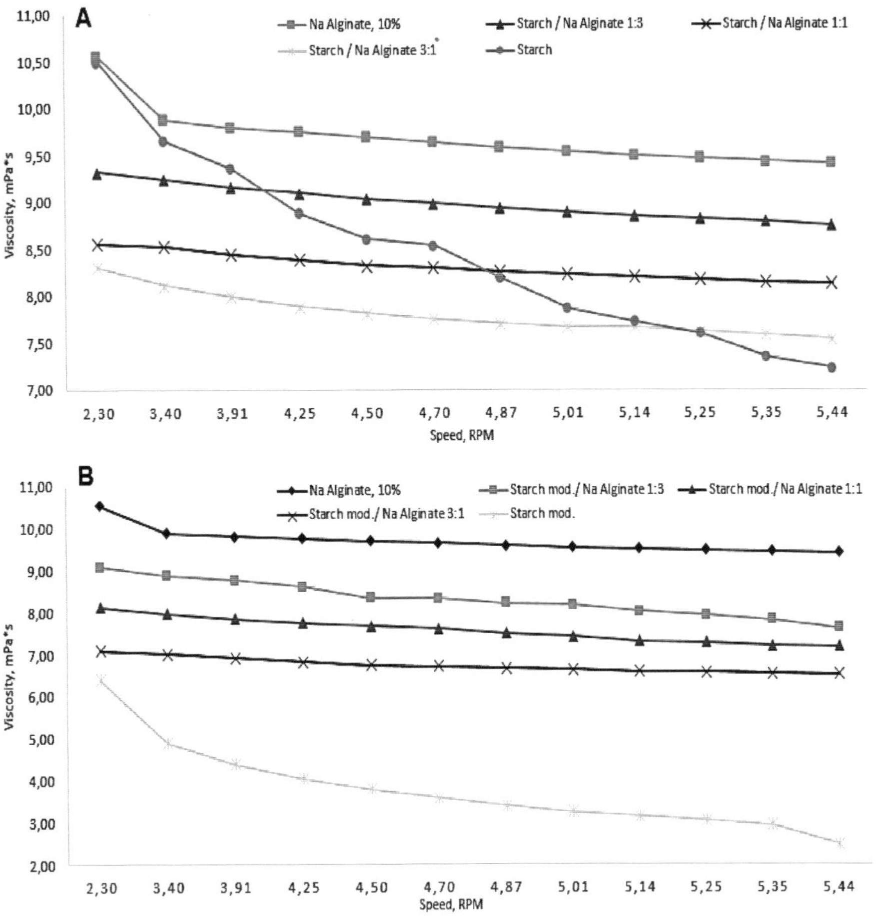

Fig.1 Dependence of changes of the viscosity of aqueous solutions on the content of starch (A) or Na-CM (B) in composition with sodium alginate.

It has been found that films containing carboxymethylated corn starch and sodium alginate show higher mechanical characteristics (fig.2) at 65% humidity, for the sample Starch mod./Sodium Alginate-1:1 stress ups to 2.0 N/mm^2, but it is less elastic. If the main indicator for the film is elasticity, then films based on starch are better than modified starch. However, films containing 3 vol. mass of carboxymethylated corn starch has an average strength of 2.1 N/mm^2 and elasticity up to 45%, which is the precursion case for the following studies determining water absorption of films and stability to their dissolution.

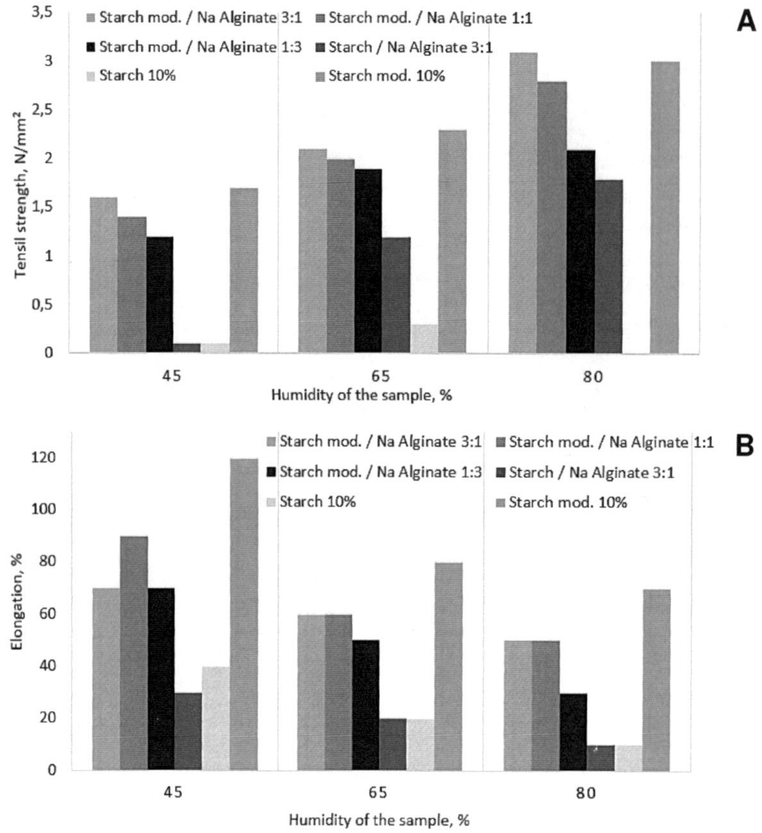

Fig. 2 Physical and mechanical properties (A – tensile strength; B – elongation) of films depending on humidity

The mechanical strength of the formed films is due the fact that alginate and carboxymethylated corn starch are hydrocolloids with an unbranched structure. This composition is characterized by stronger structure gels and higher viscosity of the solutions. The partial association of alginate and modified starch molecules results the homogeneity of the structure. Association of molecules and colloidal particles occurs due to the electrostatic interaction of charged groups, groups with high electron density, by the formation of so-called "chemical bridges" between the molecules and, also, due to the mechanical involvement of molecules. The addition of glycerol had led to an increase of the elasticity of the film but, at the same time, there was some increase of strength.

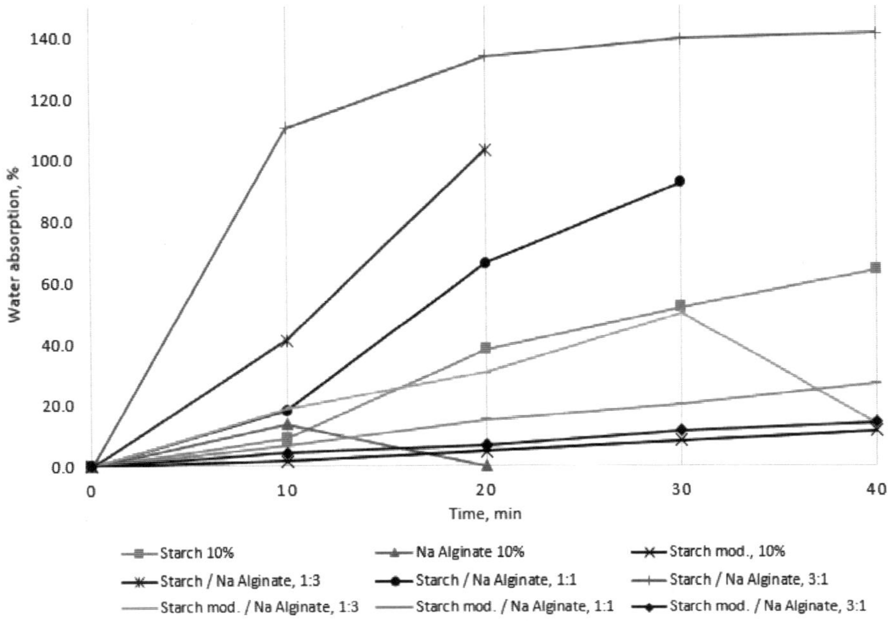

Fig.3 Dependence of water absorption of samples on the content of starch in the films

The films (fig.3) based on the pure sodium alginate and with ratio Starch mod./Sodium Alginate-1:3 are not water resistant and dissolve in 15 minutes. Films based on pure corn starch show greater water absorption, but also are not resistant to water solubility, films based on carboxymethylated corn starch show uniform swelling in water, indicating their resistance to water. Since carboxymethylated corn starch films has shown greater resistance to water solubility, this allows to develop the possibility of obtaining compositions with adjustable properties, namely prolonged water solubility.

4. Conclusions

Films based on a mixture of sodium alginate, modified starch and glycerine have the best physical and mechanical properties comparing with the starch based films. Rheological studies have shown that the values of the flow index of the two-component compositions are in the range of 0.29-0.45, and the compositions have a complex of properties predicting the possibility of their effective use for the production of films by casting method.

The addition of modified starch led to an increase of water absorption resistance of films based on sodium alginate up to 20-50% and prolonged their dissolving in time. The tensile strength of the films varies from 1.3 to 3.1 N/mm^2 and elongation varies from 30 to 130%. Thus, the use of sodium alginate in a mixture with carboxymethylated starch and glycerol has sufficient real prospects for further studies of the films and improvement of the films properties.

References

[1] Ogunsona, E., Ojogbo, E., & Mekonnen, T. (2018). Advanced material applications of starch and its derivatives. *European Polymer Journal*, 570-81.
[2] Nalamothu, N., Potluri, A., & Muppalla, M. B. (2014). Review on marine alginates and its applications. *Journal of Pharm Research*, 4(10), 4006-15
[3] Ishchanko, O., Plavan, V., Liashok, I., Kuchynska I., Resnytskiy, I. (2017, December).. Modyfikovani polisaharydy u farmacii [Modified polysaccharides in pharmacy]. Paper presented at the III International scientific and practical conference KyivPharma-2017

Pharmacology and pharmaceutical technology in active longevity (Ukraine), Kyiv (pp. 61-70). Kyiv: KNUTD. (in ukr)

[4] Ishchenko, O., Isak, V., Bessarabov, V., Plavan, V., Resnytskiy, I. (2017). Vykorystanna vodorozchynnyh polimeriv dlia otrymannia plivik medychnoho pryznachennia [Use of water-soluble polymers to obtain medical purpose films]. *Lehka Promyslovist*, 1, 30-33 (in ukr.)

[5] Winkler, H., Vorwerg, W., & Rihm, R. (2014). Thermal and mechanical properties of fatty acid starch esters. *Carbohydrate polymers*, 102, 941-949.

[6] Winkler, H., Vorwerg, W., & Wetzel, H. (2013). Synthesis and properties of fatty acid starch esters. *Carbohydrate poly*mers, 98(1), 208-216.

[7] Chi, H., Xu, K., Wu, X., Chen, Q., Xue, D., Song, C., Zhang, W., Wang, P. (2008). Effect of acetylation on the properties of corn starch. *Food Chemistry*, 106(3), 923-928.

[8] López, O. V., García, M. A., & Zaritzky, N. E. (2008). Film forming capacity of chemically modified corn starches. *Carbohydrate polymers*, 73(4), 573-81.

[9] Sasaki, T., Yasui, T., & Matsuki, J. (2000). Effect of amylose content on gelatinization, retrogradation, and pasting properties of starches from waxy and nonwaxy wheat and their F1 seeds. *Cereal chemistry*, 77(1), 58-63.

[10] Simi, C. K., & Abraham, T. E. (2008). Physicochemical rheological and thermal properties of Njavara rice (Oryza sativa) starch. *Journal of agricultural and food chemistry*, 56(24), 12105-113.

[11] Fujita, S., & Fujiyama, G. (1993). The Study of Melting Temperature and Enthalpy of Starch from Rice, Barley, Wheat, Foxtail and Proso millets. *Starch Stärke*, *45*(12), 436-41.

[12] Waliszewski, K. N., Aparicio, M. A., Bello, L. A., & Monroy, J. A. (2003). Changes of banana starch by chemical and physical modification. *Carbohydrate polymers*, *52*(3), 237-42.

[13] Vanier, N. L., El Halal, S. L. M., Dias, A. R. G., & da Rosa Zavareze, E. (2017). Molecular structure, functionality and applications of oxidized starches: A review. *Food chemistry*, *221*, 1546-59.

[14] Miyazaki, M., Van Hung, P., Maeda, T., & Morita, N. (2006). Recent advances in application of modified starches for breadmaking. *Trends in food science & Technology*, *17*(11), 591-99.

[15] Chen, Q., Yu, H., Wang, L., ul Abdin, Z., Chen, Y., Wang, J., Zhou, W., Yang, X., Khan, R.U., Zhang, H. and Chen, X. (2015). Recent progress in chemical modification of starch and its applications. *Rsc Advances*, *5*(83), 67459-74.

[16] Haroon, M., Wang, L., Yu, H., Abbasi, N.M., Saleem, M., Khan, R.U., Ullah, R.S., Chen, Q. and Wu, J. (2016). Chemical modification of starch and its application as an adsorbent material. *Rsc Advances*, *6*(82), 78264-85.

[17] Rajan, A., & Abraham, T. E. (2006). Enzymatic modification of cassava starch by bacterial lipase. Bioprocess and Biosystems Engineering, 29(1), 65-71.

[18] Rajan, A., Sudha, J. D., & Abraham, T. E. (2008). Enzymatic modification of cassava starch by fungal lipase. Industrial crops and products, 27(1), 50-59.

[19] Arijaje, E. O., Wang, Y. J., Shinn, S., Shah, U., & Proctor, A. (2014). Effects of chemical and enzymatic modifications on starch–stearic acid complex formation. Journal of agricultural and food chemistry, 62(13), 2963-72.

[20] ASTM D882-18, Standard Test Method for Tensile Properties of Thin Plastic Sheeting, ASTM International, West Conshohocken, PA, 2018.

IOP Conf. Series: Materials Science and Engineering **500** (2019) 012034 doi:10.1088/1757-899X/500/1/012034

Rheological and mechanical properties of filled with calcite concentrate polypropylene blends

S Saitarly[1], N Rezanova[1], V Plavan[1], K Beleška[2]

[1]Department of Applied Ecology, Technology of Polymers and Chemical Fibers, Kyiv National University of Technologies and Design, 2,Nemirovich-Danchenkostr.Kyiv, 01011, Ukraine svetlanasaitarly@gmail.com
[2]Department of Chemistry and Technology of Polymers, Kaunas University of Technology, K. Donelaičio str. 73, LT-44249 Kaunas, Lithuania

Abstract. It has been found that the addition of calcite concentrate 5-50% w/w allows regulation the rheological properties of polypropylene-based composition melts. It was established that the increasing of the amount of the filler up to 50% w/w in the PP based composition led to decrease of its melt viscosity by 40% comparing with initial polymer melt. This may be explained by the plasticization effect of stearic acid, which is commonly used to treat the surface of calcite concentrate. The addition of a propylene-octene block copolymer as a modifying additive makes it possible to produce high-filled materials that are more resistant to impact loading. The increasing the calcite concentrate content up to 10% w/w increases the impact strength by 10% compared with the initial polypropylene. Further increasing of calcite concentrate content up to 20 % led to decrease impact strength almost to the value for the initial polypropylene, at not critical decrease in tensile strength. Elongation of compositions is higher with the presence of block copolymer. The developed compositions with propylene-octene block copolymer as a modifying additive can be used for production of automotive details.

Introduction

Modern polymer composite materials are multicomponent systems. The addition of mineral fillers in polymers makes it possible to develop materials with required performance properties and contributes to cost reduction [1]. Products, which made from mineral-filled plastics, are widely used in the automotive industry [2], in the construction industry [3], for anticorrosive coatings of pipelines [4], in the production of medical supplies [5], sports coverings [6] and consumer goods [7]. The main problem, as a rule, is the reduction of strength and technical characteristics when the amount of filler in the compositions is increased. Srivabut et al. [8] studied the effect of filling with nano-clay, talc and calcium carbonate on polypropylene (PP) already filled with wood flour. It was reported that samples containing 7% w/w of calcium carbonate were characterized by optimal physical and mechanical properties. One of the directions of development of the materials with increased impact resistance is a modification of the filled polymers adding elastomers [9].Nguen et al.[10] have shown that the modification of PP by a metallocene ethylene-propylene elastomer leads to a significant increase of the impact strength and elongation at break, however, the modulus of elasticity and the yield stress decrease.

The addition of fillers and elastomers not only affects the performance properties of polymer material, but also it's processing ability. The main technological factor determining the way of processing is the fluidity of the composition [11]. The filling always leads to products' molding difficulties due to the increase of the viscosity of the melt [12]. Compositions [13] based on PP and ethylene-propylene rubber, which meet the technical characteristics of the bumpers and other parts of cars, characterized by high impact strength have been developed; however, the melt flow rate of such mixtures reaches only 3-6 g/10 min.

A knowing of rheological properties allows an estimation of the effect of composition components on the properties of the composition melt [14].The melts of the filled polymers can be considered as concentrated suspensions. For such systems, it is reasonable to take into account the possibility of interaction between the particles, which affects the nature of the flow. It is known that the properties of such polymer compositions are determined by the content of the filler, the size and degree of aggregation of the particles, as well as the surface characteristics of the filler [1].

The aim of the work is to study the impact of the amount of mineral filler and propylene-octene block copolymer on rheological and mechanical characteristics of extrusion-type polypropylene melts and to choose the composition which suitable for the production of materials with the required properties.

Materials and methods of the research

The filled compositions based on polypropylene (PP) of 21030 mark and blends of polypropylene with 5%w/w propylene-octene block copolymer (POBC) «Vistamaxx 6202» (ExxonMobil) containing 5-50% w/w of mineral filler have been investigated. The calcite concentrate (CC) of 1TK mark (LLC «Tekhnokom») with particle size of 2.5 micrometres (20 micrometres – 1.5%) was used as inexpensive, common and affordable filler, which contains polyolefin and calcium carbonate treatedwith stearic acid.Polymer compositions were prepared by mechanical mixing of the components and subsequent granulation in the ЧП-27x30 extruder. The temperature in the zones of the extruder was 155-210-200-200°C, screw rotation rate – 30 rpm. The viscosity (η) of the melts was measured using MB-2 capillary viscometer in the range of shear stresses $(0.1\div5.7)\cdot10^4$Pa at 230°C. The determination of melt flow index (MFI) was carried out according to GOST 11645-73 (National State Standard) using "IIRT" instrument at 230°C and load of 2.16 kgf. Flow pattern «n» was calculated using tangent tilt of the slope at a given point to the flow curve.The Charpy impact value (kJ/m^2) is calculated by dividing the fracture energy by the cross-section area of the specimen according to GOST 4647-80 (National State Standard). The determination of tensile strength and elongation at break was carried out according to GOST 11262-80 (National State Standard).

Results and discussion

The experiments have shown that the addition of (5-50) % wt. CC in the melt of PP leads to decrease in viscosity of the former throughout the range of concentrations and shear stresses (figure 1).The result obtained differs from the traditional regularities of the flow of the melts filled with solid additions, which typically tend to increase viscosity. This may be explained by the fact that the calcium carbonate particle surface is treated with stearic acid, calcium stearate or coupling agents in an amount of 0.5-3% in order to improve the rheological properties and wetting [16, 17].

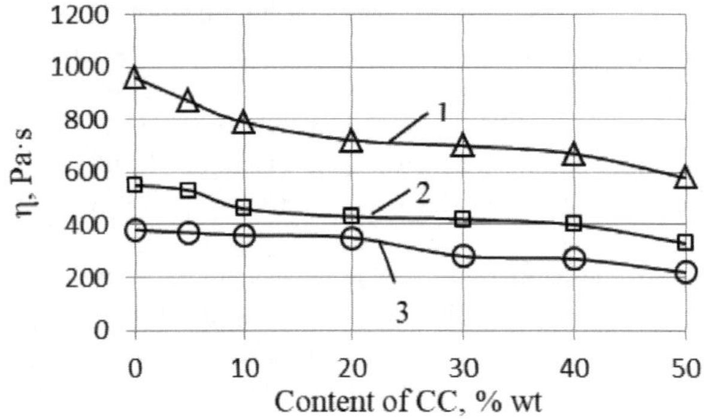

Figure1. Influence of the content of CC on the viscosity of the polypropylene melt at 230°C and shear stress of $\tau\cdot10^{-4}$Pa: 1 – 1.61; 2 – 4.2; 3 – 5.69.

As a result, calcium stearate is formed, which has a non-polar "ends" (figure 2) in the form of stearic acid chain on the surface of the particle [18].

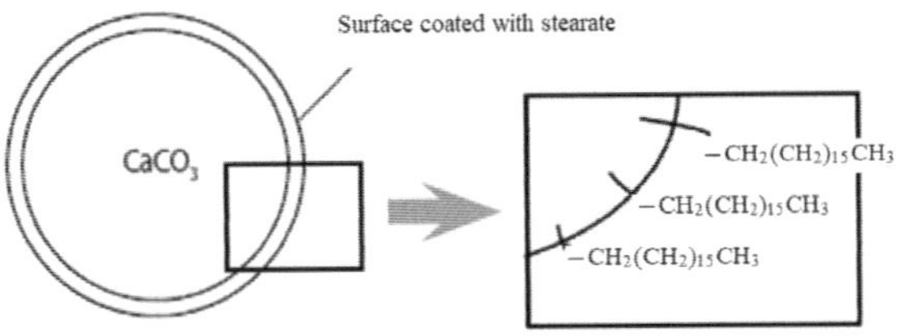

Figure2. Scheme of CC surface structure.

The formation of the described structure of filler particles improves the compatibility of the melt of hydrophobic PP and filler due to the compatibilizing action of stearic acid chains in the calcium stearate molecule $[CH_3(CH_2)_{15}CH_2OO^-]_2Ca^{2+}$ at the interface between their phases and the decrease in the hydrophilicity of calcium carbonate [19]. The viscosity of melts of PP/CC polymer suspensions is determined by two competing factors: emulsifying effect of non-polar "ends" of the acid leads to decrease of viscosity of the melt, and a solid mineral addiction provides its increase through the formation of aggregates. The predominant factor is the compatibilizing action of stearic acid chains on the surface of calcium carbonate molecules, which causes the decrease of the melt viscosity in the system. A similar effect of a significant reduction of viscosity has also been described for compositions consisting of polyvinylchloride with hydrophobizing additives based on stearic acid, the activity of which increases in the series: zinc stearate, calcium stearate, stearic acid [18]. The presence of long hydrocarbon chains that have affinity with the macromolecules of PP on the surface of the used filler provides the dispersion homogeneity in the melt and a sharp decrease of the abrasive effect of the filler on the extrusion equipment.

The studied PP compositions filled with CC, as well as the original polypropylene, demonstrate viscosity abnormality, i.e., when the stress or shear rate increases, the viscosity decreases. The behavior of such liquids is described by the polynomial equation: $\tau=\eta\cdot\gamma^n$, where n – flow index.

The addition of POBC 5% w/w into the unfilled PP reduces the viscosity of the composition(figure 3).The curves of the compositions PP+50%CC and PP+5%POBC+50%CC almost coincide, that is, the addition of copolymer into filled 50% CC polypropylene does not affect viscosity.

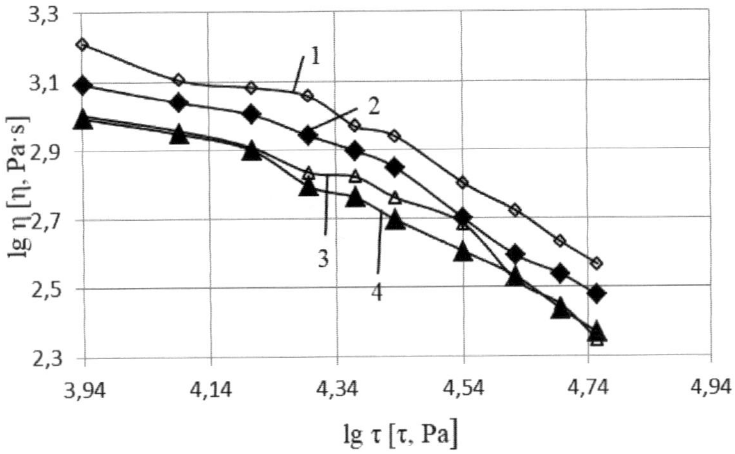

Figure 3. Dependence of the viscosity on the shear stress expressed as logarithmic coordinates for compositions: 1 – PP; 2 – PP+5%POBC;
3 – PP+50%CC; 4 – PP+5%POBC+50%CC.

It has been found that for the compositions based on polypropylene filled with CC, the degree of deviation from the Newtonian flow regime is complex: the value of *n* decreases from 2.0 (for the initial PP) down to 1.6 and then increases up to 2.3 (table 1). The described dependence is explained by the fact that the index *n*, as well as the viscosity, depends on many factors with the predominant effect of one of them in each specific case.

Table 1. Effect of calcite content of the concentrate flow mode *n* of filled compositions

Content of CC, wt.%	*n*at $\tau\cdot10^{-4}$Pa	
	5.69	1.61
0	2.0	1.5
5	2.0	1.5
10	1.6	1.2
20	1.9	1.3
30	2.0	1.4
40	2.1	1.5
50	2.3	1.5

It has been found that the increasing of content of CC up to 50% w/w in the composition leads to melt flow index increase from 3.1 g/10 min (without CC) up to 6.6 g/10 min (50%w/w of the filler). With the addition of the POBC 5% w/w as the elastomer in compositions, the curves showing the dependence of melt flow index on the content of CC are similar within the range from 3.4 g / 10 min (without CC) to 6.9 g / 10 min (50%w/w of the filler) (figure 4). It can be explained by appropriate combination of PP and POBC with its plasticizing effect. The plasticizing effect of the composition modified with the elastomer, compared to PP filled with CC50%w/w, is equal to 4.3%.

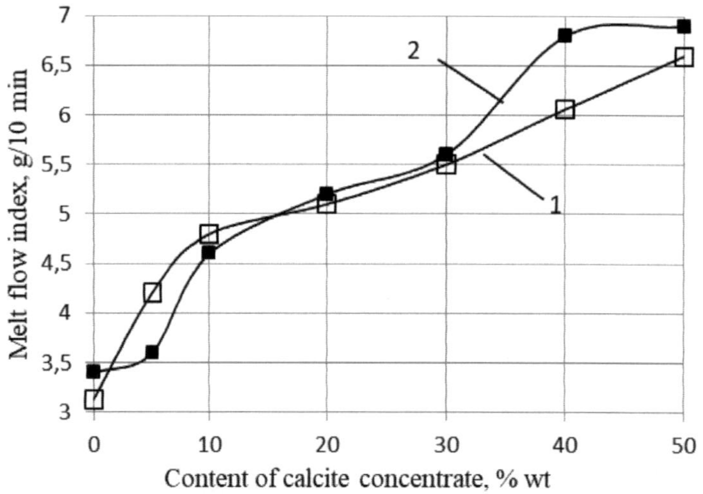

Figure 4. Influence of CC content on melt flow index in compositions: 1–PP+CC; 2 – PP+5%POBC+CC.

The impact strength index characterizes the strength of the material under impact loads. The rubber phase of the modifier combined with the polymer provides a uniform distribution and damping of the impact energy, so that the modified polymer materials withstand a higher impact load [9].The content of CC 5-10% w/w increases the impact strength probably due to the nucleation effect of PP by CC (figure 5).This effect is enhanced in the presence of POBC as an elasticizing additive.

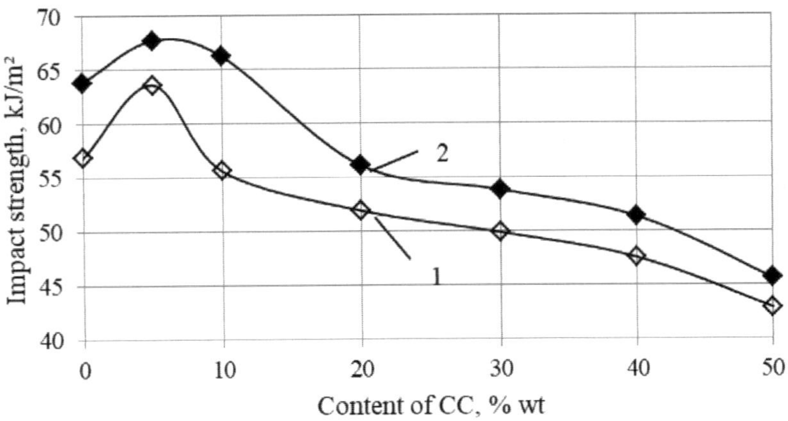

Figure 5. Dependences of the impact strength from the content of CC in composites: 1 – PP+CC; 2 – PP+5%POBC+CC

The addition of POBC5% w/w increases the elongation from 112% to 140% and practically does not change the strength of the composite (figure 6).

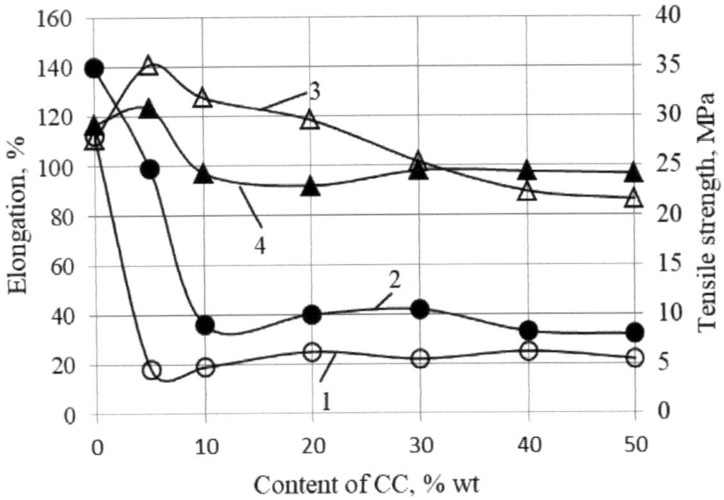

Figure 6. Dependences of the elongation (1,2) and tensile strength (3, 4) from the content of CC in composites: 1,3 – PP+CC; 2,4 – PP+5%POBC+CC

The addition of CC up to 10% w/w into compositions with 5% w/w POBC and without it leads to a decrease of elongation by 74% and 82% respectively. Further increasing of CC content does not lead to significant change of these indexes. The strength of the compositions with 5% w/w POBC with the content of CC 10-50% w/w does not change, while the strength of the composition PP+CC is reduced by 39%.

Conclusion

It has been found that the addition of calcite concentrate 5-50% w/w allows regulation the rheological properties of polypropylene-based composition melts. It was established that the increasing of the amount of the filler up to 50% w/w in the PP based composition led to decrease of its melt viscosity by 40% comparing with initial polymer melt. This may be explained by the plasticization effect of stearic acid, which is commonly used to treat the surface of calcite concentrate. The addition of a propylene-octene block copolymer as a modifying additive makes it possible to produce high-filled materials that are more resistant to impact loading. The increasing the calcite concentrate content up to 10% w/w increases the impact strength by 10% compared with the initial polypropylene. Further increasing of calcite concentrate content up to 20 % led to decrease impact strength almost to the value for the initial polypropylene, at not critical decrease in tensile strength. Elongation of compositions is higher with the presence of block copolymer. The developed compositions with propylene-octene block copolymer as a modifying additive can be used for production of automotive details.

References

[1] Xanthos M 2010 *Functional Fillers for Plastics* (*Weinheim, Germany: Wiley-VCH*)507

[2] Codolini A, Li Q M, Wilkinson A 2017 Influence of machining process on the mechanical behaviour of injection-moulded specimens of talc-filled polypropylene *Polymer Testing* **62** pp 342-347 https://doi.org/10.1016/j.polymertesting.2017.07.018

[3] Kou S-C, Poon C-S 2013 A novel polymer concrete made with recycled glass aggregates, fly ash and metakaolin. *Construction and Building Materials* **41** pp 146-151

[4] Abdou M I, Ayad M I, Diab A S, Hassan I A, Fadl A M 2017 Influence of surface modified ilmenite/melamine formaldehyde composite on the anti-corrosion and mechanical properties of conventional polyamine cured epoxy for internal coating of gas and oil transmission pipelines. *Progress in Organic Coatings* **113** pp 1-14

[5] Vinny R Sastri 2014 Polymer Additives Used to Enhance Material Properties for Medical Device. *Applications Plastics in Medical Devices* (Second Edition) Properties, Requirements and Applications pp 55–72

[6] Medvedev V P, e.a. *Kompozitsiya dlya pokryitiy* [Coatingcompositions] Patent RF, no.2190002, 2002

[7] Kryizhanovskiy V K 2005 *Tehnicheskie svoystva polimernyih materialov* (Sankt-Peterburg: Professiya) 240

[8] Srivabut C, Ratanawilai T, Hiziroglu S 2018 Effect of nanoclay, talcum, and calcium carbonate as filler on properties of composites manufactured from recycled polypropylene and rubberwood fiber *Construction and Building Materials***162** pp 450–458 http://sci-hub.tw/https://doi.org/10.1016/j.conbuildmat.2017.12.048

[9] J. G. Drobny 2014 *Handbook of Thermoplastic Elastomers* (Second Edition) A volume in Plastics Design Library ed S Ebnesajjad (Oxford: Elsevier) pp 209–218. https://doi.org/10.1016/B978-0-323-22136-8.00007-7

[10] Nguen M T, Chalaya N M, Osipchik V S 2017 Mozhifikatsiya polipropilena metallotsenovyim etilenpropilenovyim elastomerom *Uspehi v himiiihimicheskoytehnologii***11** pp 79-81

[11] Kerber M L, Vinogradov V M, Golovkin G S 2008 *Polimernye kompozitsionnye materialy: struktura, svoystva, tekhnologiya*ed. AABerlin (Sankt-Peterburg: Professiya) 560 p.

[12] Melnichenko M A, Yershova O V, Chuprova L V 2015 Vliyanie sostava napolniteley na svoystva polimernykh kompozitsionnykh materialov *Molodoy uchenyiy* **16** pp 199-202

[13] Bauman N A *Poluchenie vyisokoudaroprochnyih kompozitsiy na osnove polipropilena i etilenpropilenovogo kauchuka* Cand, Diss. [Preparation of high impact strength compositions based on polypropylene and ethylene-propylene rubber] Kazan, 2011. 153 p

[14] Sengers W G F 2005 *Rheological properties of olefinic thermoplastic elastomer blends*. Doctoral thesis (Dutch Polymer Institute)172.

[15] Arshinnikov D I, Sviderskiy V A, Nudchenko L A 2016 Sklad, struktura i dispersnist prirodnoï kreydi rodovishch Ukraïni *Visnik natsionalnogo tekhnichnogo universitetu Ukraïni «Kiïvskiy politekhnichniy institut»* **1** pp 103-107

[16] Yang Y-C, Jeong S-B, Yang S-Y, Chae1 Y-B and Kim H-S 2009 The Changes in Surface Properties of the Calcite Powder with Stearic Acid Treatment *Materials Transactions* **50** pp 695-701.

[17] Mihajlović S, Sekulić Ž, Daković A, Vučinić D, Jovanović V, StojanovićJovica 2009 Surface properties of natural calcite filler treated with stearic acid *Ceramics – Silikáty* **53** **(4)** pp 268-275

[18] Niftaliev S I, Lygina L V, Peregudov Yu S, Prokofeva L A 2014 Issledovanie reologicheskikh svoystv kompozitsiy na osnove PVKh *Vestnik Voronezhskogo gosudarstvennogo universiteta inzhenernykh tekhnologiy* **2** pp. 132-134

[19] Deshmukh G S, Pathak S U, Peshwe D R and Ekhe J D 2010 Effect of uncoated calcium carbonate and stearic acid coated calcium carbonate on mechanical, thermal and structural properties of poly(butylene terephthalate) (PBT)/calcium carbonate composites *Bulletin of Materials Science***33** pp. 277–284

IOP Conf. Series: Materials Science and Engineering **500** (2019) 012035 doi:10.1088/1757-899X/500/1/012035

The usage of modified montmorillonite to enhance the quality of leather velour

K Stadnik[1], O Okhmat[1], O Mokrousova[1,2]

[1]Department of Biotechnology, Leather and Fur, Kiev National University of Technologies and Design, Nemirovicha-Danchenko str., 2, 01011, Kiev, Ukraine,
[2]Department of Commodity Science and Customs Affairs, Kiev National University of Trade and Economics, Kioto str., 119, 02156, Kiev, Ukraine
katyshka07_08@ymail.com

Abstract. The paper considers the study of the possibility to use modified montmorillonite in velour production process in order to improve its quality. The adsorption of anionic black dye with velour derma has been investigated. A 16% increase was identified of the dye sorption by velour treated with the modified montmorillonite dispersion. The use of modified montmorillonite dispersion also provides additional formation of derma structure thereby increasing its strength by 11%.

1. Introduction

Today, leather velour is increasingly used for the production of clothing, footwear, haberdashery and gloves. This velour has a pile surface obtained by polishing. The velour surface also has a uniform and even color. The velour production capacity at enterprises is increasing year by year not only because of the substantial demand of consumers. Reducing the quality of leather raw materials as a whole, the presence of defects on the leather surface, spit mill after double glazing, are the main reasons for increasing the volume of velour production.

Typically, the technology of velour production differs from that one of grain leather only by semi-finished processes and operations. However, today, leather enterprises make adjustments to technology already at the stage of tanning processes. Such changes contribute to the formation of dense, low, uniform, elastic pile with multiple polishing. Polishing of velour is carried out several times with abrasive materials of different graininess. Too soft, loose semi-finished product is hard to grind. The result of this polishing is the presence of villus of different lengths on velour surface. The velour thus loses its nobility and its price decreases. Consequently, the additional formation of the volume of derma in tanning and post- tanning processes of velour production contributes to increasing its quality. The formation of the volume of derma is achieved through the use of both individual polymer compositions, mineral and organic tributaries, dispersions of natural minerals, and a combination of these materials [1-3].

The most common defect of velour is removing dye from the pile surface during operation of its products. The appearance of a defect may compromise production process and application of chemical materials used in the manufacture of grain leather. In order to obtain a high-quality velour, it is necessary to conduct a rinse after dying and use of sorbents for additional fixation of dyes. Montmorillonite is a silicate with properties of an effective sorbent. To increase the hydrophilicity of its surface, it is enough to use exchange sodium cations. As a result of such modification, the dispersion of montmorillonite acquires certain colloidal properties and allows forming derma structure at its various levels [1]. Most likely, the formation will contribute to additional fixation of dyes in the structure and to obtain a low and uniform pile.

The aim is to study the quality of velour, pre-treated with montmorillonite dispersion. The subject of the study is the degree of absorption of synthetic dye in the process of dying, the mechanical properties of velour and chemical interactions occurring between the dye and modified dispersion of montmorillonite.

2. Materials and Methods

Bentonite clays of the Dashukiv deposit (Ukraine) were used for the study. The content of montmorillonite in bentonite clay is 85%. Modification of montmorillonite is carried out by the introduction of sodium carbonate (5% by weight of dry mineral) into a dispersion with a concentration of mineral 100 g/l. Modification of montmorillonite with sodium carbonate leads to cation replacement of the natural metallic complex ions of the mineral with sodium ions. Such replacement contributes to the maximum level of dispersion of the mineral aggregates to its separate structural formations with a size of 0.94-1.0 nm. The presence of different particle sizes in montmorillonite modified dispersion contributes to formation of derma structure at different levels. The specific surfase of montmorilonnite is 700 - 840 m^2/g [4].

Anionic black dye (figure 1) with a pronounced green tint was used for dying.

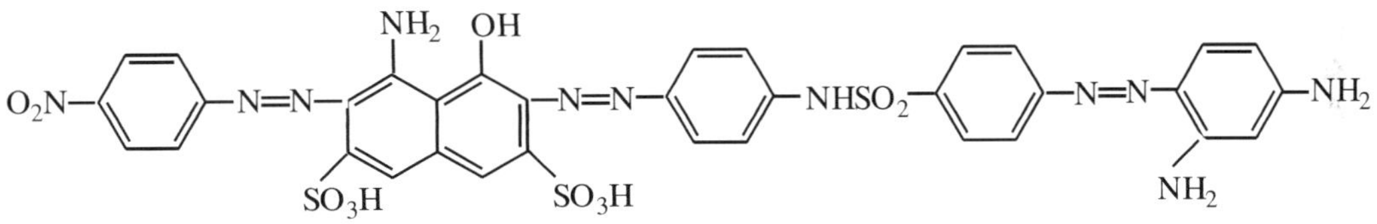

Figure 1. Structural formula of anionic black dye (Technical Specification 6-14-767).

For further research, samples of the semi-finished chrome tanning method were used, obtained by the typical method of velour production from the bottom split of cattle. The thickness of the test samples is 1.1 - 1.2 mm. Velour samples were treated with a modified dispersion of montmorillonite in the amount of 2.5% of the mineral from the mass of semi-finished product. Fixation of the modified dispersion of montmorillonite is carried out using sodium formate in the amount of 0.4% from the mass of semi-finished product. Dispersion treatment is carried out at a temperature of 30 °C. The processing time together with the fixation process is 90 minutes. Samples of the control group (option 1) are not subject to dispersion. For lubrication will used material Provol BA («Zschimmer Schwarz GMBH Co», Germany).

In order to evaluate the effectiveness of the modified dispersion of montmorillonite to intensify the velour dying, adsorption studies were conducted and the adsorption level of dyes was determined. The adsorption of dye on a semi-finished product treated with a modified dispersion of montmorillonite is based on the optical density parameters of dye solutions obtained using photoelectrocolorimetric method.

For measurements, a laboratory photovoltaic absorptive nephelometer was used [5]. The basis of the device's operation is the principle of measuring the absorption or diffusion of light of the test medium. Measurement is performed using air filter No 2 with a wavelength of 500x10 nm. The analysis is carried out by determining the light transmission rate of dye solutions at different concentrations. Based on study indicators, we construct the calibration curves of optical density (D_0) dependence on dye concentration (C_0) for the initial solutions. Upon dye interaction with the crushed weighted amount of the semi-finished product, we determine the optical density of the dye solution (D), and according to the calibration curve – equilibrium concentration of dye in the solution (C). Based on data obtained we calculate dye adsorption [mmol/g] by semi-finished product using the formula (1):

$$A = \frac{(C_i - C_f) \cdot V \cdot 1000}{S}, \tag{1}$$

where, C_i is the initial concentration dye, g/l; C_f is the concentration of dye after a certain period of time (including, the final), g/l; S is the weighted amount of semi-finished product, g; V is the volume of dye solution, l.

Quality studies of colored velour were performed in accordance with regulatory documents: Standard 939-88 - definition of physical and mechanical leather parameters; State Standard of Ukraine ISO 11640: 2005 - definition of colour resistance to mechanical impact; State Standard of Ukraine ISO 11643-2001- definition of colour resistance to organic solvents.

3. Results and Discussion

Several studies have been conducted [5] to determine the basic properties of dye solutions, which showed: dye is resistant to hard water (5 points, maximum estimate), resistant to electrolytes is estimated at 5 points. Visual observations during this study showed an increase in the brightness of the green tint when exposed to the colorant of sodium carbonate solution (the concentration of sodium carbonate solution is 100 g/l). The dye is readily soluble in water at a temperature of 45 °C.

Adsorption isotherm analysis (Figure 2) indicates a significant increase in the adsorption level for samples treated with a modified dispersion of montmorillonite. Thus, at a concentration of 0.5 mmol/l in an equilibrium solution, the adsorption level is 0.0185 mmol/g for the control sample (1), 0.0215 mmol/g for experimental option (2). An increase in the sorption of the dye by the derma will help to reduce its concentration in the dueing solution.

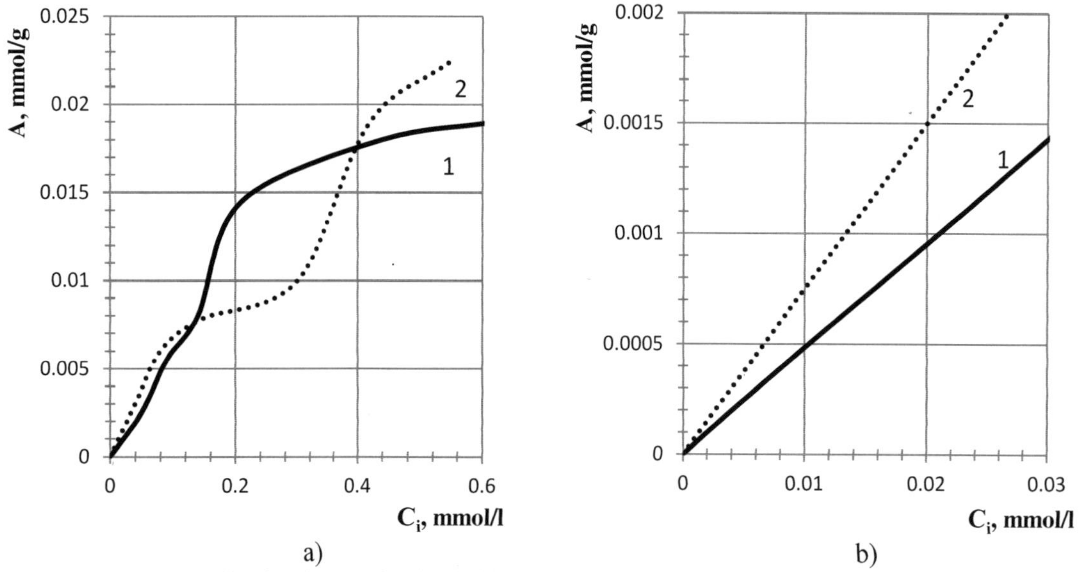

Figure 2. Adsorption isotherm of anionic black dye by the structure of the semi-finished product: total (a) and within the concentration of 0-0.03 mmol/l (b)

In order to study the quality of velour leather, a semi-finished product was coloured, made according to traditional technology (control) and treated with modified montmorillonite dispersion. Colouring of velour samples was performed using anionic black dye with consumption rate of 4% of the mass of semi-finished product. Dye fixation in derma structure was performed with a solution of formic acid, consumption rate of 0.5% of the mass of semi-finished product. Dying time – 75 minutes, temperature – 50–55 °C. After the process of dying and conduct a rince, the samles was lubrication 40 minutes at temperature 65 °C. fat consumption – 7% from weight finished product.

Adsorption studies (figure 3) for non-lubrication specimens (1) and after lubrication (2) allowed detecting that level of adsorption for non-lubrication specimens was higher by almost 30%, with the same index for samples after lubrication. This is probably due to shielding of the derm structural elements by components of fatliquoring emulsion.

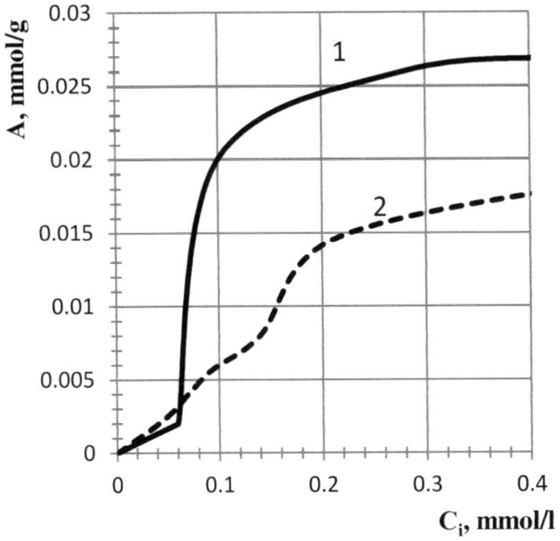

Figure 3. Adsorption isotherm of anionic black dye with structure of the semi-finished product before (1) and after (2) fatliquoring

For all samples of colored velour defined complex properties [4], which characterizes mechanical properties of velour (Table 1) and the quality of color achieved (Table 2).

Table 1. Mechanical properties of velour

Treatment options	Tensile strength at break [MPa]	The relative elongation at break [%]
1 (control)	14.1	39
2	15.5	32

Table 2. Quality indicators of velour coloration

Treatment options	Sample colour	Coloration resistance to dry friction, score	Coloration resistance to wet friction, score	Coloration resistance to organic solvents, score
1 (control)	black-green	4	2,5	2,5
2	black	5	4	5

The visual evaluation of samples showed the absence in samples treated with the modified dispersion of montmorillonite, a green tint in the resulting colour. The organoleptic evaluation showed that the samples of second group are more dense compared to samples of the control group. In this case, the samples treated with a dispersion are less dense, but not rigid. The conclusions of the organoleptic evaluation confirm the results of tests for mechanical parameters. Thus, the strength of the leather option 2 compared with the control one increased by 11%, and elongation decreased by 18%. The additional formation of derma is indicated by more uniform pile on samples of group 2 after repeated velour polishing. There is also a slight (3-4%) increase in thickness of the test samples after their dispersion treatment.

Quality tests of obtained colour showed an increase of 20% in resistance to dry friction of specimens filled with dispersion of montmorillonite. Significant difference (37.5%) is observed in parameters when determining the resistance of coloration to wet friction. The increase in these parameters indicates the formation of additional hydrogen bonds in the process of colouring. Double increase in resistance of test samples to organic solvents compared with the control group also

IOP Conf. Series: Materials Science and Engineering **500** (2019) 012035 doi:10.1088/1757-899X/500/1/012035

indicates the formation of additional bonds between the dye and derma. Hydrogen bonds formed in the system are destroyed by effect of organic solvent. Thus, most likely, we can talk about formation of covalent bonds, resistant to the specified type of processing. The formation of bonds of the aforementioned type may also indicate the disappearance of dye tint and obtaining deep black coloration in specimens of the second group.

4. Conclusions

Despite the fact, that the specific gravity of montporillonite is large, different connections are formed between the active elements of the dermal collagen, dispersion and chemical materials. This formation is shown by an increase (by 11%) of samples strength with without the softness modification.

Projections confirmed for improved quality of velour coloration. This is shown by the improvement in each parameter that characterizes the quality of leather colour. Most likely, this is due to the fact that the montmorillonite modified dispersion has anionic nature, to fix which the sodium formate was used. The sodium formate in solution recharges the system of montmorillonite and transforms it into a cationic form. The dye used is anionic in its nature, and hence, there is a probability of establishing bonds between charged groups. This was confirmed based on study results through increase in adsorption of dyes by derma, an increase in resistance of the obtained coloration to mechanical effects and organic solvents.

When implementing the technology of velour colouring the leather enterprise uses auxiliary substances for due diffusion in derma structure. Often, these are surface-active substances that reduce velour and its products resistance to water in general. The use of montmorillonite modified dispersion allows excluding these substances from the process. Treatment with the montmorillonite dispersion arranges the porous structure at the level of fibrils and primary fibers [1]. At the same time, the thickness of collagen fibers increases, derma structure adhesiveness decreases sharply and dye diffusion improves. Complete coloration of derma with dye was achieved during this study within 30 minutes of treatment.

Consequently, application of montmorillonite modified dispersion in velour production contributes to an increase in both leather colour quality and its quality as a whole.

References

[1] Marukhlenko M O, Palamar V A, Mokrousova O R 2016 Stabilizing derma collagen structure with modified dispersions of montmorillonite *IOP Conference Series: Materials Science and Engineering* **111** pp 1–8

[2] Nikonova A, Andreyeva O, Maistrenko L 2016 Investigation of the properties of polymer-titanium tanned leather *Proceedings of the 6th International Conference ICAMS Advanced Materials and Systems* pp 135–140

[3] Andreyeva O, Nikonova A, Maistrenko L 2017 Investigation on Polymer-Mineral Tanned Leather Properties *Solid State Phenomena* **267** pp 98-102

[4] Tarasovich Ju I, Ovcharenko F D 1975 Adsorbcija na glinistyh mineralah *Naukova dumka* 351

[5] Danylkovych A G, Chursyn V I 2000 Workshop on Chemistry and Technology leather and Fur *TSNYYKP* 413

IOP Conf. Series: Materials Science and Engineering **500** (2019) 012036 doi:10.1088/1757-899X/500/1/012036

Comparative characterization of aged bitumen properties

A Ābele[1], R Meri Merijs[1], R Bērziņa[1], J Zicāns[1], T Ivanova[1], V Haritonovs[2]
[1]Institute of Polymer Materials, Faculty of Materials Science and Applied Chemistry, Riga Technical University, Riga, Latvia
[2]Department of Roads and Bridges, Faculty of Civil Engineering, Riga Technical University, Riga, Latvia

Agnese.Abele_1@rtu.lv

Abstract. To provide higher rates of reclaimed bitumen uses in road industry better understanding about aging of bitumen is needed, which requires to get complete information about influence of aggressive environmental factors on bitumen. In this paper changes of bitumen structural and rheological parameters upon ageing are evaluated by using one of the most known method, RTFOT in comparison to alternative procedures, as thermally and ultraviolet light irradiation induced aging. Calorimetric properties of the raw and aged bitumens are determined by MDSC, chemical composition by FTIR and rheological parameters (viscosity, rutting resistance and fatigue cracking resistance) by DSR. The results confirm that mentioned alternative aging procedures can be used to some extent for prediction of real field-aged bitumen properties changes.

1. Introduction

Asphalts pavement service life depends on several environmental factors, including traffic load, climate and temperature, but the main factor is used binder. Most frequently used binder is bitumen; it is viscoelastic and waterproofing material derived from crude oil. As organic material bitumen, ages faster than others asphalt components. With bitumen aging generally oxidation is understood, but it also includes evaporation of volatile components, consequently, chemical structure is changed which affects other important service life properties, such as, rheological behavior. Aged bituminous materials are harder, more brittle and are characterized with higher viscosity, which reduces adhesion between pavement components [1,2].

The study of bitumen rheology and structural properties after aging is not a new field, over the years several methods have been developed for prediction of aging process in laboratory conditions, because uses of reclaimed asphalt increases [3].

Main methods for prediction of bitumen aging are rolling thin film oven test (RTFOT) and pressurized aging vessel (PAV) test for prediction of short-term and long term aging, respectively, but these methods do not include influence of ultraviolet light (UV) irradiation on the bitumen. The aim of this study is to evaluate bitumen properties after several aging procedures, including UV irradiation, and evaluate the effectiveness of the used aging methods [1,2].

2. Materials and methods

Commercial 70/100 penetration grade bitumen was used. Properties of the bitumens after different aging procedures were assessed: reclaimed asphalt pavement (RAP) derived bitumen; RTFOT aged bitumen as accords to ASTM D2872; thermally aged bitumen (70/100: 8T), obtained by mixing of bitumen at 185-188 °C for 8 hours, and UV aged bitumen (70/100:UV167 and 70/100:UV330), obtained by using QUV accelerated weathering tester utilizing UV-A 340 nm lamps with irradiance of 0.89 W/m^2 for different duration times (167 and 330 hours) at 60 °C.

Infra-red spectra were recorded by using Fourier transform infrared spectrometer Thermo Fisher Scientific, Nicolet 6700 in attenuated total reflection (FTIR-ATR) mode in the range of wave numbers from 4000 cm^{-1} to 650 cm^{-1}.

Calorimetric properties were measured by using differential scanning calorimeter DSC-3 (Mettler Toledo) utilizing modulated technique TOPEM (MDSC). Average sample mass was ca 10 mg. Test regime: heating from -90 °C to 60 °C with heating rate of 1 °C/min.

Rheological behaviour of unaged bitumen and aged bitumens was characterized by using REOLOGICA StressTech NOVA rotational viscometer NOVA (DSR) by utilizing 25 mm plate-and-plate configuration to carry out oscillatory shear tests at various temperatures (46, 58, 70, 82 °C).

3. Experimental part

Bitumen aging is very complex process, involving variety of reactions, thus making interpretation of data challenging. Consequently, MDSC analysis has been performed to determine influence of aging process on the amorphous and crystalline phases of the bitumen constituents, as it is known that heat flow in MDSC can be separated into reversing and non-reversing components, respectively attributable to the processes that can be brought to equilibrium and those, controlled by kinetics. In Figure 1 reversing and non-reversing heat flow curves of the bitumen samples are shown. Fig. 1 (a) shows very broad range of overlapping structural relaxations/glass transitions T_g for all samples over the temperature diapason of -90 – 0°C. Besides it, inexpressive endothermic peaks (e.g. for 70/100:8T around 10 °C), most probably attributable to melting of semi-crystalline constituents, can be also observed, evidently, due to the fact that melting of crystals at certain experimental conditions can contain also some reversing contribution. Non-reversing heat-flow curves, depicted in Fig. 1 (b), give more detailed information about the effects that cannot be brought to equilibrium, for instance, evaporation, oxidation, decomposition, and at certain experimental conditions crystallization. Thus, endothermic peaks, observed starting from ca -10°C can be attributed to crystallization of the saturate fractions in bitumen [4]. Plausible, due to ageing these endothermic contributions are increased, resulting in increased crystallinity of bitumens, aged by different procedures. It should, however, be mentioned that isotropization of fractions other than the saturates should be also regarded as contributors to the observed endotherms [4].

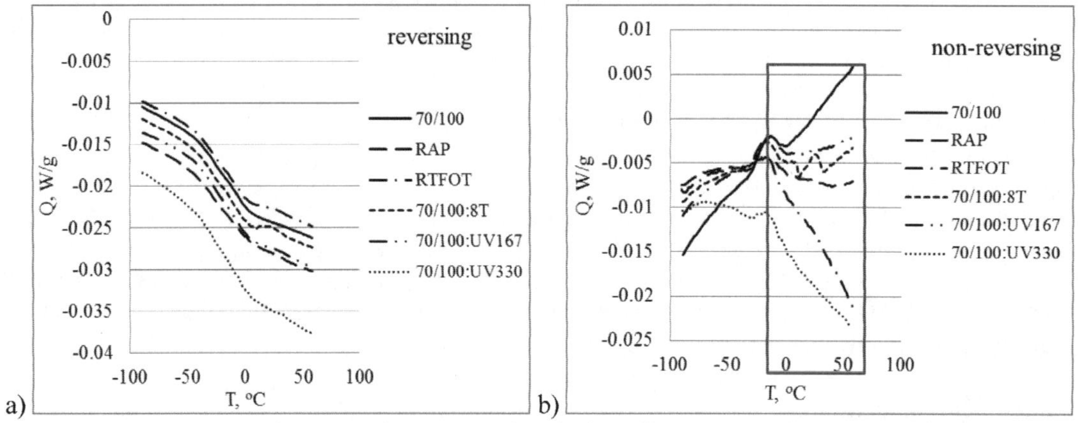

Figure 1. Reversing (a) and non-reversing (b) heat flow curves of unaged and aged bitumen

The derivative of heat capacity temperature relationships Cp_o'-T allows improved analysis of relaxation processes. From previous results, we observed broad relaxation region, but in Figure 2, multiple peaks within this region are distinguishable. Masson et.al. [4, 5] have concluded, that bitumen is characterized with four T_g that arise from different phases in bitumen, as it is known that bitumen has complex chemical composition, consisting from saturates (S), aromatics (A), resins (R) and asphaltenes (A), which ratios strongly depend on the refinement process of the crude oil [1,2]. It is generally known that decrease in aromatics content and an increase in resins and asphaltene fractions occurs upon ageing, resulting in the change of the ratio between the SARA fractions. However, these effects are hardly recognizable due to "noisy" nature of the derivative curves.

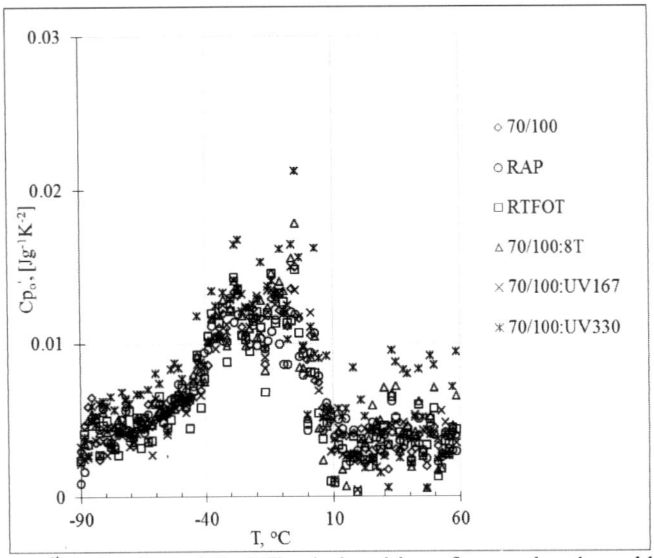

Figure 2. 1st derivative of Cp_o'-T relationships of unaged and aged bitumen

The change in chemical composition, including SARA fractions, of bitumen under aging can be also determined by FTIR analysis, because at oxidation ketones, dicarboxylic anhydrides, carboxylic acids and sulfoxides are formed, as characterized by appearance of carbonyl (C=O at 1705 cm^{-1}) and sulfoxide (S=O at 1030 cm^{-1}) groups. This method is often used for characterization of ageing process of asphalt pavement binder.

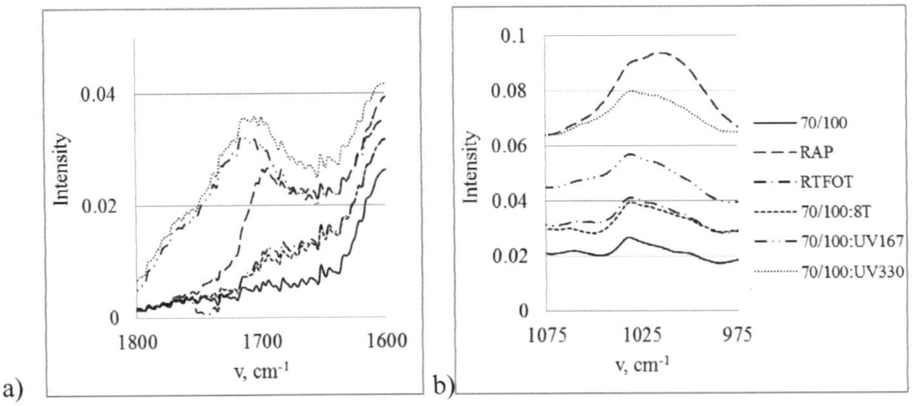

Figure 3. FTIR spectra of unaged and aged bitumens in carbonyl group (a) and sulphoxide group (b) absorption regions

In Figure 3 and Table 1 it is observed, that sulfoxide peaks intensities as well as respective sulfoxide indices of neat and aged bitumens are higher in comparison to those, characterizing carbonyl groups. According to literature S=O groups form faster than C=O groups, furthermore, the former can be present even before aging, while it depends from the used crude oil [6,7]. Carbonyl (Ico) or sulfoxide (Iso) indexes are calculated according to equations (1-2):

$$Ico = I_{1700}/I_{reference} \quad (1)$$
$$Iso = I_{1030}/I_{reference} \quad (2)$$

where I_{1700} is the amplitude of carbonyl group centred at 1700 cm^{-1}, I_{1030} is the amplitude of sulfoxide group centred at 1030 cm^{-1} and $I_{reference}$ is the amplitude of ethylene and methylene groups which are considered not to be significantly affected by aging process (1460 and 1375 cm^{-1}, respectively) [6].

Table 1. Ico and Iso of unaged and aged bitumens

	Iso	Ico
70/100	0.12	0.02
RAP	0.38	0.11
RTFOT	0.18	0.04
70/100:8T	0.18	0.04
70/100:167UV	0.26	0.14
70/100:330UV	0.33	0.14

Increase in the intensities at the absorption region of C=O and S=O groups is expected after all aging procedures. By increasing UV irradiation exposure time one can clearly see that at higher exposure times greater intensities at C=O and S=O groups absorption regions are observed, sulfoxide index almost reaches the value, characteristic for RAP. Concomitant, the values of Ico of the UV aged bitumens are higher in comparison to that of RAP derived bitumen regardless of UV irradiation exposure time.

Influence of oxidized groups on the bitumen stiffness can be various, it depends of dispersing power of non-associating components and strength of the polar molecules [7]. In our investigation it is observed, that viscosity of the bitumen derived from RAP at 46 °C is higher (above 800%) than that for unaged bitumen (see Fig.4 (a)). Viscosities of the bitumens, aged by other procedures, don`t increase so drastically, values are between unaged bitumen and RAP derived bitumen. At higher temperatures viscosity values of the investigated bitumens decrease, but keep before observed relationship.

Rheological behaviour is also characterized with rutting resistance parameter ($G^*/\sin\delta$) and fatigue cracking resistance parameter $G^*\sin(\delta)$. Plots of $G^*/\sin\delta$ versus temperature are displayed in Figure 4 (b). As already expected, critical temperature of the investigated materials is higher for aged bitumens, especially RAP derived bitumen, it is considerably stiffer than 70/100 or any other of the aged bitumens.

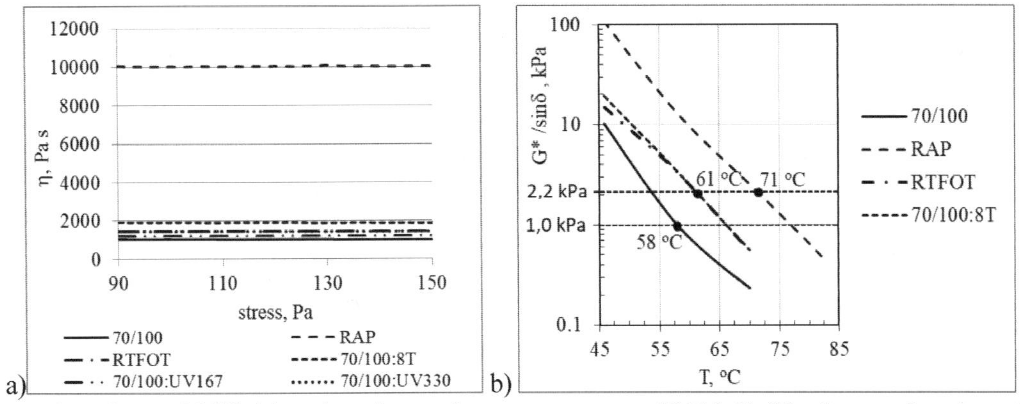

Figure 4. Viscosity at 46 °C (a) and rutting resistance parameter (G*/sinδ) (b) of unaged and aged bitumens.

Values of fatigue cracking resistance parameter G*sin(δ) of the investigated bitumens are summarized in Figure 5. Bitumen is more elastic, if the parameter is lower. According to AASHTO T 315 for bituminous materials to resist fatigue cracks, G*·sinδ for RTFOT+PAV-aged bitumen should not be greater than 5000 kPa. By applying this value to the investigated aged bitumens one can see that RAP has the lowest critical temperature in comparison to any other aged bitumen.

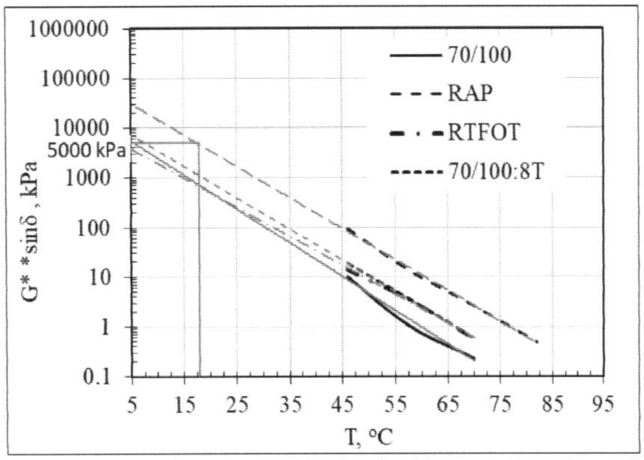

Figure 5. Fatigue cracking resistance parameter G∗sin δ of unaged and aged bitumens.

Conclusions

Results of the study show that under the influence of temperature and UV irradiation the investigated bituminous materials undergo aging, which is similar to some extent with the field aging of bitumen. Because of aging chemical structure of bitumen is affected, i.e., bitumen becomes more polar as confirmed by increasing intensities in the carbonyl (C=O) and sulfoxide (S=O) groups absorption regions. Aging of bitumen causes also increment in endothermic contributions of non-reversing heat flow curves, which could be related to increased crystallinity. Due to the changes in the investigated bitumens chemical composition, including SARA fractions, it viscosity values are increased, rutting resistance parameters increase and fatigue cracking resistance parameter increases.

Acknowledgements

Financial support in the framework of the ERDF project No 1.1.1.1/16/a/148 is greatly acknowledged.

References

[1] Little N D Allen H D and Bhasin A 2018 *Modeling and Design of Flexible Pavements and*

Materials (Springer International Publishing)

[2] Goncalves M C and Margarido F 2015 *Materials for Construction and Civil Engineering: Science, Processing, and Design* (Springer) pp 237-273

[3] http://www.eapa.org/promo.php?c=174 (accessed on 17.05.2018.)

[4] Masson J F and Polomark G M Bitumen microstructure by modulated differential scanning calorimetry *Thermochim Acta* vol 374 n 2 pp105-114

[5] Masson F Polomark G and Collins P 2005 Glass transitions and amorphous phases in SBS–bitumen blends *Thermochim Acta* vol 436 pp 96–100

[6] Marsac P Porot L Pierard N and Van den bergh W 2014 Potential and limits of FTIR methods for reclaimed asphalt characterisation *Mater Struct* vol 47 n 8 pp 1273–1286

[7] Branthaver J F et.al. *Binder Characterization and Evaluation* 1993 (Washington: Strategic Highway Research Program National Research Council) vol 2